AF325580

SOLID-STATE SCIENCE AND TECHNOLOGY LIBRARY

Volume 3

Editorial Advisory Board

Aims and Scope of the Series

The aim of this series is to present monographs on semiconductor materials processing and device technology, discussing theory formation and experimental characterization of solid-state devices in relation to their application in electronic systems, their manufacturing, their reliability, and their limitations (fundamental or technology dependent). This area is highly interdisciplinary and embraces the cross-section of physics of condensed matter, materials science and electrical engineering.

Undisputedly during the second half of this century world society is rapidly changing owing to the revolutionary impact of new solid-state based concepts. Underlying this spectacular product development is a steady progress in solid-state electronics, an area of applied physics exploiting basic physical concepts established during the first half of this century. Since their invention, transistors of various types and their corresponding integrated circuits (ICs) have been widely exploited covering progress in such areas as microminiaturization, megabit complexity, gigabit speed, accurate data conversion and/or high power applications. In addition, a growing number of devices are being developed exploiting the interaction between electrons and radiation, heat, pressure, etc., preferably by merging with ICs.

Possible themes are (sub)micron structures and nanostructures (applying thin layers, multi-layers and multi-dimensional configurations); micro-optic and micro-(electro)mechanical devices; high-temperature superconducting devices; high-speed and high-frequency electronic devices; sensors and actuators; and integrated opto-electronic devices (glass-fibre communications, optical recording and storage, flat-panel displays).

The texts will be of a level suitable for graduate students, researchers in the above fields, practitioners, engineers, consultants, etc., with an emphasis on readability, clarity, relevance and applicability.

Physics and Applications of Optical Solitons in Fibres '95

Proceedings of the Symposium
held in Kyoto, Japan, November 14 –17 1995

Edited by

Akira Hasegawa

Department of Communication Engineering,
Osaka University,
Suita, Japan

KLUWER ACADEMIC PUBLISHERS

DORDRECHT / BOSTON / LONDON

A C.I.P. Catalogue record for this book is available from the Library of Congress

ISBN 0-7923-4155-4

Published by Kluwer Academic Publishers,
P.O. Box 17, 3300 AA Dordrecht, The Netherlands.

Kluwer Academic Publishers incorporates
the publishing programmes of
D. Reidel, Martinus Nijhoff, Dr W. Junk and MTP Press.

Sold and distributed in the U.S.A. and Canada
by Kluwer Academic Publishers,
101 Philip Drive, Norwell, MA 02061, U.S.A.

In all other countries, sold and distributed
by Kluwer Academic Publishers Group,
P.O. Box 322, 3300 AH Dordrecht, The Netherlands.

Printed on acid-free paper

Printed in the Netherlands

ORGANIZER AND PROGRAM COMMITTEE

Organizer

Research Group for Optical Soliton Communications (ROSC)

Chairman:Akira HASEGAWA, Osaka University

Program Committee

Kazuro KIKUCHI, University of Tokyo

Masayuki MATSUMOTO, Osaka University

Kiyoshi NAKAGAWA, Nippon Telegraph and Telephone Corporation

Kazuo SAKAI, Kokusai Denshin Denwa Co., Ltd.

Secretariat

Support Center for Advanced Telecommunications Technology Research, Foundation (SCAT)

Tel. +81-3-3351-0540, Fax. +81-3-3351-1624

E-Mail LDE02703@niftyserve.or.jp

TABLE OF CONTENTS

Preface

This book summarizes the proceedings of the invited talks presented at the International Symposium of Physics and Application of Optical Solitons in Fibers held in Kyoto during November 14 to 17, 1995. As a result of worldwide demand for ultra high bitrate transmissions and increased scientific interests from the soliton community, research on optical solitons in fibers has made a remarkable progress in recent years. In view of these trends, and with the support of the Japanese Ministry of Posts and Telecommunications, the Research Group for Optical Soliton Communications (ROSC), chaired by Akira Hasegawa, was established in Japan in April 1995 to promote collaboration and information exchange among communication service companies, industries and academic circles in the theory and application of optical solitons. This symposium was organized as a part of the ROSC activities.

The symposium attracted enthusiastic response by worldwide researchers involved in this subject which has lead to the most intensive meeting that the editor ever attended. The reader will find the contents to be well-balanced among theory, experiment and technology. Although the evaluation of the contents shall naturally depend on the particular area of interest of the reader, the symposium has confirmed that the soliton based light wave transmission has achieved the best result in one channel, both in distance of transmission and in bitrate although in wavelength division multiplexed (WDM) systems, NRZ transmission has yet better result.

In this regard, fiber dispersion managements, which are attracting increased interests as indicated by three invited talks here, are expected, among other things, to enhance soliton WDM capability. Thus, new soliton based WDM experiments using dispersion managed fiber are awaited. This is just one additional example that research on optical solitons has room to achieve innovation and quantum jump in theory and application. The symposium has also confirmed that optical solitons and soliton phenomena have a wide range of applications not limited to optical signal transmissions, and their research is attracting scientists from a wide range of backgrounds including applied mathematics, theoretical physics, optics and communications engineers. This symposium has succeeded in bringing these scientists together under the common subject of optical soliton.

In the preparation of the final draft of the proceedings, Messrs. H. Takehara and M. Shinomiya have devoted a significant amount of their time to make the manuscript consistent and well-coordinated. On behalf of ROSC, the editor would like to thank those efforts without which the publication of the book would have faced significant delay.

The editor would also like to express his appreciation to the support of the Japanese Ministry of Posts and Telecommunications, which indicates a serious interest on soliton based communication systems by the Japanese government. The sponsorships by The Telecommunications Advancement Foundation (TAF), International Communications Foundation (ICF), and in particular Support Center for Advanced Telecommunications Technology Research, Foundation (SCAT) to which Messrs. Takehara and Shinomiya are affiliated, have enabled us to hold this successful symposium.

Kyoto, March 1996
Akira HASEGAWA

DISPERSION AS CONTROL PARAMETER IN SOLITON TRANSMISSION SYSTEMS

N. J. DORAN, N. J. SMITH, W. FORYSIAK AND F. M. KNOX
Electronic Engineering and Applied Physics
Aston University
Aston Triangle
Birmingham B4 7ET, United Kingdom

Abstract. A review is presented of the role of dispersion in managing solitons in high data rate transmission. It is shown that piecewise intra-amplifier dispersion profiling may be used to minimise the effect of periodic amplification. Jitter may be controlled by multistage dispersion compensation. Finally an investigation of periodic dispersion compensation in a generalised NLS system is presented showing the suppression of modulational instability and the existence of a new stable soliton like pulse.

1. Introduction

There is real potential for the application of solitons in high data rate transmissions systems. For such systems there are several key parameters which are required in the design for optimum operation. Of course the critical feature for all commercial systems is for cost effective operation relative to the rival NRZ approach. The most important of these parameters are: pulse duration, fibre dispersion and amplifier spacing. In addition most proposed systems include some form of soliton control [1]. There are many schemes for soliton control including fixed and sliding frequency filters and active modulation [2, 3, 4]. This article will assess the extent to which design of dispersion may be used as the primary control strategy. We will show that dispersion offers the potential for controlling solitons allowing long distance and high data rates with only the minimal additional control elements. Dispersion may be used to extend amplifier spacing and to permit operation over standard fibres.

1

A. Hasegawa (ed.), Physics and Applications of Optical Solitons in Fibres '95, 1–14.
© 1996 *Kluwer Academic Publishers. Printed in the Netherlands.*

It is first useful to review the competing requirements for dispersion in soliton systems. It seems obvious that finite dispersion is necessary for a soliton system but many of the detrimental effects are mitigated by choosing a low value for fibre dispersion. The primary limiting effects in soliton transmission are as follows:

1. Jitter: temporal jitter, predominantly Gordon-Haus jitter [5], is proportional to the long distance average dispersion and is thus minimised by operating with low dispersion. As shown in Section 3, the jitter may be reduced by post-transmission dispersion compensation.
2. Signal Power: this is proportional to dispersion, at least in a simple 'average soliton' [6] system. It is thus desirable to operate with finite dispersion to improve the SNR. For high dispersion systems (e.g. using standard fibre) the signal powers may be inconveniently high.
3. Polarisation Mode Dispersion: PMD can be eliminated in soliton systems provided $0.3\ D^{1/2} > D_p$ [7]. Thus the long distance dispersion must be chosen sufficiently large to overcome PMD.
4. Amplifier Spacing: with the average soliton model the amplifier spacing must be less than the soliton period which is proportional to the dispersion. Thus small intra-amplifier dispersion is desirable for extending spacing.
5. Soliton Interactions: the collapse of solitons depends inversely on the long distance average dispersion. Thus small long distance dispersion is desirable.
6. WDM: for wavelength multiplexing it is generally desirable for the interactions to take place over several amplifiers. Small dispersion is usually the requirement.

Thus features 1, 4, 5 and 6 would demand small dispersion and 2 and 3 require larger dispersion. It is, however, important to recognise that dispersion may be managed in the system to optimise the competing requirements. In the following we shall describe schemes for optimising the effects by varying the dispersion along the transmission path.

2. Intra-amplifier Dispersion Profiling

Here we shall consider the approach of varying the dispersion between the amplifier whilst maintaining the average dispersion at some fixed value [8, 9]. The average dispersion will determine long distance effects such as jitter and SNR; by profiling the dispersion in a finite number of discrete lengths the radiation, produced by the periodic amplification, can be minimised. The principle is straightforward. In a lossy system it is known that the effects of loss (in particular the generation of radiation) can be elimi-

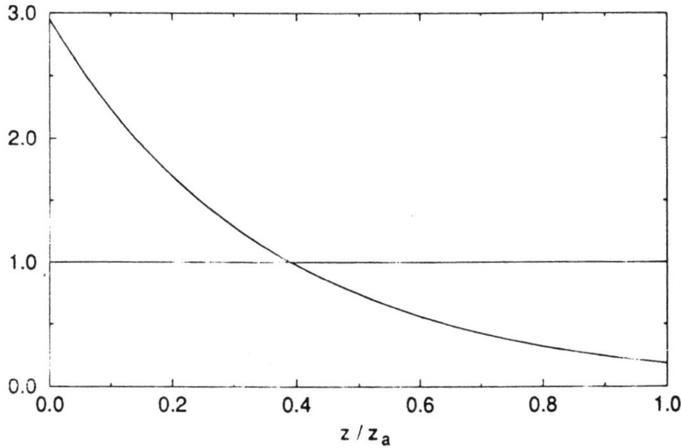

Figure 1.   Variation of the function $U(z)$ (the lossy pulse) against distance compared with lossless case.

nated by matching the power loss to a corresponding decrease in dispersion [10]. Here we examine the use of a piecewise approximation to the ideal exponential decrease in dispersion. The optimum design for n dispersion steps will define the ideal dispersions and lengths of the fibres. The design is obtained by minimising the error defined as the area difference between the ideal NLS and the lossy wave $u(z)$ as illustrated in Figure 1. For example, if we consider a two stage profiling where there are two lengths of constant (different) dispersions, the area (error) illustrated in Figure 2(b) is the sum of four regions S_1, T_1 and S_2, T_2. These are a measure of the deviation from the pure NLS. It is important to recognise that it is the area in the normalised NLS in each region which is appropriate. For n regions mathematically this procedure amounts to the minimisation of a straightforward expression in n-1 variables (the average dispersion and the total length being predefined). Generally, across the span the dispersion is decreased and the fibre lengths increase. For small inter-amplifier loss the normalised lengths are equal, but for losses greater than a few dB this is not the case. The dispersions and relative lengths depend only on the total loss (assumed to be exponential). As an example, consider a 50 km spacing with average dispersion $D_a = 1$ ps/nm/km and total loss 12 dB (i.e. 0.24 dB/km). With a four stage profiling the optimum is found to be $D_j = 2.41$, 1.51, 0.82, 0.34 and $L_j = 7.53$, 9.51, 12.91, 20.05 km. Figure 3 illustrates

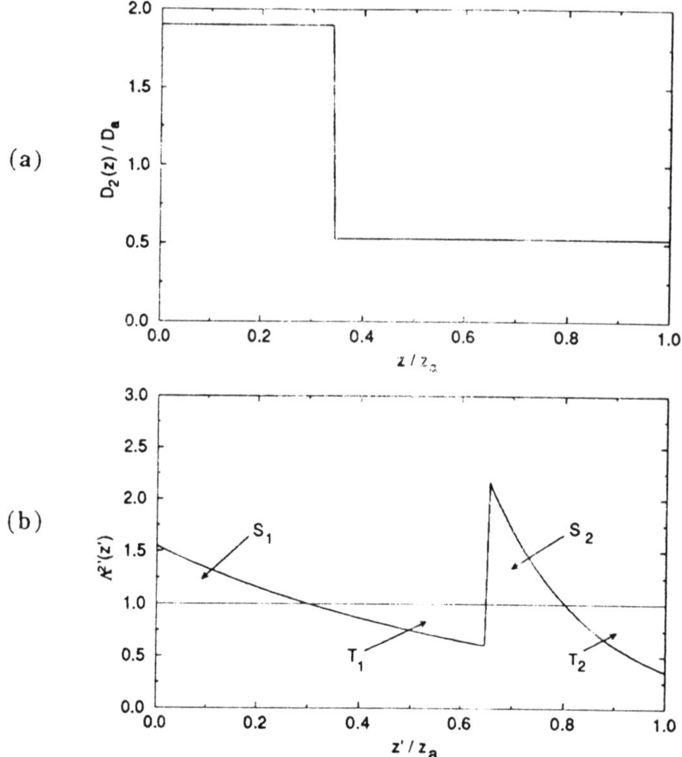

Figure 2. 2-fold dispersion variation (a) and its effect on the mismatch compared with lossless case (b).

for this example the improved stability of a 5 ps pulse over 25 amplifier spans.

The technique described above is remarkably effective. First, clear benefit will be obtained from just two fold profiling (2 stage dispersion) and unless the loss is extreme, 4 fold profiling almost completely eliminates radiation provided the optimum design is employed. Details of this approach and design parameters are presented in Reference [9].

3. Dispersion Compensation in Multiamplifier Systems

In this section we shall discuss and present a simple model for the reduction of jitter in amplified soliton systems achieved by employing a degree of dispersion compensation. In Section 2 dispersion profiling on a length scale less than the amplifier spacing (and therefore also the soliton period) was

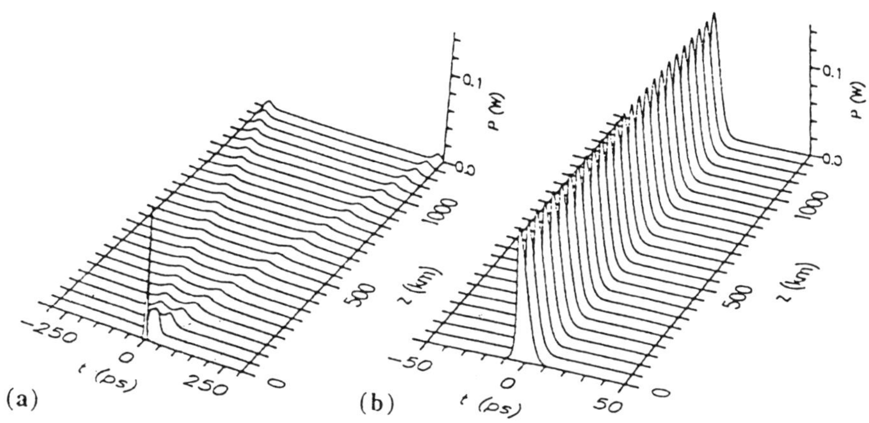

Figure 3. Pulse profiles over 25 amplifiers $z_a = 50$ km loss 12 dB, $D_a = 1$ ps/nm/km, $t = 5.62$ ps, (a) unprofiled, (b) 4 step profiling.

employed. In contrast here, we consider the possibility of periodic dispersion compensation on a length scale of many amplifiers which will reduce jitter by reducing the global system dispersion whilst maintaining the 'local' intra-amplifier dispersion. If we consider Gordon-Haus jitter alone then the temporal jitter can simply be expressed in terms of the frequency jitter (variance $< \delta\omega^2 >$) for each amplifier summed over the total amplifier chain. With an interamplifier distance z_a and an average dispersion (intra-amplifier) D the variance of the timing jitter is expressed as

$$\langle \delta t^2 \rangle = \left(\frac{\lambda^2}{2\pi c} \right)^2 \sum_{j=1}^{N} (D z_a j)^2 \langle \delta\omega^2 \rangle. \tag{1}$$

If we now consider a dispersion compensation of value D_c inserted after N amplifiers then

$$\langle \delta t^2 \rangle = \left(\frac{\lambda^2}{2\pi c} \right)^2 \sum_{1}^{N} (D z_a j + D_c)^2 \langle \delta\omega^2 \rangle. \tag{2}$$

The variance is minimised for $D_c = -\frac{N D z_a}{2}$ and is reduced by a factor of 4 [11, 12].

We can extend this analysis to dispersion compensation for a chain of $P = N x M$ amplifiers with compensation after N amplifiers performed at M

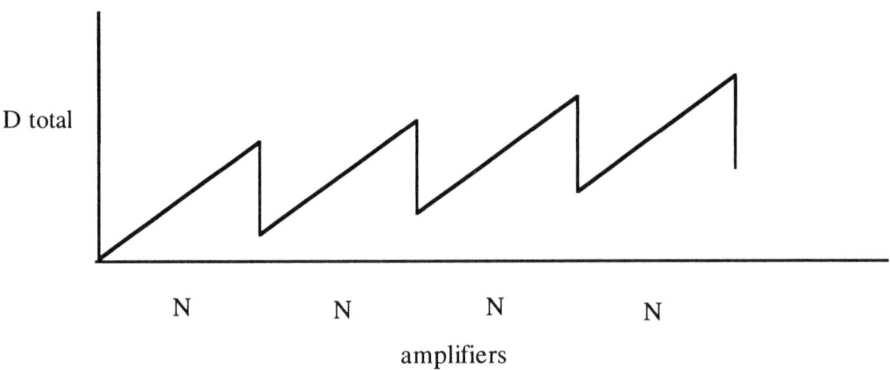

Figure 4.   Dispersion profile for a 4 fold compensation after N amplifiers.

times ($M = 1$ corresponds to the above post transmission compensation). Then (dropping the scaling terms Dz_a)

$$\langle \delta t^2 \rangle = \left(\frac{\lambda^2}{2\pi c} \right)^2 \sum_{k=1}^{M} \sum_{j=1}^{N} (N(k-1) + j + kD_c)^2 \langle \delta \omega^2 \rangle. \tag{3}$$

Figure 4 illustrates this principle for a 4 fold compensation. From Equation (3) the optimum compensation (assumed to be the same at each compensation stage) is

$$D_c = -N \left(\frac{4M - 1}{4M + 2} \right). \tag{4}$$

We can simply calculate the maximum benefit of this periodic compensation on the jitter.

The table shows that huge reductions in jitter can be obtained by a few stage dispersion compensation scheme. In addition the compensation approaches the full dispersion of the N amplifier stage.

	D_c	Reduction Factor
M = 1	$-\dfrac{N}{2}$	4
M = 2	$-\dfrac{7N}{10}$	14.5
M = 4	$-\dfrac{15N}{18}$	42.7

Of course this analysis does not take into account the pulse broadening effect of dispersion compensation and thus whether the soliton's integrity can be maintained even over this few stage scheme. We shall return to the question of soliton stability in long distance periodically dispersion compensated schemes in Section 5.

4. Solitons in Standard Fibre Systems [12]

There is much interest in upgrading the embedded fibre network to operate at 1.55 μm using fibre amplifiers to replace regenerators and operating at 10 Gbit/s. In addition higher speeds and wavelength multiplexing can be anticipated to increase flexibility and throughput. The problem is that this fibre has high dispersion 15 - 20 ps/nm/km at the desired wavelength. The solution for NRZ systems is to employ dispersion compensation schemes. The question we pose here is whether solitons could have a role to play in such systems.

We shall consider as our example a system operating at 10 Gbit/s with fibre dispersion of 15 ps/nm/km and amplifier separation of 36 km. This system has been extensively studied by us in Reference [12], here we briefly review the principle results which are directly relevant to the general discussion of dispersion as the control parameter.

In Reference [12] we show that for this typical system, because the amplifier spacing is at least a soliton period, the maximum transmission distance with no dispersion control is $\sim$ 200 km.

In order to extend the transmission distance it is necessary to include

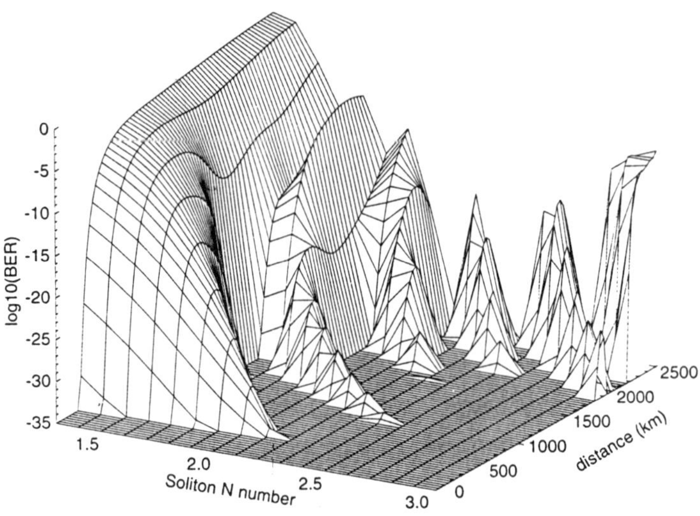

Figure 5. BER against input soliton N number and distance for 20 ps pulses and 1 ps/nm/km average dispersion.

dispersion profiling. This is, however, a different approach to that in Section 2. The principle is to include a dispersion compensating element (which may be fibre) at each amplifier stage. The dispersion is of opposite sign (i.e. normal) and reduces the average intra-amplifier dispersion. The benefit of this average dispersion reduction is two fold. First the intra-amplifier dispersion reduction means that the perturbations introduced by the periodic amplification are significantly reduced in a manner similar to that described in Section 2. Second, the average power requirements are reduced making the system specifications more practical.

Numerical studies have shown, however, that the most stable soliton power is not given by the simple average soliton prescription but rather the launch power needs to be increased by a factor of up to 3 depending on the pulse duration chosen. It is maybe not surprising, that the prescription breaks down for this system, since for the uncompensated dispersion the soliton period is less than the amplifier separation. It is, however, surprising that a significantly more stable soliton can be found by simply increasing (albeit by a significant degree) the launch power. Figure 5 shows a plot of the launch N number (as in $N sech(t)$ referred to the average dispersion (1 ps/nm/km) against calculated error rate as a function of transmission distance. For the system under consideration $N = 1.55$ would be the average soliton prescription. It can be seen that increasing this to $N \approx 2.6$ or more results in stable propagation for 2,000 km or more. Figure 6 shows a typical data stream transmitted under these conditions. Thus we can con-

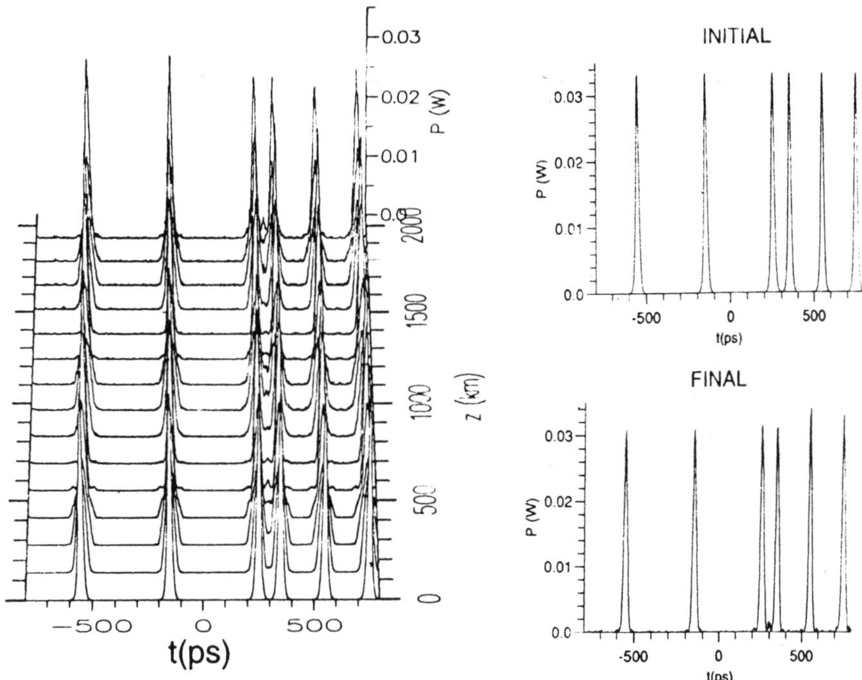

Figure 6. Pulse evolution for 1,250 km standard fibre using dispersion compensation.

clude that a degree of dispersion compensation at each amplifier stage can permit 10 Gbit/s soliton transmission with useful amplifier separation for at least 2,000 km.

5. Principle of Multistage Dispersion Compensation

The above sections have identified clear benefits which may be obtained by periodically varying the dispersion in a soliton transmission system. Here we shall present some initial results on the general principle of periodic dispersion compensation in NLS systems. In order to identify the principle which is being employed above, we shall consider a model of a lossless NLS system with periodic dispersion variations.

Figure 7 illustrates the system we are considering. Here the dispersion, $\ddot{\beta}$, is periodically negative (bright soliton supporting) and positive. There is a net average dispersion, $\ddot{\beta}_{av}$, which may be negative or zero.

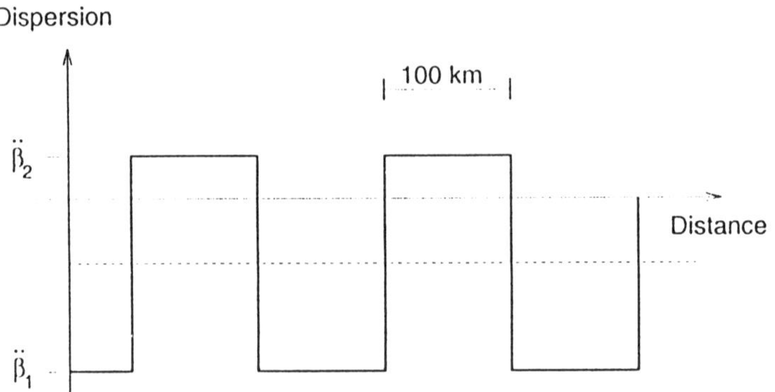

Figure 7.　Periodic dispersion map used in the calculation.

5.1. MODULATIONAL INSTABILITY IN A PERIODIC DISPERSION SYSTEM

The simplest application of this periodic dispersion is in the modification of the modulational instability gain.

We can analyse the modulation instability gain by using a periodic $\ddot\beta(z)$ in the NLS, i.e. we write

$$i\frac{\partial u}{\partial z} = \frac{1}{2}\left(\ddot\beta_{av} + \ddot\beta(z)\right)\frac{\partial^2 u}{\partial t^2} - \gamma|u|^2 u \tag{5}$$

By making a substitution, which is the solution of the linear dispersion equation of the fluctuations, a modified NLS is derived with a reduced effective nonlinearity. Using an analytic form for the $\ddot\beta(z)$ $\left(\ddot\beta_p \cos k_p z\right)$ a simple result for the modified gain spectrum is obtained. Namely, the gain spectrum is for values of Ω which give a positive real component of the eigenvalue K expressed as

$$K^2 = -\ddot\beta_{av}\Omega^2\left(\gamma P_0 + \ddot\beta_{av}\frac{\Omega^2}{4}\right) + \gamma^2 P_0^2\left(J_0^2\left(\ddot\beta_p\frac{\Omega^2}{k_p}\right) - 1\right), \tag{6}$$

where J_0 is the Bessel function. In Equation (6) the first part is the classical MI spectrum and the second term is the modification due to the periodic dispersion.

In Figure 8 we show the MI spectra calculated from Equation (6) and a numerical calculation of the same. The calculations are performed for 100

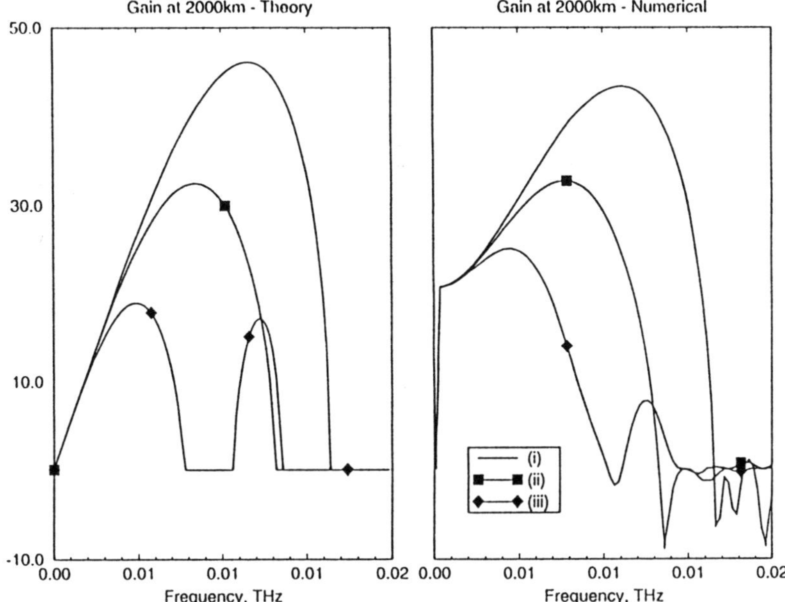

Figure 8. Modulational instability gain spectra for (i) no dispersion oscillation and (ii) alternating lengths of 100 km, -8 ps^2/km, +6 ps^2/km and -16 ps^2/km, +14 ps^2/km.

km alternating lengths with - 8 ps^2/km and +6 ps^2/km for fixed (1 mW) launch power. It can clearly be seen that the effect of dispersion modulation is to reduce the gain spectra. The peak is decreased and moves to smaller frequency shift. This is the same effect as would be obtained by reducing the average dispersion and simultaneously the power.

Thus we can conclude that dispersion oscillation will reduce the MI and its associated gain spectrum. This is really the key as to why dispersion management is so effective in reducing detrimental nonlinear effects in NRZ systems.

5.2. PULSE PROPAGATION IN PERIODIC DISPERSION SYSTEMS

The critical question here is whether there are stable pulse like solutions to the periodic dispersion oscillating problem discussed in the previous section. We have conducted a numerical investigation of the system illustrated in Figure 7 with varying amplitudes of dispersions in the two sections each now of 10 km.

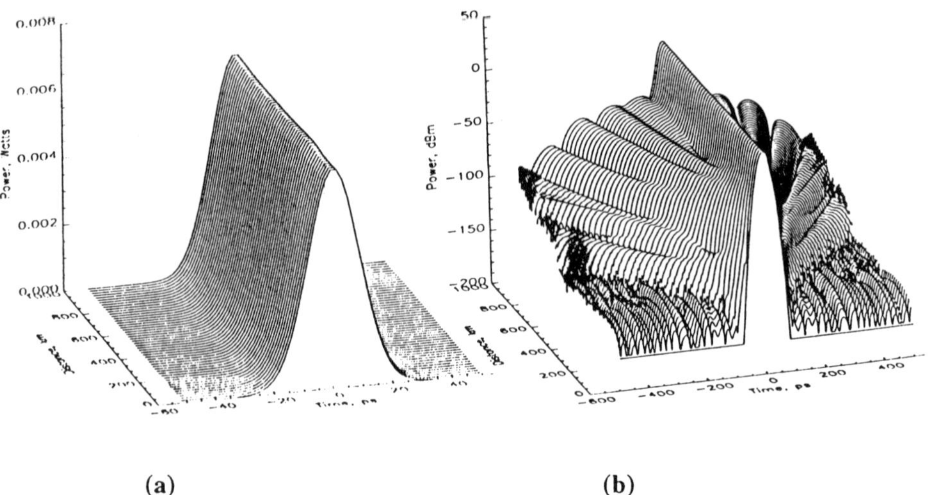

(a) **(b)**

Figure 9. Pulse evolution (a) linear, (b) logarithmic for alternating lengths 10 km, -32 ps^2/km, 30 ps^2/km, pulse width 20 ps.

We have discovered that there are indeed stable pulse solutions. For example, for the case of $\ddot{\beta} = +32$, -30 and pulse durations of 20 ps (FWHM). Figure 9 shows the temporal and spectral evolution of the pulses. Although the log plots show a small degree of radiation these pulses are stable for what would be many soliton periods in the average dispersion (-1 ps^2/km).

There are, however, several important points as to how these stable pulses were obtained. First, it is vitally important that the initial dispersion step is a half of the length of all the others. The radiation is much reduced by this approach. Second, the pulse energy must be chosen to be significantly above the soliton energy for the average dispersion. This second point is illustrated in Figure 10 where we show the numerically obtained stable pulse energy as a function of peak to peak dispersion excursion. Clearly the increase in energy increases monotonically with the increase in dispersion amplitude. Third, the stable pulse is not simply sech like in shape. In fact the pulse is well described by a simple Gaussian shape. Figure 9 shows the typical shape of a Gaussian rather than the standard *sech* of common solitons.

We can conclude this section and the work as a whole by observing that there are stable soliton-like solutions of the periodically dispersion vary-

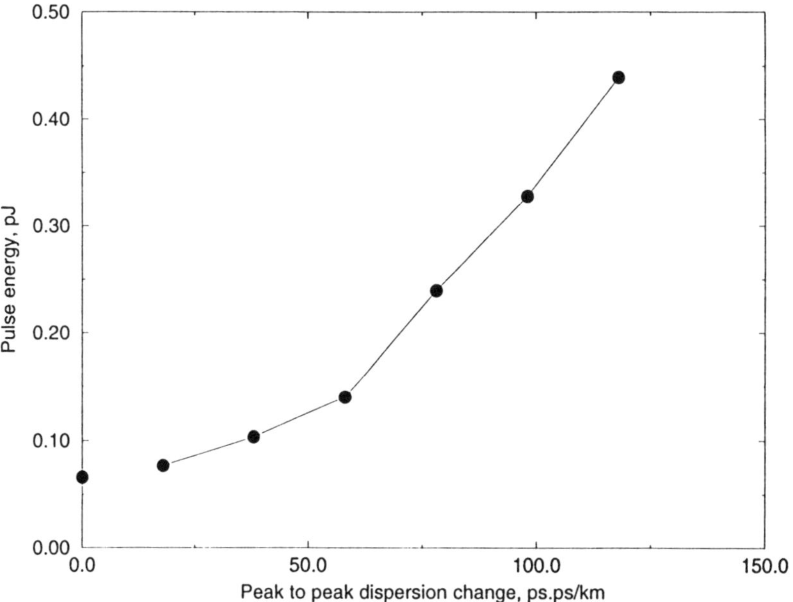

Figure 10. Stable pulse energy versus dispersion change for an average of -1 ps^2/km and 20 ps pulses.

ing system. This is consistent with the experiments in Reference [13]. We have discovered by numerical investigation that these pulses are Gaussian in shape and have energies well above that of the soliton of the average dispersion. Thus it should be possible to operate stable pulse (soliton) systems with periodic dispersion compensation whilst maintaining the pulse energy well above that dictated by the global average dispersion. Thus very low long distance dispersion averages and finite pulse energies are possible without significant pulse distortion. It is necessary to operate with more Gaussian like pulses and with an initial 1/2 dispersion step for reducing radiation generation effects. We have not included the combined effect of periodic loss and gain here, but we do not anticipate that this inclusion would substantially alter our present conclusions.

References

1. For a review of soliton control see: Hasegawa, A. and Kodama, Y.: *Solitons in Optical Communication.* (O.U.P., Oxford, 1995).
2. Mecozzi, A., Moores, J. D., Haus, H. A. and Lai, Y.: Soliton transmission control, *Opt. Lett.* **16** (1991), 1841-1843.
3. Mamyshev, P. V. and Mollenauer, L. F.: Stability of soliton propagation with sliding-

frequency guiding filters, *Opt. Lett.* **19(24)** (1994), 2083-2085.

4. Kubota, H. and Nakasawa, M.: Soliton transmission control in time and frequency domains, *IEEE J. Quantum Electron.* **29(7)** (1993), 2189-2197.

5. Gordon, J. P. and Haus, H. A.: Random walk of coherently amplified solitons in optical fiber transmission, *Opt. Lett.* bf 11 (1986), 665-667.

6. Blow, K. J. and Doran, N. J.: Average soliton dynamics and the operation of soliton systems with lumped amplifiers, *IEEE Photon. Technol. Lett.* **3** (1991), 369-371.

7. Mollenauer, L. F., Smith, K., Gordon, J. P. and Menyuk, C. R.: Resistance of solitons to the effect of polarisation dispersion in optical fibres, *Opt. Lett.* **14** (1989), 1219-1221.

8. Tajima, K.: Compression of soliton broadening in nonlinear optical fibres with loss, *Opt. Lett.* **12** (1987), 54-56.

9. Forysiak, W., Knox, F. M. and Doran, N. J.: Average soliton propagation in periodically amplified systems with stepwise dispersion profiled fibre, *Opt. Lett.* **19(3)** (1994), 174-176.

10. Forysiak, W., Knox, F. M., and Doran, N. J.: Stepwise dispersion profiling of periodically amplified soliton systems, *IEEE J. Lightwave Technol.* **12(8)** (1994), 1330-1337.

11. Forysiak, W., Blow, K. J., and Doran, N. J.: Reductiono of Gordon-Haus jitter by post-transmission dispersion compensation, *Electron. Lett.* **29(13)** (1993), 1225-1226.

12. Knox, F. M., Forysiak, W., and Doran, N. J.: 10 Gbit/s soliton communication systems over standard fibre at 1.55 mm and the use of dispersion compensation, *IEEE J. Lightwave Technol.* **13(10)** (1995), 1955-1963.

13. Edagawa, N., Morita, I., Suzuki, M., Yamamoto, S., Taga, H. and Akiba, S.: 20 Gbit/s 8100 km straight line single channel soliton based transmission experiment using periodic dispersion compensation, *21st European Conference on Optical Communication (ECOC'95)*, Brussels, Postdeadline (1995), 983-986.

DISPERSION MANAGEMENTS ON SOLITON TRANSMISSION IN FIBERS WITH LUMPED AMPLIFIERS

Y. KODAMA
Department of Mathematics
Ohio State University
Columbus, OH 43210 USA

S. KUMAR
Department of Communication Engineering
Osaka University,
Suita, 565 Japan

AND

A. HASEGAWA
Department of Communication Engineering
Osaka University,
Suita, 565 Japan

Abstract. Radiations of continuous spectra and collision induced time jitters at lumped amplifiers in adiabatic soliton transmission systems can be reduced to negligible level by means of properly designed dispersion management schemes.

1. Introduction

Transmission properties of optical solitons are distinctively different between nonadiabatic regime (dispersion distance $z_0 \gg$ loss distance γ^{-1}) and adiabatic regime ($z_0 \ll \gamma^{-1}$). In the former regime, stable soliton transmission is possible by means of the guiding center soliton [1, 2] with properly enhanced initial amplitude [3, 4] and a choice of amplifier spacing $z_a \ll z_0$. In the adiabatic regime, when the soliton goes through a lumped amplifier, its amplitude jumps and non-soliton components are generated [5]. In addition, collision between soliton induces frequency shifts in WDM systems.

15

A. Hasegawa (ed.), Physics and Applications of Optical Solitons in Fibres '95, 15–26.
© *1996 Kluwer Academic Publishers. Printed in the Netherlands.*

As a result, long distance ($z \gg z_0$) and high-bit rate ($z_a \geq z_0$) soliton transmission in adiabatic regime with periodic lumped amplifiers is considered to face difficulty. As a solution to this problem, a system having amplifiers with distributed gain has been proposed and successful transmission has been demonstrated [6, 7]. Forysiak ct al. [8, 9] have proposed an alternative idea of using a stepwise dispersion profile between amplifiers and have numerically shown that the ultra short (5 ps) soliton can propagate stably with the usual amplifier spacing by using stepwise dispersion decreasing fibers.

In this paper, we present a theoretical investigation of the radiations emitted by a soliton propagating in stepwise and comb-type dispersion managed systems, using perturbed inverse scattering transform method and show that adiabatic soliton transmission is possible.

We further show that the dispersion management scheme reduces the frequency shift due to soliton collisions in wavelength division multiplexing (WDM) systems with lumped amplifiers. The scheme is effective even when collision distance is shorter than the amplifier spacing where the collision induced frequency shift has been considered to be intolerable [10].

2. A System with Varying Dispersion and Lumped Amplifiers

The complex amplitude q of a light wave in a fiber with a variable dispersion $d(Z)$ in the presence of lumped amplifiers at $Z = nZ_a$, $n = 0, 1, \cdots, N$ (the distance Z is normalized by z_0, $Z = z/z_0$) is described by the nonlinear Schrödinger equation [11]

$$i\frac{\partial q}{\partial Z} + \frac{d(Z)}{2}\frac{\partial^2 q}{\partial T^2} + |q|^2 q = -i\Gamma q + i\alpha \sum_{n=0}^{N} \delta(Z - nZ_a)q(Z - 0), \quad (1)$$

where $\Gamma (= \gamma z_0)$ describes the fiber loss and α represents the gain of the amplifiers, $\alpha = \exp(\Gamma Z_a) - 1$. We first transform q to a new variable u to eliminate the oscillating part due to the periodic amplification [1],

$$q = a_1(Z)u, \quad (2)$$

where $a_1(Z)$ is given by

$$a_1(Z) = a(0)\exp[-\Gamma(Z - nZ_a)], \quad \text{for} \quad nZ_a < Z < (n+1)Z_a, \quad (3)$$

with $a(0)$ defined by

$$a(0) = \sqrt{\frac{2\Gamma Z_a}{1 - \exp(-2\Gamma Z_a)}}. \quad (4)$$

With this change of variable, a periodic shift of the soliton eigenvalues is absorbed in the function $a_1(Z)$. Then u satisfies

$$i\frac{\partial u}{\partial Z'} + \frac{1}{2}\frac{\partial^2 u}{\partial T^2} + \frac{a_1^2(Z)}{d(Z)}|u|^2 u = 0, \tag{5}$$

where the new coordinate Z' is defined by

$$Z' = \int_0^Z d(Z)\,dZ. \tag{6}$$

It is clear from Equation (5) that if we choose

$$d(Z) \propto a_1^2(Z), \tag{7}$$

Equation (5) gives the ideal nonlinear Schrödinger equation for $u(Z', T)$ and an exact soliton solution emerges.

Although some attempt of producing fibers with continuously decreasing dispersion has been made [12], fabrication of a fiber having a precise dispersion variation given by $a_1^2(Z)$ for an extended distance is impractical especially when we allow a sufficiently large Z_a so that ΓZ_a may become much larger than unity. Consequently we first consider a fiber with stepwise decreasing dispersion $d(Z) = D_m$, the constant group velocity dispersion for $Z_m < Z < Z_{m+1}$, with $m = 0, 1, \cdots, M - 1$, and $Z_0 = 0, Z_M = Z_a$, as shown in Figure 1. The number M is the steps (divisions) of the fiber within an amplifier spacing. D_m may be chosen to provide dispersion corresponding to the average of the ideal dispersion,

$$\begin{aligned}
D_m &= \frac{1}{\Delta_m}\int_{Z_m}^{Z_{m+1}} a^2(0)\exp[-2\Gamma Z]\,dZ \\
&= \left[\frac{a(0)}{a_m(0)}\right]^2 \exp[-2\Gamma Z_m].
\end{aligned} \tag{8}$$

Here, $\Delta_m \equiv Z_{m+1} - Z_m$, and $a_m(0)$ is defined as $a(0)$ in Equation (4) with replacing Z_a by Δ_m, i.e.

$$a_m(0) \equiv \sqrt{\frac{2\Gamma\Delta_m}{1 - \exp(-2\Gamma\Delta_m)}}\,. \tag{9}$$

Equation (5) then reduces to

$$i\frac{\partial u}{\partial Z'} + \frac{1}{2}\frac{\partial^2 u}{\partial T^2} + a_M^2(Z)|u|^2 u = 0, \tag{10}$$

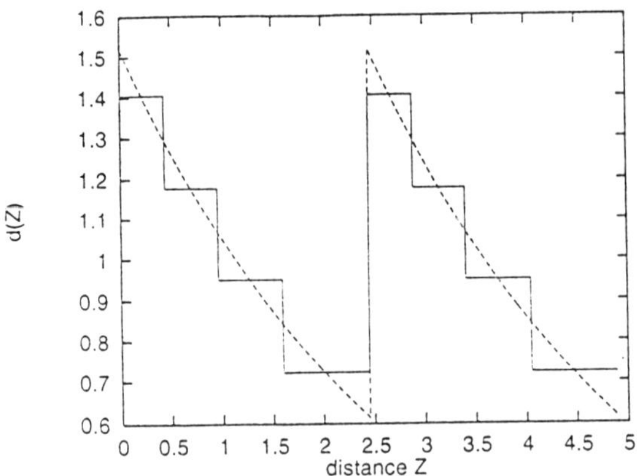

Figure 1. Step-wise dispersion profile with fiber sections equally spaced in Z' (solid line) and ideal exponential dispersion profile (broken line).

where $a_M(Z) = a_m(0) \exp[-\Gamma(Z - Z_m)]$ for $Z_m < Z < Z_{m+1}$. We note that, if $Z' \approx Z$, Equation (10) is equivalent to the expression for u with amplifier spacing reduced to Δ_m. As M the number of steps increases, a_M becomes unity and converges to the case of ideal dispersion. This fact indicates that the present scheme may also be applicable to nonadiabatic regime where the guiding center soliton emerges if $Z_a/M \ll Z_0(= 1)$ even if $Z_a \geq Z_0$. However Z' and Z are essentially different in the dispersion management discussed here, and we use the perturbed inverse scattering method [13] to study the soliton propagation in a dispersion managed system.

3. Perturbation Methods

Let us first write Equation (10) in the form,

$$\frac{\partial u}{\partial Z'} - i\frac{1}{2}\frac{\partial^2 u}{\partial T^2} - i|u|^2 u = \epsilon R[q, q^*], \tag{11}$$

where the perturbation term is given by

$$\epsilon R[q, q^*] = i(a_M^2(Z) - 1)|u|^2 u. \tag{12}$$

If $a_M(z)$ is close to unity, the term (12) describes the small perturbation due to varying dispersion and lumped amplifications. The solution $u(T, Z')$ of Equation (11) can be given by a sum of soliton and radiation, $u = u_s + u_r$, where u_s and u_r may be expressed by [14]

$$u_s(T, Z') = \eta \operatorname{sech}(\eta T)\, e^{i\eta^2 Z'/2}, \tag{13}$$

$$u_r(T, Z') = \frac{-1}{\pi} \int_{-\infty}^{\infty} \{r(Z'; \xi)\psi_1^2(T, Z'; \xi) + r^*(Z', \xi)\psi_2^2(T, Z'; \xi)\}\, d\xi. \tag{14}$$

As a perturbative solution, we here take ψ_1 and ψ_2 to be the wave functions for the one-soliton solution (13),

$$\psi_1 = \frac{\eta}{2(\xi + i\eta/2)} \operatorname{sech}(\eta T)\, e^{i\xi T + i\eta^2 Z'/2}, \tag{15}$$

$$\psi_2 = \left[1 - \frac{i\eta}{2(\xi + i\eta/2)} \operatorname{sech}(\eta T)e^{-\eta T}\right] e^{i\xi T}, \tag{16}$$

and $r(Z'; \xi)$ is the reflection coefficient describing the spectral components of the radiation on the bases of ψ_1^2 and ψ_2^2.

Using the perturbed inverse scattering transform [13], the reflection coefficient $r(Z'; \xi)$ can be shown to evolve in accordance with

$$\frac{d}{dZ'}\left[\hat{r}\, e^{2i\xi^2 Z'}\right] = i\left(\frac{\xi + i\eta/2}{\xi - i\eta/2}\right)^2 e^{2i\xi^2 Z'} \int_{-\infty}^{\infty} (R^*\psi_1^2 - R\psi_2^2)dT. \tag{17}$$

Here the function $\hat{r}$ is related to r in (14) with $|\hat{r}| = |r|$ which we only need to estimate the radiation energy due to the perturbation of (12). The total energy of radiation is given by

$$W_r(Z') = \frac{2}{\pi} \int_0^{\infty} \ln\left[1 + |r(Z'; \xi)|^2\right]\, d\xi. \tag{18}$$

4. Estimation of the Radiation

Using Equations (12),(15) and (16), we find from Equation (17) that $r(Z'; \xi)$ at the end of the transmission line $Z' = L \equiv NZ_a$ is evaluated as

$$|r(L; \xi)| = \pi\nu(\xi)\operatorname{sech}\left(\frac{\pi Z_a \xi}{\eta}\right) \left|\frac{\sin[\nu(\xi)L]}{\sin[\nu(\xi)Z_a]}\right| |B(\nu(\xi))|, \tag{19}$$

where $\nu(\xi) = \xi^2 + \eta^2/4$, and the Fourier component $B(\nu)$ is given by

$$\begin{aligned}
B(\nu) &= \frac{1}{Z_a} \int_0^{Z_a} e^{i2\nu Z'}[a_M^2(Z) - 1]dZ', \\
&= \frac{1}{Z_a} \int_0^{Z_a} e^{i2\nu Z'}[a_1^2(Z) - d(Z)]dZ, \\
&= \frac{a^2(0)}{Z_a} \left|\sum_{m=0}^{M-1} e^{i2\nu Z'_m} \exp[-\Gamma(Z_{m+1} + Z_m) + i\nu D_m Z_m] \right. \\
&\quad \left. \times \left[\frac{\sinh\{(i\nu D_m - \Gamma)\Delta_m\}}{i\nu D_m - \Gamma} - \frac{\sinh(\Gamma\Delta_m)}{\Gamma\Delta_m}\frac{\sin(\nu D_m \Delta_m)}{\nu D_m}\right]\right| \tag{20}
\end{aligned}$$

Y. KODAMA ET AL.

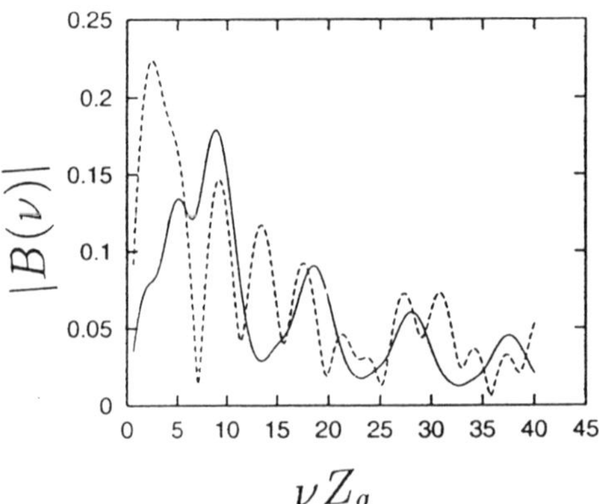

Figure 2. Fourier component $B(\nu)$ versus νZ_a. $Z_a = 2.45$, $M = 3$, $\Gamma = 0.7$, for equal spacing in Z' (solid line) and equal spacing in Z (broken line).

where $Z'_m = Z'(Z_m)$. The above expression for $B(\nu)$ is valid for arbitrary Z_m, and D_m as given by Equation (8). From Equation (19), we see that resonance due to periodic amplifications occurs at $\nu(\xi)Z_a = n\pi$, for $n = 1, 2, \cdots$, and the resonance is *partially* suppressed by the dispersion management described by $|B(\nu)|$.

We analyse two types of dispersion management schemes: (1) equal spacing in Z, i.e. $Z_m = mZ_a/M$ and (2) equal spacing in Z'. For the latter case, the length Z_m and dispersion D_m of the mth section of the fiber are so chosen that

$$Z'(Z_{m+1}) - Z'(Z_m) = Z_a/M, \quad m = 0, 1, \ldots M - 1. \tag{21}$$

Using Equation (21), Z_m and D_m are found to be

$$Z_m = -\frac{1}{2\Gamma}\ln\left[1 - \frac{m\{1 - \exp(-2\Gamma Z_a)\}}{M}\right], \tag{22}$$

$$D_m = \frac{Z_a}{M(Z_{m+1} - Z_m)}. \tag{23}$$

The Fourier component $B(\nu)$ is plotted in Figure 2 for both the cases with $M = 3$. As can be seen, the main resonance at $\nu = \pi/Z_a$ is suppressed in both the cases. In particular in the case of equal spacing in Z', we see the further suppression of main resonance and the peak of $B(\nu)$ shifts to the higher resonance at $\nu = M\pi/Z_a$.

From Equations (18),(19) and (20), for large M radiation energy for the case of equal spacing in Z' is given by

$$W_r(L) \approx \left(\frac{1}{M}\right)^4 (I_1 + I_2L), \tag{24}$$

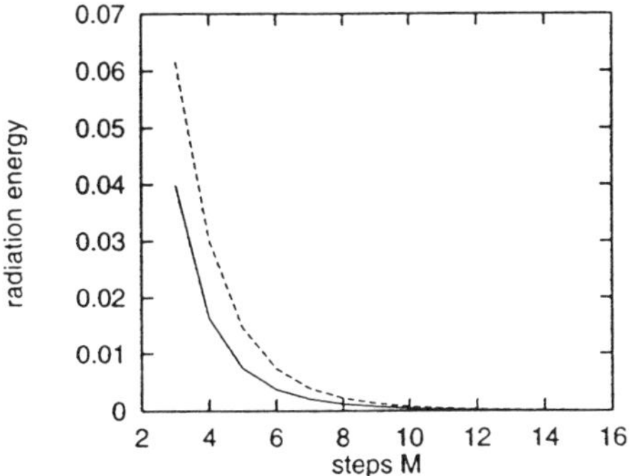

Figure 3.   Plots of radiation energy $W_r(L)$ versus steps M, $Z_a = 2.45$, $(Z_0 = 1)$, $\Gamma = 0.7$, for equal spacing in Z' (solid line) and equal spacing in Z (broken line).

where I_1 and I_2 are given for large L by

$$I_1 \approx \frac{\pi Z_a^4}{72} \int_0^{\xi_r(Z_a)} \frac{\nu^4(\xi)\mathrm{sech}^2(\pi\xi/\eta)}{\sin^2[\nu(\xi)Z_a]} |J(\nu)|^2 \, d\xi, \tag{25}$$

$$I_2 \approx \frac{\pi^6 a^4(0)}{144(\Gamma Z_a)^2} \frac{\mathrm{sech}^2[\pi\xi_r(Z_a)/\eta]}{\xi_r(Z_a)} |J(\pi/Z_a)|^2, \tag{26}$$

with

$$J(\nu) = \int_{\exp(-2\Gamma Z_a)}^1 \frac{\exp[i\nu a^2(0)p/\Gamma]}{p} \, dp, \tag{27}$$

and

$$\xi_r(Z_a) = \sqrt{\frac{\pi}{Z_a} - \frac{\eta^2}{4}}. \tag{28}$$

Equation (24) shows the improvement factor of $O(M^{-4})$. Here I_2 is evaluated only at the first resonant point $\xi = \xi_r(Z_a)$, assuming $Z_a < 4\pi/\eta^2$, which avoids the one-soliton resonance [2]. Note that I_2 may be negligible for small Z_a. $Z_a \ll Z_0(= 1)$ implies the guiding center regime, and it makes the resonant point ξ_r to move away from the soliton spectrum. In this case $(I_2 \approx 0)$, the radiation energy is bound and does not increase as the number of amplifiers.

Figure 3 shows the energy of radiation as a function of number of steps M, obtained by Equation (18), and shows M^{-4} dependence for large M as given by Equation (24). As can be seen, energy emitted as radiation in the case of equal spacing in Z', is less compared to the case of equal spacing in Z. This is due to the fact that the suppression of the first resonance is better in the case of equal spacing in Z'.

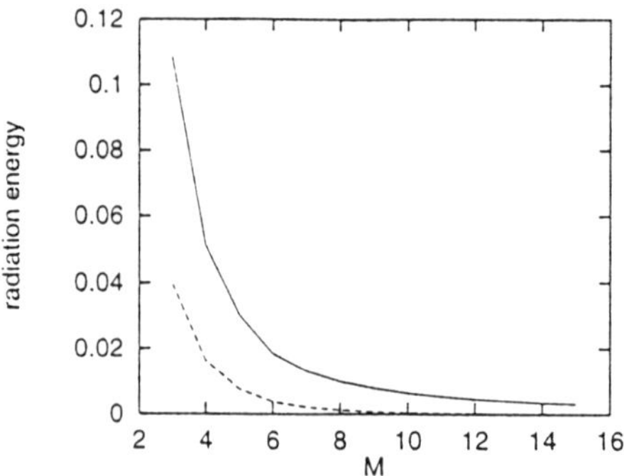

Figure 4.   Radiation energy $W_r(L)$ versus M, for comb-type (solid line) and stepwise (broken) dispersion profiles with equal spacing in Z', $Z_a = 2.45$, $\Gamma = 0.7$ and $D_\epsilon = 0.01$.

From the practical implementation point of view, it is advantageous to use comb-type dispersion profile, as given by

$$
\begin{aligned}
d(Z) &= D_\epsilon \quad \text{for} \ \ Z_m < Z < Z_{m+1} - \delta_m \\
&= D_c \quad \text{for} \ \ Z_{m+1} - \delta_m < Z < Z_{m+1}.
\end{aligned}
\tag{29}
$$

Here, D_ϵ and D_c are so chosen that,

$$
(\Delta_m - \delta_m)D_\epsilon + D_c \delta_m = Z_a/M.
\tag{30}
$$

As $D_\epsilon \to 0$ and $\delta_m \to 0$ with $D_c \delta_m \to Z_a/M$, the Fourier component $B(\nu)$ is given by

$$
\begin{aligned}
|B(\nu)| &= \frac{a^2(0)}{\Gamma Z_a} \left| \sum_{m=0}^{M-1} e^{2i\nu m Z_a/M} \sinh(\Gamma \Delta_m) \exp[-\Gamma(Z_{m+1} + Z_m)] \right. \\
&\quad \times \left. \left[1 - \frac{\sin(\nu Z_a/M)}{\nu Z_a/M} e^{i\nu Z_a/M} \right] \right|.
\end{aligned}
\tag{31}
$$

Figure 4 shows that comb-type dispersion management is worse than stepwise dispersion management. However, comb-type scheme is easy to install and may be upgradable by choosing a proper design of parameters, D_ϵ, D_c and δ_m.

5. Wavelength Division Multiplexing

We now consider the frequency shift $\Delta K(Z')$ of a soliton in one channel induced by a collision with another channel with a frequency separation

ΔB. The frequency shift due to a collision may be obtained using the adiabatic perturbation technique [14]

$$\frac{d\Delta K}{dZ'} = -a_M^2(Z) \int_{-\infty}^{\infty} \operatorname{sech}^2\left(T - \frac{T_0(Z')}{2}\right) \frac{\partial}{\partial T}\operatorname{sech}^2\left(T + \frac{T_0(Z')}{2}\right) dT. \tag{32}$$

With $T_0(Z') \approx -\Delta B Z'$, from Equation (32) total frequency shift at $Z = \infty$ (not including a shift due to the initial overlap), $\Delta K(\infty)$ is found to be

$$\begin{aligned}
\Delta K(\infty) &= \frac{1}{\Delta B} \int_{-\infty}^{\infty} dZ' \, [a_M^2(Z) - 1] \\
&\times \frac{d}{dZ'} \int_{-\infty}^{\infty} \operatorname{sech}^2\left(T + \frac{\Delta B}{2}Z'\right) \operatorname{sech}^2\left(T - \frac{\Delta B}{2}Z'\right) dT.
\end{aligned} \tag{33}$$

By writing $a_M^2(Z) - 1$ in Fourier series, i.e.

$$a_M^2(Z) - 1 = \sum_{n=-\infty}^{\infty} B_n e^{i2\pi n Z'/Z_a}, \tag{34}$$

we obtain

$$\Delta K(\infty) = \operatorname{Im}\left\{ \frac{16\pi}{Z_a(\Delta B)^2} \sum_{n=1}^{\infty} n B_n \left[\frac{\pi^2 n/Z_a \Delta B}{\sinh(\pi^2 n/Z_a \Delta B)} \right]^2 \right\}, \tag{35}$$

where $B_n = B(-\pi n/Z_a)$ defined in Equation (20).

Note here that if we choose $d(Z) = a_1^2(Z)$, B_n vanishes and therefore, $\Delta K(\infty) = 0$ (ideal dispersion management).

We have numerically evaluated the total frequency shift $\Delta K(\infty)$ from Equation (35), for various values of the number of steps M between amplifiers for the case where the fiber is divided into equal lengths in Z (the real distance) and that in Z' (where Z_l is given by Equation (22)) for the choise of $Z_a = 2.45$, $\Gamma = 0.185$ and $\Delta B = 5$. Figure 5 shows the result. We see that if the fiber sections are of equal length in Z', for M larger than 6, $\Delta K(\infty)$ is practically reduced to zero.

We have carried out the numerical simulation with the following parameters of fiber: wavelength $= 1.56$ μm, pulse width $= 5$ ps, loss rate 2γ $= 0.0461$ km^{-1} (0.2 dB/km), dispersion $k'' = 1$ ps^2/km, which corresponds to the dispersion distance $z_0 = 8.16$ km, $A_{eff} = 25$ μm^2, nonlinear coefficient $n_2 = 3.18 \times 10^{-16}$cm^2/W, the amplifier spacing $z_a = 20$ km. The channel separation is 0.281 THz. We have neglected the amplifier noise and higher order terms in the nonlinear Schrödinger equation, by assuming that their effects can be reduced to a low level by using filters. Figure 6 shows

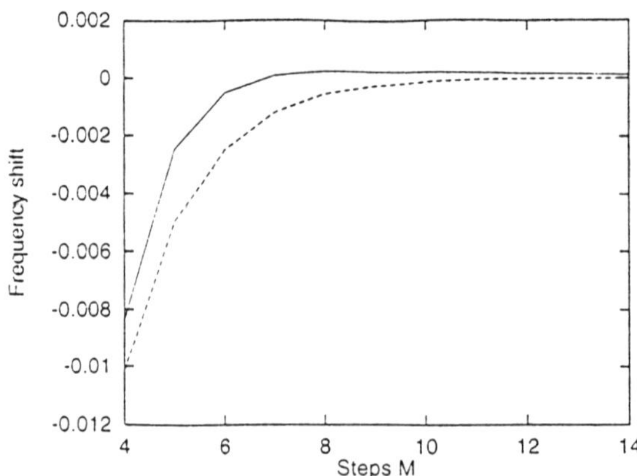

Figure 5. Frequency shift $\Delta K(\infty)$ versus steps M for the case of fiber sections equally spaced in Z' (solid line) and that in Z (broken line).

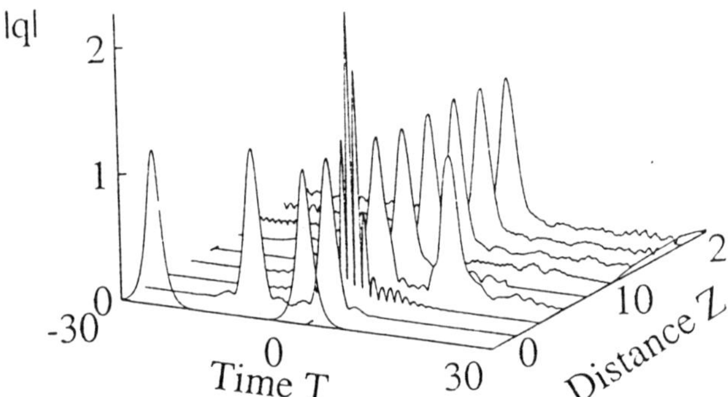

Figure 6. Absolute amplitude $|q|$ as a function of distance Z and time T with no dispersion management.

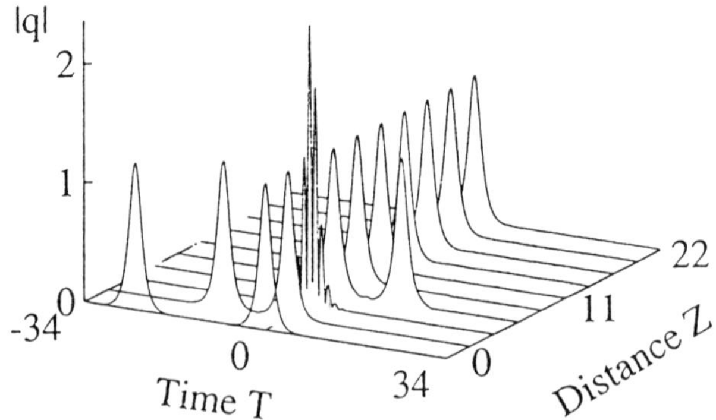

Figure 7. Same as Figure 6, but with dispersion management. Fiber sections are equally spaced in Z', $M = 4$.

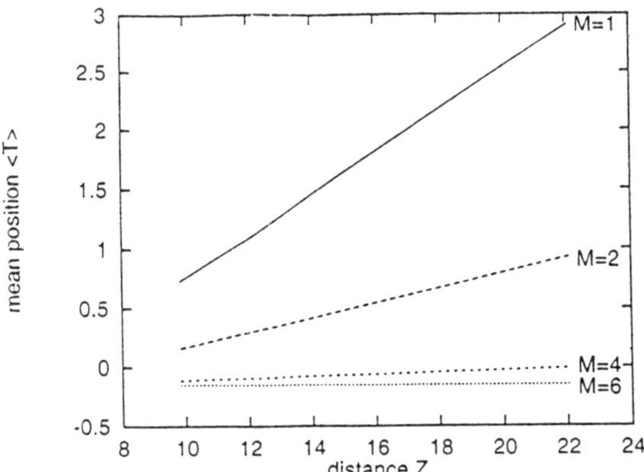

Figure 8.   Mean position of the faster soliton versus distance for different choice of M. Fiber sections are equally spaced in Z'.

the soliton collision without dispersion management. Due to periodic amplifications, dispersive waves are emitted by solitons. In addition, collision induced shift of soliton's mean position can be seen. Figure 7 shows the soliton collision with 4 fiber sections between amplifiers with fiber lengths and dispersion as given by Equations (22) and (23). As can be seen, both radiation and collision induced temporal shift of soliton are reduced. The mean position of the faster soliton is evaluated and plotted in Figure 8. We see that $M \geq 4$ is sufficient in reducing the effect of collisions within a tolerable level.

6. Conclusions

We have shown that with appropriate choice of the stepwise or comb-type dispersion profiles, adiabatic soliton transmission is possible even with lumped amplifiers. This result applies both for nonadiabatic (guiding center soliton) and adiabatic (high bit rate) regimes. We have also shown that with the dispersion management scheme, the frequency shift due to collisions in amplifiers can be reduced even when the amplifier spacing is comparable to or longer than the collision distance. We note that the present dispersion management scheme involves only passive elements and therefore, easier to implement than a fiber with distributed gain. If the dispersion management scheme and the amplifier spacing is designed for a particular pulse width and channel separation, the same transmission link may be used over a wide range of pulse widths and channel separation.

7. Acknowledgements

The authors appreciate the suggestion of comb-type dispersion management by Dr.Toda and Mr.Sawada. S. Kumar is supported by a fellowship from the Ministry of Education, Science, Sports and Culture while the collaboration between Y. Kodama and A. Hasegawa is supported by International Science Research Program - Joint Research of the Ministry of Education, Science, Sports and Culture.

References

1. Hasegawa, A. and Kodama, Y.: Guiding center soliton in optical fibers, *Opt. Lett.* **15** (1990), 1443-1445.
2. Hasegawa, A. and Kodama, Y.: Guiding center soliton, *Phys. Rev. Lett.* **66** (1991), 161-164.
3. Nakazawa, M., Suzuki, K., Kubota, H., Yamada, E. and Kimura, Y.: Dynamic optical soliton communication, *IEEE J. Quantum Electron.* **26** (1990), 2095-2102.
4. Mollenauer, L. F., Evangelides, S. G. and Haus, H. A.: Long distance soliton propagation using lumped amplifiers, *J. Lightwave Tech.* **9** (1991), 194-197.
5. Kodama, Y. and Hasegawa, A.: Amplification and reshaping of optical solitons in glass fiber-II, *Opt. Lett.* **7** (1982), 339-341.
6. Simpson, J. R., Shang, H. T., Mollenauer, L. F., Olsson, N. A., Becker, P. C., Kranz, K. S., Lemaire, P. J. and Neubelt, M. J.: Performance of a distributed erbium-doped dispersion shifted fiber amplifier, *IEEE J. Lightwave Tech.* **9** (1991), 228-233.
7. Nakazawa, M., Kimura, Y. and Suzuki, K.: Gain-distribution measurements along an ultra long erbium-doped fiber amplifier using optical time-domain reflectometry, *Opt. Lett.* **15** (1990), 1200-1202.
8. Forysiak, W., Knox, F. M. and Doran, N. J.: Average soliton propagation in periodically amplified systems with stepwise dispersion profiled fiber, *Opt. Lett.* **19** (1994), 174-176.
9. Forysiak, W., Knox, F. M. and Doran, N. J.: Stepwise dispersion profiling of periodically amplified soliton systems *J. Lightwave Technol.* **12** (1994), 1330-1335.
10. Mollenauer, L. F., Evangelides, S. G. and Gordon, J. P.: Wavelength division multiplexing with solitons in ultra long distance transmission using lumped amplifiers, *J. Lightwave Tech.* **9** (1991), 362-367.
11. Hasegawa, A. and Kodama Y.: Guiding-center soliton in fibers with periodically varying dispersion, *Opt. Lett.* **16** (1991), 1385-1387.
12. Stentz, A. J., Boyd, R. W. and Evans, A. F.: Dramatically improved transmission of ultrashort solitons through 40 km of dispersion-decreasing fiber, *Opt. Lett.* **20** (1995), 1770-1772.
13. Kaup, D. J. and Newell, A. C.: Solitons as particles, oscillators, and in slowly changing media: a singular perturbation theory, *Proc. R. Soc. London Ser. A* **361** (1978), 413-446.
14. Hasegawa, A. and Kodama, Y.: *Solitons in Optical Communications*, Oxford University Press, Oxford.

A DISPERSION-ALLOCATED SOLITON AND ITS IMPACT ON SOLITON COMMUNICATION

HIROKAZU KUBOTA
Senior Research Engineer
NTT Access Network System Laboratories
Tokai-mura, Ibarakiken, 319-11 Japan
MASATAKA NAKAZAWA
Group Leader
NTT Access Network System Laboratories
Tokai-mura, Ibarakiken, 319-11 Japan

Abstract. The dispersion allocation technique is very effective for realizing soliton transmission system. This technique made it possible to undertake a soliton communication field trial very easily using the conventional fiber cable already installed for commercial systems. We have succeeded in a field demonstration of soliton transmission at 20 Gb/s over 2,000 km and 10 Gbit/s over 2,500 km using part of the Tokyo metropolitan optical loop network.

1. Introduction

The advent of erbium-doped fiber amplifiers (EDFAs) has led to numerous interesting reports on high-speed soliton transmission [1]. Here EDFAs are used as lumped amplifiers, with which the concept of the average soliton, or dynamic soliton has been developed [2]-[6]. This technique has a wide range of applicability from L_a (amplifier spacing) $\ll Z_{sp}$ (soliton period) to $L_a \sim Z_{sp}$, depending on the maximum required transmission distance. It is also well known that bright solitons can withstand dispersion fluctuation within the anomalous GVD region when the soliton period Z_{sp} is much longer than the dispersion change-over period [4].

It has long been believed that a bright optical soliton will only propagate in an anomalous GVD region. We have reported that a bright soliton can exist and

A. Hasegawa (ed.), Physics and Applications of Optical Solitons in Fibres '95, 27–36.

propagate in a normal GVD region as long as the average dispersion D_{ave} throughout the amplifier spacing is properly set so that it is anomalous [7, 8].

In this paper we show detailed analyses of the dispersion-allocated soliton and point out that a small change in the initial soliton amplitude can control how the soliton stably propagates throughout a fiber over long distances.

2. Dispersion-allocated Soliton

Here we consider a sech-type bright soliton. The nonlinear Schrödinger equation with a position dependent dispersion, $\beta(\xi)$, is given by

$$(-i)\frac{\partial u}{\partial \xi} = \frac{1}{2}\beta(\xi)\frac{\partial^2 u}{\partial \tau^2} + |u|^2\, u.$$

(1)

Assuming that u can be described in the form

$$u = U(\xi,\tau)e^{if(\xi,\tau)}$$

(2)

and that $U(x, t)$ and f satisfy the following equations.

$$U(\xi,\tau) = 2\eta\, \text{sech}(2\eta\tau)\,,$$

(3)

$$\frac{\partial f}{\partial \tau} = 0,$$

(4)

one obtains

$$U\frac{\partial f}{\partial \xi} = \frac{1}{2}\beta(\xi)\frac{\partial^2 U}{\partial \tau^2} + U^3,$$

(5)

Thus, the total nonlinear phase change for a propagation disrtance Z is given by

$$f(Z) = \int_0^Z (\frac{\partial f}{\partial \xi})d\xi$$

$$= 2\eta^2\int_0^Z \beta(\xi)d\xi + (2\eta)^2\text{sech}^2(2\eta\tau)\left[Z - \int_0^Z \beta(\xi)d\xi\right].$$

(6)

In order for u to behave as a soliton, f should be time independent and therefore the coefficient of $\text{sech}^2(2\eta\,\tau)$ should be zero.

$$\frac{1}{Z}\int_0^Z \beta(\xi)d\xi = 1.$$

(7)

Which means that stable soliton exists by defining D_{ave} as :

$$D_{ave} = \frac{1}{L}\int_0^L D(z)dz \;.$$

(8)

It is permissible to have a random change in the GVD value between normal and anomalous as long as D_{ave} is anomalous.

Each fiber section is allowed to have large normal or anomalous dispersion as long as D_{ave} is slightly anomalous over the soliton period. The average soliton period, $Z_{sp(ave)}$, is expressed as

$$Z_{sp(ave)} = 0.322\,\frac{\pi^2 c}{\lambda^2}\,\frac{\tau^2}{D_{ave}},$$

(9)

where t is the input pulse width. $Z_{sp(ave)}$ can be extended much further than L_a, even though we use highly normal or highly anomalous GVD fibers which usually result in short soliton periods in each section. By choosing a small D_{ave}, L_a can be easily extended to a distance comparable to that in a conventional IM/DD system.

For a pulse width of τ, the soliton peak power in the present system is given by

$$P_{N=1(ave)} = 0.776\,\frac{\lambda^3}{\pi^2 cn_2}\,\frac{D_{ave}}{\tau^2}\,A_{eff}.$$

(10)

When the loss is included, the dispersion-allocated soliton amplitude at the input, U_0, is the same as that of an average soliton. A significant difference between the present soliton and a conventional average soliton is that it can propagate even in a normal GVD region. Here we analyze how much perturbation can be allowed for stable soliton propagation. Assuming that D_i is much larger than D_{ave}, the dispersion average soliton behaves just like a dispersive wave in each region. Thus, the pulse broadening $\Delta\tau$ $(=\tau_{out} - \tau_0)$ due to D_i is

$$\frac{\Delta\tau}{\tau_0} = \frac{1}{2}\frac{(k_i'' l_i)^2}{\tau_0^4} = \frac{1}{2}\left(\frac{D_i l_i}{D_{ave} Z_0}\right)^2.$$

(11)

An important requirement for the present soliton is that this broadening should be much less than the original pulse width. Thus, a requirement of

$$\left(\frac{D_i l_i}{D_{ave} Z_0}\right) < 1$$

(12)

is obtained. This is quite reasonable because if the broadening is not negligible to the original pulse width, the perturbation to the soliton becomes large and eventually the soliton pulse becomes a mere dispersive wave. A soliton is stable when the product of dispersion and fiber length is smaller than $D_{ave} Z_{0(ave)}$. Here $Z_{0(ave)}$ is the normalized distance which has a relationship of $Z_{sp(ave)} = (\pi/2)Z_{0(ave)}$.

Figure 1 shows how the pulse width of this new soliton changes as it propagates down a dispersion allocated fiber. Here L_a is 90 km and • indicates amplifier position. The fiber consists of a 30 km fiber with a normal GVD of +2.0 ps/km/nm and a 60 km fiber with an anomalous GVD of -1.3 ps/km/nm, giving an average anomalous GVD of -0.2 ps/km/nm. The soliton width is 20 ps. The typical feature of the present soliton is that the pulse width changes alternately following the sign of the GVD, but it eventually becomes a steady-state pulse due to the nonlinear phase change over the average soliton period. As the average dispersion is set at D_{ave} = - 0.2 ps/km/nm, without the nonlinearity, the pulse simply disperses out. However, the pulse width remains near its original value when the input peak intensity is set at that of a conventional average soliton.

We have already shown that a stable soliton exists in the present dispersion-allocated technique.[7,8] There was no distortion in the waveform when D_{ave} was negative (anomalous). When D_{ave} was made positive (normal) with D_{ave} = +0.2 ps/km/nm, nonsoliton transmission occured where the pulse was distorted and eventually became rectangular. This pulse was the positively chirped nonlinear pulse used for pulse compression with a grating pair [9]. We also investigated change in the pulse width as a function of propagation distance. When the dispersion perturbation was large, the pulse width of the present soliton deviated slightly from that of a uniform average soliton as it propagated, but it was entirely different from a linear pulse. When a small perturbation was applied to the system, stable soliton pulse propagation was achieved.

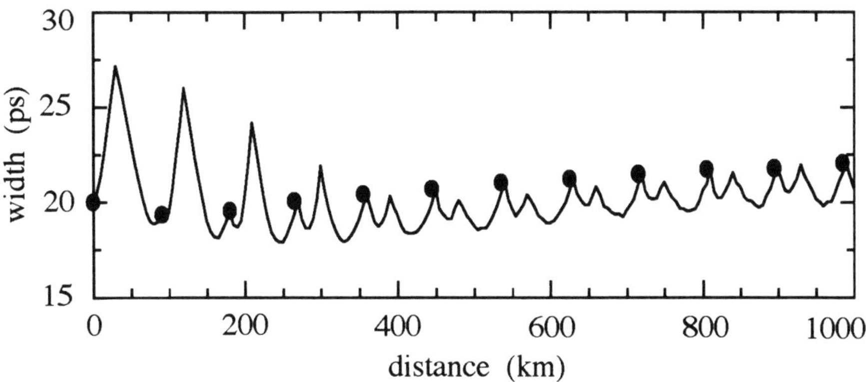

Figure 1.   Change in the pulse width of a dispersion-allocated soliton as a function of propagation distance. The filled circles show the pulse width at each amplifier position.

As the dispersion-allocated soliton power is small, it does not undergo a large nonlinear phase change, but it behaves like a linear pulse such that the pulse broadening due to normal GVD can be compensated for by anomalous GVD. However, the nonlinear phase change is gradually accumulated during this process and eventually balances with the average dispersion. Thus, the dispersion-allocated soliton has the nature of a pure soliton over $Z_{sp(ave)}$.

It is important to note that the nonlinear phase rotation in each section is slightly modified because the dispersive wave decreases its amplitude during propagation. Therefore, the initial amplitude U_0 is modified to a slightly bigger value $U_0 A$ to compensate for the decrease in the nonlinear phase shift of the average soliton. Here A is a correction factor which is close to unity. For simplicity, we use a lossless Gaussian pulse and neglecting the noncommuting nature of nonlinearity and dispersion. Let the intensity of a Gaussian pulse at $t = 0$ be I_G. In the propagation of the pulse $0 < z < l_{\mathrm{p}}$, the average intensity $\overline{I_{Gp}}$ is given by

$$\overline{I_{Gp}} = \frac{1}{l_p}\int_0^{l_p}\left\{1 - \frac{1}{2}\left(\frac{4k_p'' z}{\tau_0}\right)^2\right\}dz$$

$$= 1 - \frac{8}{3}\left(\frac{k_p'' l_p}{\tau_0^2}\right)^2 .$$

$$(13)$$

As the sign of the dispersion changes in the next fiber section, the slightly broadened soliton in the first stage recovers to its original intensity. It should be noted here that the average dispersion ($k_P^{''} l_P + k_n^{''} l_n$) does not have to be zero in order to retain the soliton property, but must be smaller than $k_P^{''} l_P$. Thus one obtains $\overline{I_{Gn}} \cong \overline{I_{Gp}}$. That is, the correction factor A is

$$A \cong 1 + \frac{4}{3}\left(\frac{D_p l_p}{D_{ave} Z_0}\right)^2 \quad \text{for} \quad \left(\frac{D_p l_p}{D_{ave} Z_0}\right) < 1. \tag{14}$$

Figure 2 shows how the change in the pulse width is modified when the correction factor A changes. L_a is 90 km, and the fiber consists of a 30 km fiber with a GVD of + 0.5 ps/km/nm (normal) and a 60 km fiber with a GVD of -0.4 ps/km/nm (anomalous). The correction factor A in this case is calculated to be $A = 1.03$ from Equation (14). When A is equal to unity, a pulse broadening of 1 ps is observed. However, when A is increased to 1.02-1.05, this broadening is suppressed, and a more uniform propagation is realized. A slight pulse compression occurs in the early stage of the transmission, which is due to the strong accumulation of nonlinear phase change. An over correction using a factor of $A = 1.1$ considerably reduces the transmission quality.

These results prove that new soliton can exist in a dispersion-allocated fiber system. This offers us the enormous advantage of being able to construct an actual soliton transmission line using various fiber combinations.

3. Field Demonstration of Soliton Transmission Using the Dispersion-allocated Soliton Technique

We have shown that as long as the average GVD is in the anomalous region, a soliton can propagates even in the fibers with normal GVD. This dispersion-allocation technique made it possible to undertake a soliton communication field trial very easily using the conventional fiber cable already installed for commercial systems.

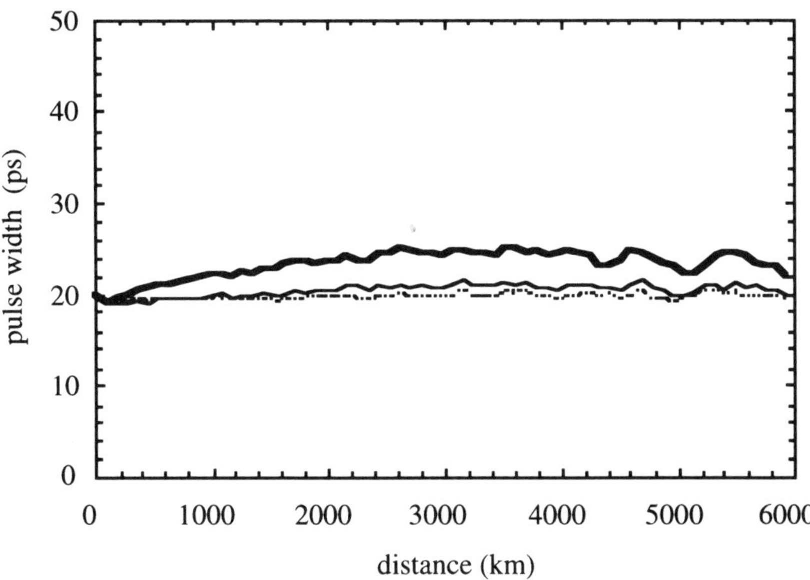

Figure 2. Change in the soliton pulse width vs. propagation distance with a different correction factor A. The 90 km repeater span consists of a 30 km fiber with + 0.5 ps/km/nm (normal) and a 60 km fiber with -0.4 ps/km/nm (anomalous).

A conventional fiber cable usually consists of many fiber segments with different GVDs. However, by properly setting the signal wavelength, a conventional fiber cable is capable of handling soliton communication. With this technique, we have succeeded in transmitting 10-20 Gb/s soliton signals over 2,000 km in the Tokyo metropolitan optical network [10]. The location of the field test is shown in Figure 3(a). A total of four repeater terminals were used which were located at Mito, Utsunomiya, Sano and Maebashi. The distances between the Mito and Utsunomiya terminals (MU), Utsunomiya and Sano terminals (US) and Sano and Maebashi terminals (SM) were 70 km, 50 km and 50 km, respectively. The transmitter and receiver were placed at the Mito terminal and the total transmission distance was extended by repeatedly cascading these terminal spans as shown in Figure 3(b). The average fiber losses of the MU, US and SM sections were 18. 5 dB, 14.3 dB and 14. 3 dB, respectively.

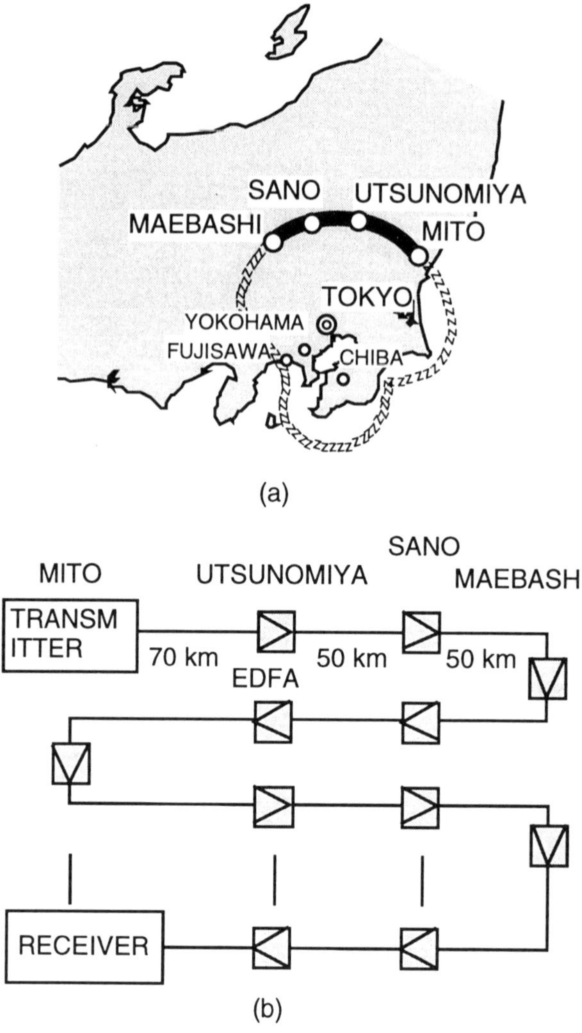

Figure 3. Experimental field transmission of solitons at 10 and 20 Gb/s using part of the Tokyo metropolitan optical loop network with a radius of 100 km.   (a) Location of field test   (b) Soliton transmission route

The optical soliton source for the field test was a regeneratively mode-locked fiber laser operating at 10 GHz with a pulse width of 10.3 ps and a spectral width of 0.27 nm. The pulse was modulated at 10 Gb/s with a 2^{23}-1 pseudorandom binary sequence using a LiNbO3 modulator and was multiplexed in the time domain to a 20 Gb/s signal using 3 dB fiber couplers. The EDFAs consisted of a 1.48 μm InGaAsP LD for pumping, a WDM coupler, an isolator, erbium fiber, a 3 nm optical filter and an APC circuit.

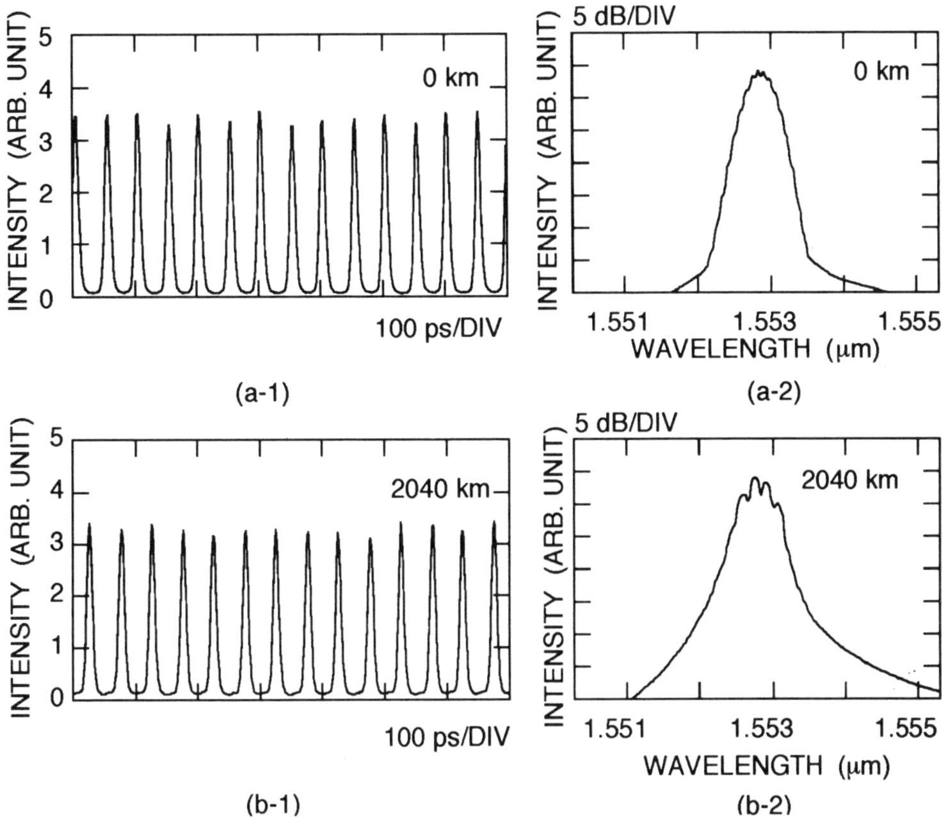

Figure 4 20 Gb/s PRBS input and output waveforms and corresponding spectra (a-1)Input psuedo random soliton waveform, (a-2) Spectrum, (b-1) Output psuedo random soliton waveform, (b-2) Spectrum.

The input and output waveforms (2,000 km) of a pseudorandom soliton pulse and their spectra are shown in Figure 4. (a-1) and (a-2) are the input waveform and its spectrum and (b-1) and (b-2) are the output waveform and its spectrum. It is surprising to note that a 10 ps input pulse at 20 Gb/s remains unchanged even after a transmission of 2,000 km through the installed conventional dispersion-shifted fiber cable.

The 2:1 demultiplexing of the 20 Gb/s transmitted signal to 10 Gb/s signals (CH. 1 and CH. 2) was achieved by using a polarization-insensitive electro absorption (EA) modulator. The EA modulator was a strained multi quantum well type and the fiber to fiber insertion loss was approximately 10 dB [11].

The dispersion allocation technique is very effective even for soliton transmission. This technique made it possible to undertake a soliton communication field trial very easily using the conventional fiber cable already installed for commercial 2.4 Gb/s IM/DD systems. We have succeeded in a field demonstration of soliton transmission at 20 Gb/s over 2,000 km using part of the Tokyo metropolitan optical loop network with a radius of 100 km. Further extension of the error-free transmission distance will be possible through proper dispersion allocation which can be realized by adding dispersion compensation fibers.

4. Acknowledgment

The authors express their sincere thanks to K. Ishihara and Y. Koyamada of NTT Access Network Systems Laboratories for their encouragement.

References

1. Nakazawa, M, Kimura, Y. and Suzuki, K.; Soliton amplification and transmission with Er^{3+}-doped fibre repeater pumped by GaInAsP laser diode, *Electron Lett.* **25** (1989), 199-200.
2. Kubota, H and Nakazawa, M.; Long-distance optical soliton transmission with lumped amplifiers, *IEEE, Quantum Electron.* **26** (1990), 692-700.
3. Nakazawa, M., Suzuki, K. and Kimura, Y.; 3.2-5 Gbit/s, 100 km error-free soliton transmissionwith erbium amplifiers and repeaters, *IEEE Photon Tech. Lett.* **2** (1990), 216-219.
4. Hasegawa A. and Kodama, Y.; Guiding-center soliton, *Phys. Rev. Lett.* **66** (1991), 161-164.
5. Mollenauer, L. F., Evangelides, S. G. and Haus, H. A.; Long distance soliton propagation using lumped amplifiers and dispersion-shifted fiber, *IEEE J. Lightwave Tech.* **9** (1991), 194-197.
6. Blow, K. J. and Doran, N. J.; Acerage soliton dynamics and operation of soliton systems with lumped amplifiers, *IEEE Photo. Tech. Lett.* **3** (1991), 369-371.
7. Nakazawa, M. and Kubota, H.; Optical soliton communication in a positively and negatively dispersion-allocated optical fiber transmission line, *Electron Lett.* **31** (1995), .216-217.
8. Nakazawa, M. and Kubota, H.; Analyses of the dispersion-allocated bright and dark solitons, *Jpn. J. Appl. Phys.* **34** (1995), L889-L891.
9. Tomlinson, W. J., Stolen, R. H. and Shank, C. V.; Compression of optical pulses chirped by self-phase modulation in fibers, *J. Opt. Soc. Am. B.* **2** (1984), 139-149.
10. Nakazawa, M., Kimura, Y., Suzuki, K., Kubota, H., Komukai, T., Yamada, E., Sugawa, T., Yoshida, E., Yamamoto, T., Imai, T., Sahara, A., Nakazawa, H., Yamauchi, O. and Umezawa, M.; Field demonstration of soliton transmission at 10 Gbit/s over 2000 km in Tokyo metropolitan optical loop network, *Electron Lett.* **31** (1995), .992-994.
11. Kondo, S., Wakita, K., Noguchi, Y and Yoshimoto, N. (May 1995) *IPRM'95*, pp.61-64, Japan.

NRZ SOLITON TRANSMISSION SCHEME

M. HAELTERMAN AND M. BADOLO
Service d'Optique et Acoustique
Université Libre de Bruxelles
50, av. F.D. Roosevelt, CP 194/5 B-1050 Brussels, Belgium

AND

A. P. SHEPPARD
Optical Science Centre
Australian National University
Canberra, Australia

Abstract. A new type of optical solitary wave is shown to exist in Kerr media. These solitary waves consist of localized structures separating uniform domains of orthogonal modes of the electromagnetic field and can be viewed as optical domain walls. The fundamental nature of these new solitary waves is revealed by their link with intrinsic instabilities of light propagation in Kerr media. We suggest application of the new solitary waves to optical fiber communications. We show that their topology is adapted to the non-return-to-zero modulation format and that their fundamental features offer significant potential advantages with respect to the usual return-to-zero NLS soliton transmission scheme.

1. Introduction

The use of optical solitons in fiber communications was suggested more than 20 years ago as a natural mean to avoid the detrimental effects of fiber chromatic dispersion and Kerr nonlinearity [1]. Solitons in fibers are bright or dark pulses in which dispersion and nonlinearity balance in such a way that their shape is maintained as they propagate through the fiber links. They thus appear as being ideal information carriers for data transmissions based on the return-to-zero (RZ) modulation format [1]. However, soliton propagation in actual fiber links is subject to several detrimental

37

A. Hasegawa (ed.), Physics and Applications of Optical Solitons in Fibres '95, 37–52.

effects which limit the transmission capacity. The main limiting factors in soliton-based transmission systems are soliton interactions and the Gordon-Haus time jitter. These effects are inherent to the fundamental dynamical properties of the nonlinear Schrödinger (NLS) solitons. As a consequence, the bit-rates obtained up to now with soliton-based systems are not higher than those achieved with the conventional non-return-to-zero (NRZ) transmission scheme [2, 3]. From this viewpoint it appears natural to address the question of the existence of other types of solitary waves which would not have the fundamental drawbacks of NLS solitons and could therefore lead in a natural way to the possibility of increasing the fiber links capacity.

The aim of the present communication is to show the existence of such a new solitary wave in single-mode optical fibers and to discuss its potential advantages as regards applications to optical communications. The paper is organized as follows: In Section 2 we discuss the phenomenon of cross-phase modulation (CPM) inherent to the simultaneous propagation of two waves belonging to different optical modes. We show that CPM is responsible for the existence of a dark-type compound solitary wave characterized by a domain-wall structure. In Section 3 we discuss the fundamental nature of this new domain-wall solitary wave by considering its links with intrinsic instabilities of light propagation in Kerr media. In Section 4 we consider application of the optical domain walls to fiber communications. We discuss the condition of existence of the domain walls in single-mode fibers. Finally we analyze the results of preliminary numerical studies of a new NRZ transmission scheme based on domain-wall solitons. Section 5 is devoted to our conclusions. The theory presented in this communication has been the object of several preliminary reports [4]-[7].

2. Optical Domain Walls

In this section we consider the basic properties of light propagation in dispersive Kerr media in the presence of two orthogonal modes of the electromagnetic field. We first briefly discuss the physical conditions in which CPM occurs between such two modes. Taking the example of orthogonal polarization components, we show by means of simple mathematics that CPM is responsible for the existence of domain-wall solitons. Note that for the sake of simplicity, we use here the colloquial term *soliton* for what should be more rigorously called a stationary solitary wave.

2.1. CROSS-PHASE MODULATION

Cross-phase modulation is a phenomenon of common occurrence in non-linear optics. It is characteristic of the simultaneous propagation of several waves belonging to different optical modes. It refers to each mode modulating the phase of the other. From a mathematical point of view it is represented by incoherent coupling terms between the NLS equations describing the nonlinear propagation of both modes. For instance, in bulk isotropic Kerr media CPM occurs between waves of orthogonal polarizations provided that nonlinear birefringence takes place. In this case the dimensionless equations governing propagation of the two modes are of the following form [8].

$$iU_z - \sigma\frac{1}{2}U_{tt} + \frac{1}{2}[(1 - B)|U|^2 + (1 + B)|V|^2]U = 0,$$

$$iV_z - \sigma\frac{1}{2}V_{tt} + \frac{1}{2}[(1 - B)|V|^2 + (1 + B)|U|^2]V = 0, \qquad (1)$$

where $U(t, z)$ and $V(t, z)$ are the envelopes of the circular polarization components, B is the nonlinear birefringence coefficient related to the Kerr susceptibility tensor by $B = \chi^{(3)}_{xyyx}/\chi^{(3)}_{xxxx}$ [9] and σ is the sign of chromatic dispersion. Since the early work of Berkhoer and Zakharov, these equations were shown to exhibit much richer dynamics than the single NLS equation [10]. These authors showed, in particular, that CPM leads to an extension of the phenomenon of modulational instability to normally dispersive fibers. Since then many studies have been reported on the dynamics of the coupled NLS equations. Special attention has been paid to the soliton dynamics in these equations. Some studies have shown that the nonlinear coupling between the two optical modes can lead to the formation of bound pairs of solitons belonging to the two modes. These bound pairs of solitons have been called vector solitons. Stationary bound pairs of orthogonally polarized bright and dark solitons were shown to propagate in birefringent Kerr media [11]-[13]. We have shown that such solutions exist in isotropic Kerr media in the form of rotating polarized solitary waves [14, 15]. We shall see in the following section that CPM can also originate bound pairs of semi-infinite kink waves in normally dispersive Kerr media.

CPM, as described by Equations (1), does not only occur between circularly polarized waves. It can take place between linearly polarized waves in highly birefringent media [8], between waves of different frequencies [8], between orthogonal modes of a nonlinear waveguide and between counter-propagating waves [16]. The theory of the following section is developed from Equations (1) for circularly polarized waves in dispersive isotropic Kerr media. In principle, it can however be applied to all physical conditions involving CPM. In section 4 we apply the theory to dual frequency light signals in single-mode fibers.

2.2. DOMAIN-WALL SOLITONS

Let us now consider the soliton solutions of the coupled NLS Equations (1). These equations are relatively complex, they are non-integrable and the search for their solutions requires elaborate techniques. In order to provide here a simple picture of the problem we consider numerical solutions in association with an analogy to mechanics.

The soliton solutions of Equations (1) have the form $U = u(t) \exp(i\beta z)$, $V = v(t) \exp(i\beta z)$. The envelopes u, v and the propagation constant β are real quantities. Substituting these expressions into Equations (1) yields a set of two coupled ordinary differential equations.

$$
\begin{aligned}
u'' &= -2\sigma\beta u + \sigma(1 - B)u^3 + \sigma(1 + B)v^2 u, \\
v'' &= -2\sigma\beta v + \sigma(1 - B)v^3 + \sigma(1 + B)u^2 v,
\end{aligned}
\tag{2}
$$

where the primes denote time-derivatives. It proves convenient to analyze this set of equations as being the equation of motion of a unit mass in the plane (u, v) in a 2-D conservative potential V. This potential is given by

$$
V = \sigma\beta(u^2 + v^2) - \sigma\frac{1}{4}(u^2 + v^2)^2 + \sigma\frac{B}{4}(u^2 - v^2)^2.
\tag{3}
$$

If we restrict the motion of the unit mass to the u-axis, V reduces to the 1-D potential $V = \sigma\beta u^2 - \sigma(1 - B)u^4/4$. The separatrices of this potential represent the usual soliton solutions of the scalar NLS equation. Considering anomalous dispersion ($\sigma = -1$), the potential exhibits a minimum surrounded by two zeroes. One of them being located at the origin $u = 0$, which is also a maximum of V, and the other at $u = [(1 - B)/2\beta]^{1/2}$. In the mechanical analog, the separatrix trajectory which starts and ends at the origin, passing through the minimum, represents the bright soliton solution of the scalar NLS. It is easy to show in this case that the amplitude of the motion of the unit mass follows an evolution in $sech(t)$. Considering the two dimensions of the problem ($u \neq 0, v \neq 0$) in the anomalous dispersion regime, the potential V exhibits four minima on the u- and v-axes separated by saddle points on the bissecting lines ($u = \pm v$). Straight line separatrices along the axes and the bissecting lines represent the circularly and linearly polarized bright NLS solitons, respectively. We have studied more elaborate separatrix trajectories of this potential in our work on soliton interactions [14, 15]. We have shown in particular that separatrices exist which represent bound states of orthogonally polarized NLS solitons. In such bound states the interaction force between adjacent solitons is suppressed [14].

Here we restrict our analysis to the normal dispersion regime which leads to more intriguing features. As can be seen in Equation (3), when $\sigma = +1$, the 1-D potential along the u-axis exhibits two maxima on each

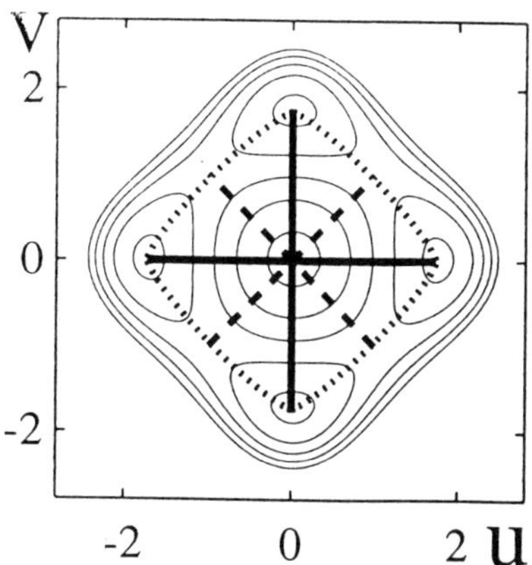

Figure 1. Contour plot and separatrices of the potential V in the (u,v) plane for $B = 1/3$ and $\beta = 1$.

side of the origin which is now a minimum of zero potential energy. The separatrix trajectory which links these maxima, passing through the origin, represents the usual dark soliton solution of the scalar NLS equation. It is easy to show in this case that the amplitude of motion of the unit mass follows an evolution in $\tanh(t)$. The full, two-dimensional potential V is shown in Figure 1 for $\beta = 1$ in the case where $B = 1/3$ corresponding to the nonlinearity of silica. Note that Equations (2) can be normalized with respect to the propagation constant β. For the sake of simplicity we can therefore set $\beta = 1$ without loss of generality. The potential V exhibits four maxima on the u- and v-axes and four saddle points on the bisecting lines ($u = \pm v$). These fixed points represent stationary cw solutions of Equations (1). The separatrices which link pairs of opposite maxima and pairs of opposite saddle points represent circularly and linearly polarized dark NLS solitons, respectively.

At a first glance, we see in Figure 1 that the potential also possesses separatrix trajectories which link adjacent maxima. They are represented by dotted lines in Figure 1, they are calculated by means of a standard shooting technique. These separatrices correspond to a new type of solitary-waves. Since they link pairs of adjacent maxima they represent localized structures of the field separating semi-infinite cw domains of orthogonal eigenpolarizations. This is illustrated in Figure 2 where we plotted the envelopes $u(t)$ and $v(t)$ corresponding to the separatrix of the first quadrant. Figure 2

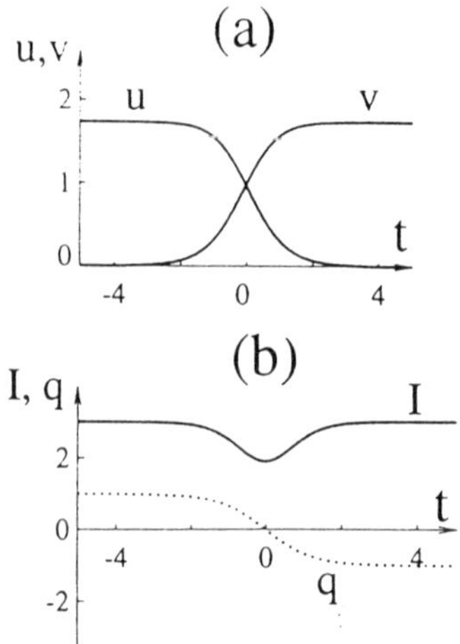

Figure 2.   (a) Envelopes $u(t)$ and $v(t)$ of the domain-wall soliton. (b) Its intensity $I(t)$ and ellipticity $q(t)$ profiles.

also shows the intensity and ellipticity profiles of the new solitary wave. The values $q = \pm 1$ represent opposite circular polarizations while $q = 0$ represents linear polarization. The new wave can be seen as a dark soliton where the phase variations are simply replaced by ellipticity variations of the polarization state. By analogy with the theory of ferromagnetism, the new soliton solution can be interpreted as a polarization domain wall (the atomic spin being replaced here by the polarization state of the electromagnetic field). Notice that, contrary to linearly polarized waves, cw circularly polarized waves are stable in isotropic normally dispersive Kerr media. This suggests that the domain-wall soliton is a stable structure of the field. We have checked the stability of the domain wall by direct numerical integration of Equations (1). Our results have confirmed the robust nature of the new soliton.

3. CPM-Induced Instabilities and Domain-Wall Solitons

In this section we briefly discuss some fundamental aspects of the optical domain-wall soliton. We show the links which exist between the soliton and the intrinsic instabilities of light propagation in Kerr media. These links reveal the fundamental nature of the new soliton.

3.1. FAST-AXIS INSTABILITY AND POLARIZATION DOMAIN WALLS

It is easy to generalize the above theory to birefringent Kerr materials. Assuming low birefringence, one can neglect the difference in group velocity of both linear polarization components. In this approximation the equations governing propagation of the circularly polarized components in normally dispersive Kerr media are [17].

$$iU_z - \frac{1}{2}U_{tt} + \frac{1}{2}[(1 - B)|U|^2 + (1 + B)\,V|^2]U + \Delta V \;=\; 0,$$

$$iV_z - \frac{1}{2}V_{tt} + \frac{1}{2}[(1 - B)|V|^2 + (1 + B)\,U|^2]V + \Delta U \;=\; 0, \qquad (4)$$

where Δ is the scaled linear birefringence: $\Delta = \Delta n/n_2$, where $\Delta n = n_y - n_x$, and n_2 is the nonlinear index coefficient. It is well known that the nonlinear birefringence leads to an instability of the linear polarization state oriented along the fast axis of the birefringent medium [18, 19]. This instability relates to the concepts of spontaneous symmetry breaking, multistability and nonlinear eigenpolarization [20]. More precisely, when the intensity of the light wave reaches a threshold value, the linear polarization switches, due to arbitrarily small fluctuations, to right- or left-handed polarization states. This instability is associated with the existence of two new stable elliptic polarization states which are mutually symmetric with respect to the fast axis and constitute two eigenpolarizations of the birefringent Kerr medium. This phenomenon is illustrated in Figure 3 where we plotted the ellipticity of the polarization state as a function of the optical power P. We see that above the critical power $P_c = \Delta/B$, two elliptic polarization states appear while the linear polarization becomes unstable. The new states are called eigenpolarization in the sense that they constitute stationary states of the polarization of light propagating in the birefringent Kerr material. Note that when the Kerr material is isotropic $(\Delta = 0)$, the critical power becomes zero and the eigenpolarizations become the circular polarization components $(q = \pm 1)$ for which the theory of section 2 applies.

This description of fast axis instability does not include dispersion. It can therefore not account for possible fast temporal variations of the field. Taking into account time t and dispersion, one can address the question of the simultaneous formation of domains of different eigenpolarizations. If such domains were formed they would be separated by localized structures of the field in which the polarization state switches from one eigenpolarization to the other, as suggested by the arrow in Figure 3. We shall see in the following that such localized structures exist and constitute a generalization to birefringent media of the domain-wall soliton presented above.

Looking for soliton solutions of Equations (4) in the form

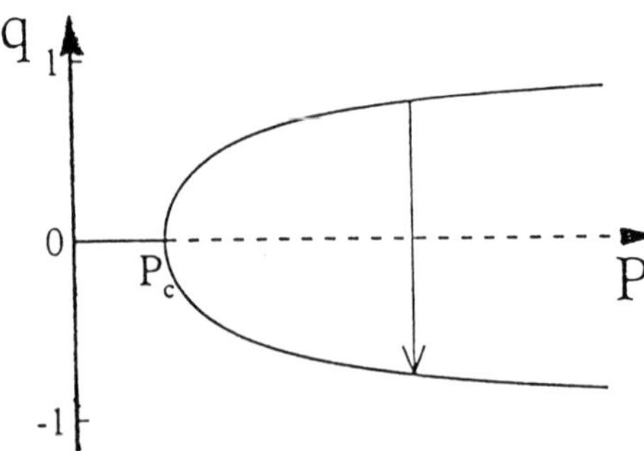

Figure 3. Illustration of the fast axis instability in birefringent Kerr media. The curves show the ellipticity degree q of the nonlinear eigenpolarizations as a function of the optical power P. The dotted line shows the unstable linear eigenpolarization. The arrow suggests the existence of localized structures separating domains of opposite elliptic eigenpolarizations.

$$U = u(t)\exp(i\beta z), \; V = v(t)\exp(i\beta z)$$

and using the mechanical analog picture leads to the following expression of the potential

$$V = \beta(u^2 + v^2) - \frac{1}{4}(u^2 + v^2)^2 + \frac{B}{4}(u^2 - v^2)^2 + 2\Delta uv. \qquad (5)$$

The contour lines of this potential are shown in Figure 4 for $B = 1/3$ and different values of Δ. The scaling is chosen here so that the x-polarized eigenpolarization has a power equal to unity, that is, $\beta = 1 - \Delta$ so that the saddle point of the first quadrant is located in $(1/\sqrt{2}, 1/\sqrt{2})$. With this scaling an increase of the power is represented by a decrease of Δ and the fast axis instability condition reads $\Delta < B$.

As shown in Figure 4, when Δ increases from zero, the maxima of the potential move away from the u- and v-axes. This change in position represents the fact that the eigenpolarizations of a birefringent Kerr material are elliptic. When Δ is increased further (i.e., in practice, when the power decreases) the maxima merge with the saddle points. The merging is obtained for $\Delta = B$, which naturally corresponds to the fast axis instability threshold $P = \Delta/B$ or, in other terms, to the bifurcation point of Figure 3. We see in this way that the existence of elliptic eigenpolarizations is always associated with the presence of saddle points between the four maxima of the potential. It is therefore always possible to find separatrix trajectories between adjacent maxima. These separatrices are shown by

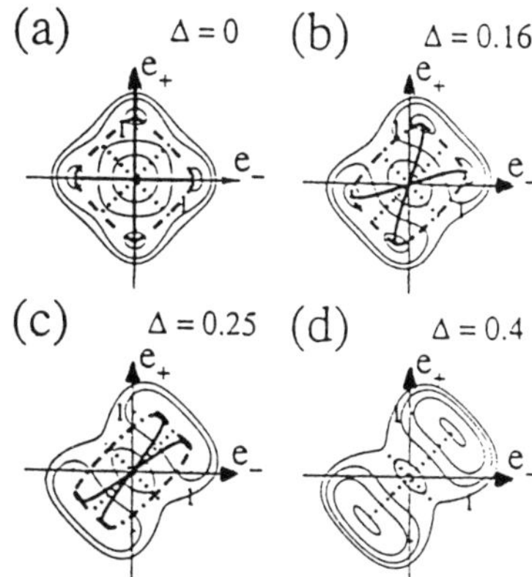

Figure 4. Contour lines of the normalized potential V given in Equations (4) for $B = 1/3$, $\beta = 1 - \Delta$, and different values of the birefringence parameter Δ. The dotted lines are the separatrices corresponding to the linearly polarized dark solitons. The solid lines correspond to a generalization of the circularly polarized dark NLS solitons to birefringent media, the so-called "elliptically polarized dark solitons". The dashed and dot-dashed lines are the separatrices of the domain-wall solitons.

dot-dashed lines in Figure 4. The corresponding solitary waves consist of localized structures separating two semi-infinite uniform domains of different nonlinear eigenpolarizations. They are the optical domain-wall solitons of birefringent Kerr media.

3.2. EXTENDED MODULATION INSTABILITY AND DOMAIN WALLS

In this section we deal with the phenomenon of polarization modulational instability (PMI) in the normal dispersion regime. We show that the domain-wall soliton is associated with PMI in the normal dispersion regime in the same way as the bright NLS soliton is associated with scalar modulational instability. A detailed analysis of this problem has been given in Ref. [5]. We only give here a shortened and simplified description of the theoretical developments.

Since the pioneering work of Berkhoer and Zakharov [10] it is well known that CPM leads to an extension of the phenomenon of modulational instability to the normal dispersion regime. This can be easily verified from a linear stability analysis of Equations (1). Let us consider the linearly po-

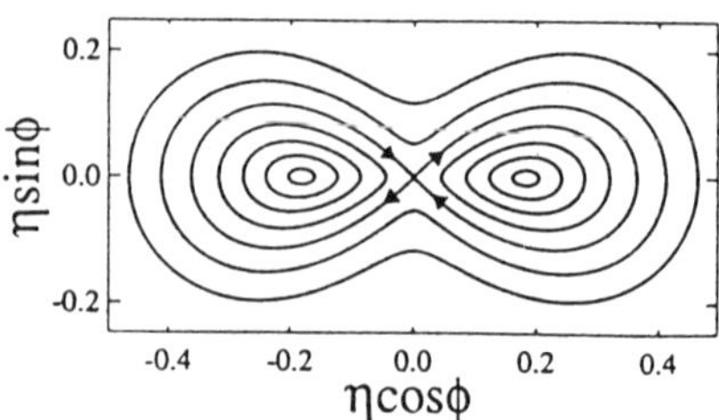

Figure 5. Representation of the dynamics of PMI in the plane (η, ϕ). The origin represents the cw linearly polarized wave. The arrows show its stable and unstable directions. This figure reveals the existence of stationary time-periodic solutions associated with PMI.

larized cw solutions of Equations (1): $U = V = A\exp(iA^2 z)$. The stability of this solution against the growth of periodic perturbation is examined by introducing the ansatz $U, V = (A + e_\pm \exp(iA^2 z))$ where $e_\pm$ are first-order perturbations of the form $e_\pm = \epsilon_\pm \exp(\lambda z)\cos(\Omega t)$. Among the four eigenvalues λ, only one is potentially unstable in the normal dispersion regime $(\sigma = +1)$, it reads $\lambda = \Omega[BA^2 - \Omega^2/4]^{1/2}$. The corresponding eigenvector has the form $v = (\epsilon_+, \epsilon_-) = (c, -c)$, where c is a complex number. This indicates that PMI induces the exponential growth of periodic perturbations of opposite sign in the two circular polarization components. In other words, PMI is responsible for the formation of two π out-of-phase periodic patterns in U and V.

The dynamics of these periodic waves in the nonlinear stage of evolution of PMI can be described in a good approximation by means of a truncated Fourier expansion of the field envelopes. The waves U and V then take the form $U, V = U_0(z) \pm \sqrt{2}U_1(z)\cos(\Omega t)$. Substituting these expressions into Equations (1) one obtains two coupled ordinary differential equations in U_0 and U_1. These equations can be reduced further by introducing the normalized sideband power $\eta = |U_1|^2/(|U_0|^2 + |U_1|^2)$ and the relative phase $\phi = Arg(U_0) - Arg(U_1)$. Figure 5 shows the dynamics of PMI in the plane (η, ϕ) in polar coordinates. The origin, which is an unstable hyperbolic fixed point, represents the initial unstable linearly polarized cw field. The corresponding homoclinic orbit surrounds two elliptic fixed points which reveal the existence of stationary time-periodic solutions of the coupled NLS Equations (1).

These solutions are equivalent to the cnoidal waves of the scalar NLS equation. We calculated numerically these peculiar solutions by means of

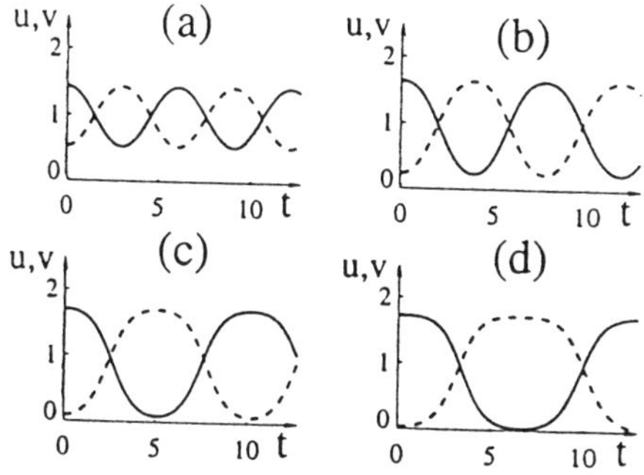

Figure 6. Envelopes of the stationary periodic solutions of Equations (1) with $B = 1/3$ obtained by numerical integration. The amplitudes a and periods T of the "cnoidal" waves are: (a) a = 1.43, T = 6.2; (b) a = 1.62, T = 7.8; (c) a = 1.69, T = 11; and (d) a = 1.73, T = 13.

the above mentioned shooting technique. They are shown in Figure 6 for different amplitudes. We verify that, as their amplitude increases, their period increases without bond. Figure 6(d) shows that, in this limit, the "cnoidal" wave consists of an alternation between quasi-cw domains of orthogonal polarization states. Each domain appears to be separated from its neigbours by localized structures in which the polarization switches from one circular polarization to the other. Naturally, these structures are nothing but the domain-wall solitons studied in Section 2. This analysis clearly shows that the domain-wall soliton is the soliton associated with polarization modulational instability in the normal dispersion regime.

4. Color Domain Walls: Application to Optical Data Transmissions

Polarization domain-wall solitons can in principle propagate in single-mode fibers and could be considered for applications to data transmissions. However, in practice, optical fibers exhibit residual random birefringence which destroys the polarization domain walls [21]. The use of these new solitons would require considerable and probably unrealistic progress in the design and manufacturing of low birefringence fibers. In view of this difficulty and considering the universality of the CPM process, we investigated other physical situations in which domain-wall solitons can exist. We consider here the use of two fields of different frequencies instead of two orthogonal polarization components.

4.1. COLOR DOMAIN WALLS IN W-TYPE FIBERS

In Section 2 we have shown that the coupled NLS equations admit domain-wall soliton solutions in the normal dispersion regime. Similar equations also describes simultaneous propagation of two fields of different frequencies. In this case the ratio between CPM and SPM is equal to 2, which corresponds to $B = 1/3$ as considered above. Due to group velociy dispersion, however, one must take into account the difference between the group velocities of both frequencies. This leads to an additional term in Equations (1) which therefore no longer possess domain-wall soliton solutions. To overcome this difficulty we consider here, among other solutions, the use of especially designed W-type single-mode fibers.

W-type fibers, characterized by a depressed inner cladding, have been developed for telecommunication purposes to get shifted and flattened dispersion. As shown in Figure 7 their dispersion characteristic exhibits two zero-dispersion wavelengths located near the absorption windows of Silica. As a result, the group velocity characteristic is S-shaped and possesses two extrema at the zero-dispersion wavelengths. This is illustrated in Figure 7(b) where it is easy to see that it is possible to find, in a given range, pairs of wavelengths corresponding to identical group velocities and located in the normal dispersion regions. If such a choice of wavelengths pair is made, the propagation of the two fields is ruled by coupled NLS equations similar to Equations (1) and domain-wall solitons are solutions of the problem. We call them "Color Domain-Wall" solitons. The numerical example of Figure 7 has been obtained with the following fiber parameters: core radius $a = 3.4\mu$m, inner cladding radius $a' = 4.42\mu$m, 13.5% GeO_2 mole concentration in the core, 9.1% P_2O_5 mole concentration in the cladding and pure Silica in the inner cladding [7].

This way of obtaining CPM with equal group velocities in not unique. One can easily imagine other physical systems which exhibit this feature. For instance, the one could use fiber gratings to compensate for the group velocity difference in standard single-mode fibers. Linear polarization components in highly birefringent polarization-preserving fibers are subject to CPM [8]. Equal group velocities could be obtained in that case by adjusting the frequencies of both components.

4.2. NRZ-SOLITON TRANSMISSION SCHEME

Let us now consider application of the dual frequency domain walls to optical data transmissions. The topology of the domain-wall solitons is adapted to NRZ transmission scheme which offers the advantage of a narrower modulation bandwidth as compared to the usual RZ soliton scheme. Information

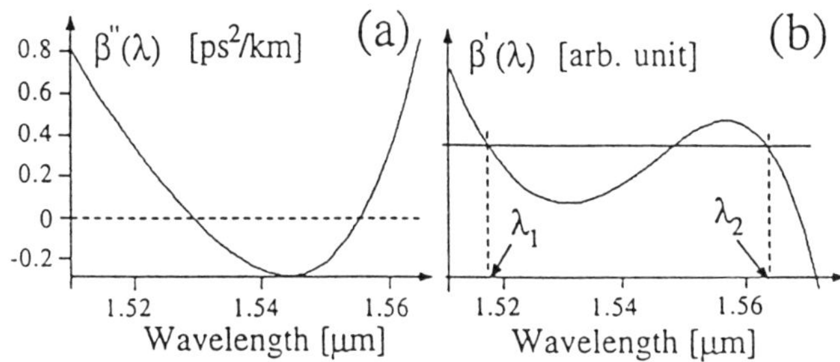

Figure 7.   Wavelength dependence of (a) the dispersion coefficient β'' and (b) the inverse group velocity β' for the fiber considered in the text. The wavelengths λ_1 and λ_2 correspond to identical group velocities.

would be coded by associating one wavelength with bits 0 and the other wavelength with bits 1. By using the domain-wall soliton to perform the transitions between 0's and 1's, one obtains distortion-free transmission in the NRZ format. As we shall see, potential advantages of this dual frequency NRZ soliton scheme are the absence of mutual interactions between solitons as well as the absence of timing jitter due to amplified spontaneous emission.

Figure 8 shows two adjacent domain-wall solitons before and after propagation in a 6,000 km long fiber link. Their separation is of 50 ps which corresponds to a 20 Gb/s repetition rate. The fiber parameters are those of Figure 7. The fiber loss of 2 dB/km is assumed to be compensated exactly by a distributed gain. The amplified spontaneous emission noise has been introduced following the procedure of Ref.[22]. No filtering is applied. The Gordon-Haus time jitter for bright NLS solitons with the same parameters would be of 6 ps [23]. The soliton width being of 10 ps, as required for a 20 Gb/s repetition rate, this value is excessive and implies the use of pulse control techniques along the line. As illustrated in Figure 8(b), which is representative of all our simulations, we verified that such timing jitter does not exist with color domain walls. The uncertainty on the position of the domain walls is only due to the fast erratic modulations of the noisy field envelopes and no significant jitter can be measured.

We also verified that domain walls exhibit no mutual interactions. In other words, the distance between adjacent domain walls does not vary during propagation. This property as well as the absence of Gordon-Haus effect suggest that the use of domain-wall solitons, instead of NLS solitons, could lead to a significant increase of the fiber links capacity. These remarkable stability properties come from energy conservation laws of the coupled

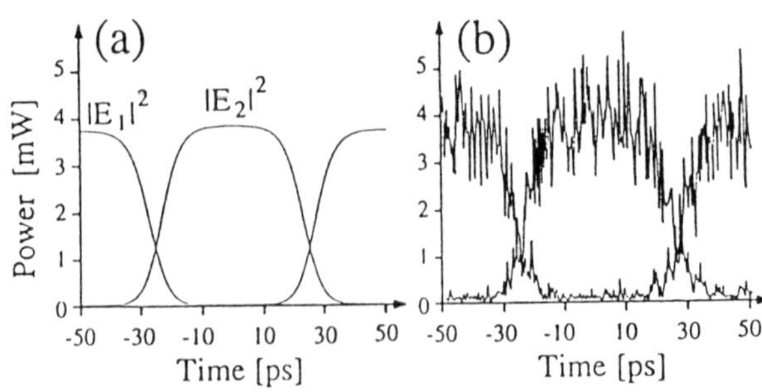

Figure 8.   (a) Power profiles of two adjacent color domain walls given in real units corresponding to the fiber of Figure 7(b) The same domain-wall solitons after propagation over 6000 Km in an amplified fiber link.

NLS equations. This can be easily explained from Figure 7(a) where we see that the distance between the domain walls is determined by the total energy in the central field E_2. Since this energy is a conserved quantity of the coupled NLS equations, the domain wall separation remains unchanged during propagation. Somehow, the two fields which compose the train of solitons lock each other. This simple analysis shows that, unlike NLS solitons, domain-wall solitons are stable as regard their position in time.

Of course, in practice, this property is only verified if the two fields are equally amplified. Amplification of two fields of different frequencies can be performed with rare earth doped fiber amplifiers provided that the two frequencies are sufficiently close to one another. Our numerical example above shows that this is in principle possible. The pair of wavelengths, as shown in Figure 7, are inside the Erbium gain bandwidth. However, in order to keep constant positive dispersions and equal group velocities in a W-type fiber with such a characteristic, the tolerance on the variations of the fiber parameters must be set very small. To avoid this technical difficulty, other amplification techniques could be used such as the simultaneous use of Erbium and Praseodymium amplifiers or the use of Raman amplifiers. These techniques allow for the amplification of well separated wavelengths for which the tolerance on the parameters of the fiber would be much less severe.

5. Conclusions

On the basis of simple mathematical treatments, we have shown the existence of a new type of optical soliton in Kerr media: the domain-wall soliton. This soliton consists of a nonlinear dispersive localized structure of the field separating domains of orthogonal modes of the electromagnetic field. The fundamental nature of the new soliton has been revealed by its links with intrinsic instabilities of light propagation in Kerr media. We have shown that, under certain conditions, single-mode fibers can propagate dual frequency domain-wall solitons. The properties of these solitons appear to be well suited to NRZ-type optical data transmissions This new transmission scheme offers several advantages with respect to the usual RZ-NLS soliton scheme such as the absence of both mutual interaction and Gordon-Haus jitter. Naturally, the difficulty in generating and amplifying coded dual frequency domain wall trains, the requisite fiber manufacturing accuracy as well as the inaptitude for wavelength multiplexing constitute drawbacks with respect to NLS soliton-based systems. However, in view of the novel remarkable features of the domain-wall solitons, we believe that the efforts required to overcome these difficulties deserve consideration.

6. Acknowledgments

This work was partially supported by the Belgian National Funds for Scientific Research and the Attraction Pole Program of the Belgian Government (grant P3-047).

References

1. Hasegawa, A.: *Optical Solitons in Fibers*, Springer-Verlag, Berlin, (1989).
2. Bergano, N. S. et al.: *in Conference on Optical Fiber Communications*, OSA Technical Digest (Optical Society of America, Washington D. C., 1995), postdeadline paper PD19.
3. Nyman, B. M. et al.: *in Conference on Optical Fiber Communications*, OSA Technical Digest (Optical Society of America, Washington D. C., 1995), Postdeadline paper PD21.
4. Haelterman, M. and Sheppard, A. P.: *Opt. Lett* **19** (1994), 96.
5. Haelterman, M. and Sheppard, A. P.: *Phys. Rev* **E49** (1994), 3389.
6. Haelterman, M.: *Electron. Lett* **31** (1995), 741.
7. Haelterman, M. and Badolo, M.: *Opt. Lett.* (1995), 2285.
8. Agrawal, G. P.: *Nonlinear Fiber Optics*, Academic Press, New-York, (1995) 243-252.
9. Maker, P. D. and Terhune, R. W.: *Phys. Rev* **A137** (1965), 801.
10. Berkhoer, A. L. and Zakharov, V. E.: *Sov. Phys.* **JETP 31** (1987), 486.
11. Christodoulides, D. N. and Joseph, R. I.: *Opt. Lett.* **13** (1988), 53.
12. Tratnik, M. V. and Sipe, J. E.: *Phys. Rev.* **A38** (1988), 2011.
13. Kivshar, Yu. S. and Turitsyn, S. K.: *Opt. Lett.* **18** (1993), 337.
14. Haelterman, M., Sheppard, A. P. and Snyder, A. W.: *Opt. Lett.* **18** (1993), 1406.

15. Haelterman, M. and Sheppard, A. P.: *Phys. Lett.* **A194** (1994), 191.
16. Haelterman, M., Sheppard, A. P. and Snyder, A. W.: *Opt. Commun.* **103** (1993), 145.
17. Wabnitz, S.: *Phys. Rev.* **A38** (1988), 2018.
18. Daino, B., Gregori, G. and Wabnitz, S.: *J. Appl. Phys.* **58** (1985), 4512.
19. Winful, H. G.: *Opt. Lett.* **11** (1986), 33.
20. Zhedulev, N. I.: *Sov. Phys. Usp.* **32** (1989), 357.
21. Haelterman, M.: *Electron. Lett.* **30** (1994), 1510.
22. Hamaide, J. P., Emplit, Ph. and Haelterman, M.: *Opt. Lett.* **16** (1991), 1578.
23. Gordon, J. P. and Haus, H. A.: *Opt. Lett.* **11** (1986), 665.

PERTURBED SOLITONS IN BIREFRINGENT FIBRES

J. N. ELGIN
Department of Mathematics
Imperial College
London SW7 2BZ, United Kingdom

Abstract. A Perturbation Theory based on Inverse Scattering Theory is described, which finds useful application to the study of pulse propagation down a birefringent optical fibre. The Theory is described, then applied to the problem of the generation of soliton shadow's.

1. Introduction

When an ultrashort pulse propagates down an anomalously dispersive birefringent optical fibre, complex features develop which require explanation. The object of this article is to outline a perturbation theory based on inverse scattering theory which admits useful application to the study of many of these observed features. The theory is an extension of one previously published [1] which has application to the isotropic (i.e. non-birefringent) case. After describing the method, it is then applied in Section 3 to the problem of the generation of soliton shadow's.

The governing equations for pulse propagation down an anomalously dispersive birefringent fibre are

$$iq_{1x} + i\delta q_{1t} + i\beta q_2 - q_{1tt} - 2q_1 \left(|q_1|^2 + (1-\sigma)|q_2|^2 \right) = iF_1,$$

$$iq_{2x} - i\delta q_{2t} - i\beta q_1 - q_{2tt} - 2q_2 \left(|q_2|^2 + (1-\sigma)|q_1|^2 \right) = iF_2. \tag{1}$$

Here, q_1 and q_2 are the (complex) pulse envelopes in each polarization mode, a suffix denotes a partial derivative, and all quantities appear in normalised form. The parameter δ is one-half the group velocity difference for pulses in each mode, β is the birefringence parameter which allows for a twisting of the polarization axis with distance down the fibre, and $(1-\sigma)$

53

A. Hasegawa (ed.), Physics and Applications of Optical Solitons in Fibres '95, 53–58.
© *1996 Kluwer Academic Publishers. Printed in the Netherlands.*

is the cross-phase modulation coefficient. The remaining terms on the left hand side of Equations (1) have their usual meanings: that is, q_{itt} and $q_i|q_i|^2$, with $i = 1, 2$ in turn, are terms which describe the effects of second-order dispersion and self-phase modulation, respectively. The remaining terms F_1 and F_2 on the right hand sides correspond to appropriate choices of further perturbations, such as third-order dispersion, the soliton self-frequency shift, gain or loss in the fibre, and so on. Later, we will rearrange Equations (1) so that the (perturbing) terms containing the parameters δ, β and σ are taken over to the right hand side and thereafter considered as special choices of F. When δ, β, F_1 and F_2 are set to zero, Equations (1) are integrable whenever σ takes one of two values: (i) $\sigma = 1$, which trivially uncouples the system leaving two independent integrable equations. This case is not pursued here. (ii) $\sigma = 0$, this case was first studied by Manakov [2], who showed that the coupled system was integrable. In particular, it has the exact single vector-soliton solution

$$\mathbf{q}(x, t) = q_s(x, t) \begin{bmatrix} \cos\theta \\ \sin\theta \end{bmatrix}$$

with

$$q_s = 2\eta_1 e^{-2i\xi_1 t + 4i(\xi_1^2 - \eta_1^2)x} \mathrm{sech} 2\eta_1 (t - 4\xi_1 x). \tag{2}$$

The parameters ξ_1 and η_1 characterise the soliton, and θ is the projection angle of the pulse onto each polarization mode. The addition of perturbations, such as permitting non-zero values for any of the constants δ, σ or β, or some appropriate choice for F_i, modify the vector soliton solution in three ways:

i) The soliton parameters, which are constants of the motion in the unperturbed case, may now vary with propagation distance x down the fibre.

ii) A background radiation field will be generated. Not only will this have many of the features typically found in the scalar case - such as resonances [3]-[5] - but also some additional features not found there. In particular, it is known that a soliton in one polarization mode can cast a "shadow" in the complementary mode [6]. Moreover, this in turn can profoundly affect the collision properties of two solitons in orthogonal polarization modes, making such collisions inelastic [7]. The shadow is a non-dispersive localised pulse of radiation which travels with the parent soliton, unlike the usual dispersive radiation modes. The main objective of this article is to outline a theory which can usefully be applied to a study of the generation of such shadows.

iii) Possible fission of the vector soliton into two daughter pulses, one in each polarization mode. This can happen when the parameters δ and σ become suitably large [7]

2. Perturbation Theory

A perturbation theory appropriate to the study of Equations (1) is now briefly described; full details are presented elsewhere [8],[9]. Equations (1) are first rearranged so that terms containing the parameters δ, β and σ are incorporated into F, leaving an integrable form on the left hand side. The scattering problem associated with the integrable vector nonlinear Schrö dinger equation (NLS) is [2]

$$\begin{aligned}
u_t + i\zeta u &= q_1 v + q_2 w, \\
v_t - i\zeta v &= -q_1^* u, \\
w_t - i\zeta w &= -q_2^* u
\end{aligned} \tag{3}$$

where $*$ denotes complex conjugate, $\zeta = \xi + i\eta$ is a complex eigen-parameter, and $[u, v, w]$ are components of a three-component spinor. An adjoint scattering problem is also required; this is again defined by Equations (3) on replacing ζ with $-\zeta$, q_i with q_i^* ($i = 1, 2$) and $[u, v, w]$ with $[\hat{u}, \hat{v}, \hat{w}]$. Unlike the scalar case, there is no simple relationship between the adjoint and the direct scattering problems. Special linearly independent sets of solutions - the Jost functions - are now defined such that

$$\phi^{(l)} \sim \begin{bmatrix} \delta_{l1} e^{-i\zeta t} \\ \delta_{l2} e^{i\zeta t} \\ \delta_{l3} e^{i\zeta t} \end{bmatrix}, t \to -\infty; \quad \psi^{(l)} \sim \begin{bmatrix} \delta_{l1} e^{-i\zeta t} \\ \delta_{l2} e^{i\zeta t} \\ \delta_{l3} e^{i\zeta t} \end{bmatrix}, t \to +\infty, \tag{4}$$

where δ_{lm} is the Kroenecker delta. Corresponding definitions hold for the Jost function solutions $\hat{\phi}^{(1)}$, $\hat{\phi}^{(2)}$, $\hat{\phi}^{(3)}$ and $\hat{\psi}^{(1)}$, $\hat{\psi}^{(2)}$, $\hat{\psi}^{(3)}$ to the adjoint problem, so that $\hat{\psi}^{(3)} \sim [0, 0, e^{-i\zeta t}]$ as $t \to +\infty$, for example. Hereafter, we denote by $\phi_j^{(i)}$ the jth component of spinor $\phi^{(i)}$, with similar definitions for the other spinor components.

The scattering data S_{ij} is defined by the relationships

$$\phi^{(i)} = \sum_{j=1}^{3} S_{ji} \psi^{(j)}. \tag{5}$$

Knowledge of the input pulse $\mathbf{q}(0, t)$ to the fibre then determines all $S_{ji}(x = 0)$ uniquely. Subsequently, at any propagation distance x down the fibre, knowledge of $S_{11}, S_{21}, S_{31}, \Delta_{11}, \Delta_{21}$ and Δ_{31}, where Δ_{ij} is the cofactor of the element S_{ij}, and of ζ_l and $\bar{\zeta}_m$, the (complex) zeros of $S_{11}(\zeta)$ and $\Delta_{11}(\zeta)$ in the upper and lower halves of the complex $\zeta-$ plane, respectively, are sufficient to reconstruct $\mathbf{q}(x, t)$. Moreover, it is easily shown that $\Delta_{ij}(\zeta) = S_{ij}^*(\zeta^*)$, so that $\bar{\zeta}_l = \zeta_l^*$.

56   J. N. ELGIN

We now state, without proof, the main results of the perturbation theory. For any choice of $\mathbf{F} = [F_1, F_2]^T$, not necessarily small, the following are true [8],[9]:
i)

$$C_{n,x} = \int_{-\infty}^{\infty} [\mathbf{F}^T, -\mathbf{F}^\dagger](2iL)^n \begin{bmatrix} \mathbf{q}^* \\ -\mathbf{q} \end{bmatrix} dt. \tag{6}$$

Here $C_n(n = 0, 1, 2, ...)$ are the infinity of conserved functionals associated with the integrable vector NLS equation, T and $\dagger$ denote transpose and transpose conjugate respectively, $\mathbf{q} = [q_1, q_2]^T$, and L is an integro-differential operator associated with a bilinear form of the scattering problem. The structure of L is not discussed here; suffice it to say that its appropriate application in Equations (6) determines how each functional C_n evolves under the action of perturbation F.
ii)

$$S_{ij,x} = S_{ij,x}^{(0)} - \int_{-\infty}^{\infty} [F^\dagger, F^T]\chi^{(ji)} dt, \tag{7}$$

where the (mixed) bilinear form $\chi^{(pq)}$ is defined by

$$\chi^{(pq)} = \left[\phi_1^{(p)} \hat{\psi}_2^{(q)}, \phi_1^{(p)} \hat{\psi}_3^{(q)}, -\phi_2^{(p)} \hat{\psi}_1^{(q)}, -\phi_2^{(p)} \hat{\psi}_1^{(q)} \right]. \tag{8}$$

Here, $S_{ij,x}^{(0)}$ denotes evolution of the appropriate element of the scattering data for the integrable case, and $\phi^{(i)}$ and $\hat{\psi}^{(j)}$ are Jost function solutions to the scattering problem and its adjoint, discussed previously.
iii) Using a completeness theorem [9] for the bilinear forms, it can be shown that

$$\begin{bmatrix} \mathbf{q} \\ \mathbf{q}^* \end{bmatrix} = -\sum_{j=1}^{N} \Big(\gamma_{21}(\zeta_i)\Psi^{(21)}(\zeta_i) + \gamma_{31}(\zeta_i)\Psi^{(31)}(\zeta_i)$$
$$+ \gamma_{12}(\zeta_i^*)\Psi^{(12)}(\zeta_i^*) + \gamma_{13}(\zeta_i^*)\Psi^{(13)}(\zeta_i^*) \Big)$$
$$+ \frac{1}{\pi}\int_{-\infty}^{\infty} \left(\frac{S_{21}}{S_{11}}\Psi^{(21)} + \frac{S_{31}}{S_{11}}\Psi^{(31)} - \frac{S_{21}^*}{S_{11}^*}\Psi^{(12)} - \frac{S_{31}^*}{S_{11}^*}\Psi^{(13)} \right). \tag{9}$$

The bilinear form $\Psi^{(ij)}$ is defined as for $\chi^{(ij)}$, but $\psi^{(i)}$ now replaces $\phi^{(i)}$. The number N indicates the number of zeros of S_{11} in the upper half of the complex $\zeta-$ plane, ζ_i denotes these zeros (so that $S_{11}(\zeta_i) = 0$) and $\gamma_{21}(\zeta_j) = 2iS_{21}(\zeta_j)/S'_{11}(\zeta_j)$ for bound state ζ_j, where prime denotes derivative with respect to ζ. The discrete sum represents both the soliton and *the shadow* contribution to the total field, whereas the integral represents the contribution from the continuum modes.

3. Generation of The Soliton Shadow

The above theory is now applied to the phenomenon of shadow generation. Consider the case when the input pulse to the fibre corresponds to the single soliton $\mathbf{q}_s(0, t)$. We choose F to correspond to the optical birefringence term, so that $F = [-\delta q_{1t}, \delta q_{2t}]$ in all the above expressions. The nature of the calculation is necessarily perturbative, so that δ is assumed to be a suitably small parameter. Then, to leading order, appropriate solitonic expressions can be used for $\phi_j^{(i)}$ etc. in quantities such as $\chi^{(ij)}$ and $\Psi^{(ij)}$, and the soliton components q_{1s} and q_{2s} can be used in the expression for F. The following statements are then true:

i) When n is zero or one, the integral on the right hand side of Equation (6) vanishes so that c_0 and c_1 are conserved quantities (actually, this is true for any value of the parameter δ, and for arbitrary choice of q_1 and q_2, requiring only that these vanish at $|t| \to \infty$). Then to $O(\delta)$ the soliton parameters ξ_1 and η_1 and the birefringence angle θ are constants of the motion.

ii) It is useful to introduce the following combinations of scattering data:

$$\begin{aligned}
b_{\parallel} &= S_{21}\cos\theta + S_{31}\sin\theta, \\
b_{\perp} &= -S_{21}\sin\theta + S_{31}\cos\theta.
\end{aligned} \tag{10}$$

Equation (7) is now evaluated using solitonic expressions for F and $\chi^{(ij)}$, as discussed above. We leave the soliton parameter η_1 arbitrary, but choose $\zeta_1 = \frac{1}{4}\delta\cos2\theta$. This increases the (relative) soliton velocity uniformly from $+\delta$ to $-\delta$ as θ is increased from 0 to $\pi/2$. Using Equations (10), Equations (7) become for real $\zeta = \xi$

$$\begin{aligned}
b_{\parallel,x} &= -4i\xi^2 b_{\parallel} - 2i\xi\delta\cos2\theta b_{\parallel}, \\
b_{\perp,x} &= -4i\xi^2 b_{\perp} - 2i\xi\delta\cos2\theta b_{\perp} - i\epsilon\sin2\theta(\xi - i\eta_1)\tilde{q}_s^*,
\end{aligned} \tag{11}$$

where $\tilde{q}_s$ is the (Fourier) transform of the soliton pulse. Inverting Equations (10), then substituting into the integral contribution to Equation (9), one finds that $b_{\parallel}$ and $b_{\perp}$ are respectively responsible for generating radiation in the vector modes $[\cos\theta, \sin\theta]^T$ and $[-\sin\theta, \cos\theta]^T$- i.e. parallel to, and perpendicular to, the soliton pulse. Since the equation for $b_{\parallel}$ is homogeneous, and since $b_{\parallel}(x = 0) = 0$ for soliton input, no radiation is present having the same polarization as the soliton pulse. The equation for $b_{\perp}$ is inhomogeneous; radiation is generated by the soliton into the orthogonal mode, even though $b_{\perp}(x = 0) = 0$. Note that the source term vanishes whenever $\theta = 0, \pi/2$.

Consider next the contribution of the discrete terms to $\mathbf{q}(x, t)$. Using again the inverse of the relations (10) (and their conjugate) in Equation (9), the discrete contribution in each mode can be found. In mode $[cos\theta, sin\theta]^T$, it is simply q_s, the soliton pulse, as expected. In the orthogonal mode $[-sin\theta, cos\theta]^T$, one gets

$$\gamma_\perp \psi_1^{(1)} \left[\begin{array}{c} \hat{\psi}_2^{(3)}cos\theta - \hat{\psi}_2^{(2)}sin\theta \\ \hat{\psi}_3^{(3)}cos\theta - \hat{\psi}_3^{(2)}sin\theta \end{array} \right], \tag{12}$$

where $\gamma_\perp$ satisfies the evolution equation

$$\gamma_{\perp,x} = -i\gamma_\perp + \delta cos2\theta\gamma_\perp + i\delta e^{-4i(\xi_1^2-\eta_1^2)x}sin2\theta \tag{13}$$

subject to $\gamma_\perp = 0$ at $x = 0$. The Jost function components $\psi_1^{(1)}$, etc. are evaluated at the bound state eigenvalue $\zeta = \xi_1 - i\eta_1$. It is easy to show that the shadow Equation (12) is indeed in the orthogonal mode.

A more detailed study of the properties of the generated shadow is in progress and results will be published elsewhere [8], [9].

References

1. Elgin, J. N.: Perturbations of Optical Solitons, *Phys. Rev.* **A.47** (1993), 4331-4341.
2. Manakov, S. V.: On the theory of two-dimensional stationary self-focussing of electromagnetic waves, *Sov. Phys. JETP* **38** (1974), 248-253.
3. Elgin, J. N.: Soliton propagation in an optical fibre with third order dispersion, *Optics Letters* **17** (1992), 1409-1410.
4. Elgin, J. N., Brabec, T. and Kelly, S. M. J.: A perturbative theory of soliton propagation in the presence of third order dispersion, *Opt. Commun.* **114** (1995), 321-328.
5. Elgin, J. N. and Kelly, S. M. J.: Spectral modulation and the growth of resonant modes associated with periodically amplified solitons, *Optics Letters* **18** (1993), 787-789.
6. Kaup, D. J. and Malomed, B. A.: Soliton trapping and daughter waves in the Manakov model, *Phys. Rev. A.* **48**, (1993), 599-604.
7. Malomed, B. A. and Wabnitz, S.: Soliton annihilation and fusion from resonant inelastic collisions in birefringent optical fibres, *Optics Letters* **16** (1991), 1388-1390.
8. Elgin, J. N. and Docksey, R. G.: Perturbed solitons in birefringent fibres. In preparation for submission to Phys. Rev. A. (1995).
9. Elgin, J. N. and Docksey, R. G.: Closure of bilinear forms of the Manakov eigenstates. In preparation for submission to SIAM. (1995).

PARAMETRIC COOLING OF SOLITONS

V. S. GRIGORYAN
"IRE-POLUS Co.", Russian Academy of Sciences
Vvedenskogo sq.1, Fryazino 141120, Moscow region, Russia
A. HASEGAWA AND A. MARUTA
Department of Communication Engineering
Faculty of Engineering, Osaka University
2-1 Yamada-Oka, Suita, Osaka 565, Japan

Abstract. Self-induced locking of soliton time position at equidistant points in the presence of noise is predicted. The effect is based on four-wave parametric interaction occurring when injected monochromatic controlling wave co-propagates with the soliton train. The effect permits the self-adjusted optical control of solitons by light wave.

1. Introduction

One basic problem in high bit rate soliton communication and storage systems is the soliton's time position jittering. This jittering is caused by the soliton frequency diffusion which originates either from amplified spontaneous emission in repeaters [1] or from soliton-soliton interactions. Use of filtering along with amplification in repeaters permits only partially suppress the time position jittering [2,3]. Without filtering mean-square time position increases as distance cubed, whereas with filtering it increases linearly in distance. Recently big progress was achieved in experiment of M.Nakazawa with the help of active modulation of losses along with filtering and amplification of solitons [4]. Modulator provides a clocking mechanism by which an ordering of solitons occurs, whereby each of solitons tends to occupy a time interval when the modulator has a maximum transmission. However, in practical use of this method for a communication line, when a number of modulators is needed, a serious problem of mutual synchronization of all those modulators arises. In addition, the bit rate is limited by the modulator electronics. The way-out is a passive mechanism of

A. Hasegawa (ed.), Physics and Applications of Optical Solitons in Fibres '95, 59–73.
© *1996 Kluwer Academic Publishers. Printed in the Netherlands.*

ordering or a mechanism of self-ordering of solitons which does not require electronics.

In present paper we predict the effect of parametric locking of soliton time position on the base of four-wave parametric interaction (FWPI) in Kerr nonlinear medium. The following speculations can give us a clue. Let us consider a four-wave parametric interaction having the frequency scheme

$$\omega_1 + \omega_2 = 2\omega_s, \tag{1.1}$$

where ω_s is the soliton frequency, $\omega_{1,2}$ are frequency of controlling waves. We assume the controlling waves to have fixed spectra. It is well known that, in parametric wave interactions, the phases of interacting waves are tightly coupled with each other. One of the examples of such a couple is known as phase locking effect. If the amplitudes of controlling waves are much more than the soliton amplitude then the phase of the soliton wave will be locked and the energy will flow from the controlling waves to the soliton wave. In the presence of linear losses and filtering at the soliton frequency one can occur the energy balance. As was shown in our recent paper [5] this situation leads not only to suppression of frequency (the phase time derivative) jittering but also to the soliton position locking too. The latter is caused by pushing of the soliton to the maximum of parametric amplification determined by the closest maximum of phased grating induced by the controlling waves.

From practical standpoint the controlling waves are more desirable to have amplitudes smaller than the soliton amplitude. In that case there is no locking effect of soliton phase. However the soliton phase behavior is still determined by parametric grating, induced by controlling waves, according to a more complicated low. If in some way we can keep the parameters of controlling wave (i.e. using narrow-band filters and injecting the new controlling waves continuously) then we can control the soliton phase dependence in distance z and time t space. As follows from geometric optics approximation the gradient of phase in z, t space is proportional to the soliton time position derivative with respect to z. The time position will be constant if the gradient is zero. Bellow we will show that the soliton position rest point (fixed by phased grating, induced by controlling waves) proves to be asymptotically stable for rather general conditions of FWPI using narrow-band filtering for weak controlling waves. One of the controlling wave ω_2 we choose as the monochromatic radiation, the second wave ω_1 is generated from the FWPI. Hence the phased grating is self-induced and the mechanism considered below relates to a passive one, and, besides, it does not require electronics.

2. Basic Equations

Slowly varying normalized amplitudes of soliton waves q_s and controlling waves $q_{1,2}$ satisfy the equations

$$i\left(\frac{\partial}{\partial z}+\frac{1}{v_1}\frac{\partial}{\partial t}\right)q_1+\frac{1}{2}d_1\frac{\partial^2 q_1}{\partial t^2}+r_1\left[q_1|q_1|^2+2q_1\left(|q_s|^2+|q_2|^2\right)+q_s^2 q_2^* \exp(-i\delta kz)\right]=$$

$$ig_1 q_1+\int_{-\infty}^{t}\chi_1(t-t')q_1(t')dt',$$

$$(2.1)$$

$$i\left(\frac{\partial}{\partial z}+\frac{1}{v_2}\frac{\partial}{\partial t}\right)q_2+\frac{1}{2}d_2\frac{\partial^2 q_2}{\partial t^2}+r_2\left[q_2|q_2|^2+2q_2\left(|q_s|^2+|q_1|^2\right)+q_s^2 q_1^* exp(-i\delta kz)\right]=$$

$$ig_2 q_2+\int_{-\infty}^{t}\chi_2(t-t')q_2(t')dt',$$

$$(2.2)$$

$$i\left(\frac{\partial}{\partial z}+\frac{1}{v_s}\frac{\partial}{\partial t}\right)q_s+\left(\frac{1}{2}-iI\right)\frac{\partial^2 q_s}{\partial t^2}+q_s|q_s|^2+2q_s\left(|q_1|^2+|q_2|^2\right)+$$

$$2q_s^* q_1 q_2\, exp(i\delta kz)=ig_s q_s+\vec{F}(z,t),$$

$$(2.3)$$

where $d_j=k_j''/k_s''$, $r_j=\omega_j/\omega_s, j=1,2$, wave vector mismatch $\delta k=k_1+k_2-2k_s$, $g_j=g_{ampl.}^j-\gamma_{loss.}^j$ is the effective gain for j-th wave, $\chi_{1,2}$ is complex linear susceptibility caused by narrow-band filtering of controlling waves respectively, I determines a wide-band filtering of the soliton wave, the Langevin force term $\vec{F}(z,t)$, simulating noise and describing spontaneous emission in amplifying medium, is assumed to be delta-correlated stochastic process.

2.1. PROBLEM STATEMENT AND EQUATION FOR SOLITON WAVE

Suppose there is some soliton train in fiber. We inject in fiber cw monochromatic four-wave mixing with the scheme (1). Furthermore, we use comb-like filtering radiation q_2; the field q_1 is parametrically generated there as a product of (Fabry-Perot etalon - type) with frequency period $\Omega=2\pi/T_o$ (where T_o is average interval between adjacent solitons) to select narrow equidistant spectral components from the q_1 spectrum. We suppose that q_1 field remains weak everywhere in z due to typically large wave mismatch, i.e. field scale is following

$$|q_1|<<|q_2|<|q_s|. \qquad (2.4)$$

ω_1 frequency is chosen such that this wave has a group velocity the same that of at the soliton frequency: $v_1 = v_s = v$. This choice is possible if the carrier frequency of q_2 is selected to be lower than the zero-dispersion frequency. Under these conditions one can derive the equation (A.9) (in Appendix A) for soliton wave. Rewriting it with small terms gathered in right-hand side (where small parameter ε is extracted) we have the equation for the soliton wave q

$$i\frac{\partial q}{\partial z} + \frac{1}{2}\frac{\partial^2 q}{\partial \tau^2} + q|q|^2 = \varepsilon\left[-q^*\mathcal{P}\exp(i\Delta kz) + igq + iI\frac{\partial^2 q}{\partial \tau^2} + \varepsilon F(z,\tau)\right],$$

$$(2.5)$$

where

$$q = q_s \exp\left(-i2|q_2|^2 z\right), \quad \Delta k = \eta_o^2 - 4|q_2|^2,$$

η_o is average soliton amplitude in the soliton train; g and I are excess gain and filtering constants respectively; $\mathcal{P}$ is defined by amplitudes p_m which are proportional to $|q_2|^2$ according to Equations (A8). In the case of practical interest q_2 is injected periodically in z. In that case one can show that under some reasonable conditions g, I and $|q_2|^2$ may be treated as relative averaged in z constants and soliton wave q as the averaged soliton wave in the sense similar to that of [6]. The Langevin noise term $F(z,\tau)$ is such that

$$\langle F(z,\tau)F^*(z',\tau')\rangle = Q_F\delta(z-z')\delta(\tau-\tau'). \qquad (2.6)$$

For simplicity bellow we will consider the case when comb-filter consists of only three peaks: one central and two sidebands, i.e. in $\mathcal{P}$ term of Equation (A.8) $p_m = 0$ when m>1:

$$\mathcal{P} = p_o + 2p_1 \cos(\Omega\tau) \ .$$

3. Equation for Parameters of Soliton

We employ the soliton perturbation theory and following to [7] we are looking for a solution of Equation (2.5) in the form

$$q = \eta \sec h(v)\exp(i\varphi),$$

where

$$\varphi = \xi(\tau - \theta) + \delta, \quad v = \eta(\tau - \theta),$$

η, θ, ξ and δ are functions of z which satisfy the following set of equation

$$\frac{d\eta}{dz} = \varepsilon\{2\eta[p_o\Phi(x_1)\sin 2\delta_1 - p_1\Phi(x_2)\sin(\Omega\theta - 2\delta_1) +$$

$$p_1\Phi(x_3)\sin(\Omega\theta + 2\delta_1) + g - I(\xi^2 + \tfrac{1}{3}\eta^2)] + \varepsilon\Gamma_\eta(z)\}, \tag{3.1}$$

$$\frac{d\xi}{dz} = -\varepsilon\{\tfrac{2\eta}{\pi}[p_o x_1\Phi(x_1)\sin 2\delta + p_1 x_2\Phi(x_2)\sin(\Omega\theta - 2\delta) +$$

$$p_1 x_3\Phi(x_3)\sin(\Omega\theta + 2\delta)] + \tfrac{4}{3}I\xi\eta^2 - \varepsilon\Gamma_\eta(z)\}, \tag{3.2}$$

$$\frac{d\theta}{dz} = \xi + \varepsilon\{-\tfrac{\pi}{\eta}[p_o Y(x_1)\cos 2\delta_1 - p_1 Y(x_2)\cos(\Omega\theta - 2\delta_1) +$$

$$p_1 Y(x_3)\cos(\Omega\theta + 2\delta_1)] + \varepsilon\Gamma_\theta(z)\}, \tag{3.3}$$

$$\frac{d\delta}{dz} = \tfrac{1}{2}(\eta^2 + \xi^2) + \varepsilon\{-\tfrac{\pi}{\eta}\xi[p_o Y(x_1)\cos 2\delta_1 - p_1 Y(x_2)\cos(\Omega\theta - 2\delta_1) +$$

$$p_1 Y(x_3)\cos(\Omega\theta + 2\delta_1)] + p_o H(x_1)\cos 2\delta_1 + p_1 H(x_2)\cos(\Omega\theta - 2\delta_1) +$$

$$p_1 H(x_3)\cos(\Omega\theta + 2\delta_1) + \varepsilon\Gamma_\eta(z)\},$$

$$\tag{3.4}$$

where

$$\delta_1 = \delta - \frac{1}{2}(\eta_o^2 - 4|q_2|^2)z, \qquad\qquad \Phi(x) = \frac{x}{\sinh x},$$

$$Y(x) = \frac{\sinh x - x\cosh x}{\sinh^2 x},$$

$$H(x) = \frac{x^2\cosh x}{\sinh^2 x}, \quad x_1 = \pi\xi/\eta, \quad x_2 = \pi(\Omega - 2\xi)/2\eta,$$

$$x_3 = \pi(\Omega + 2\xi)/2\eta,$$

$$\Gamma_\eta(z) = -\mathbf{Re}\left[i\int_{-\infty}^{\infty}\frac{F(z,\tau)}{\cosh v}\exp(-i\varphi)dv\right],$$

$$\Gamma_\xi(z) = -\mathbf{Im}\left[i\int_{-\infty}^{\infty}\frac{\tanh v\,F(z,\tau)}{\cosh v}\exp(-i\varphi)dv\right],$$

$$\Gamma_\theta(z) = -\frac{1}{\eta^2}\mathbf{Re}\left[i\int_{-\infty}^{\infty}\frac{v F(z,\tau)}{\cosh v}\exp(-i\varphi)dv\right], \tag{3.4a}$$

$$\Gamma_\delta(z) = \xi\Gamma_\theta(z) - \frac{1}{\eta}\mathbf{Im}\left[i\int_{-\infty}^{\infty}\frac{1 - v\tanh v}{\cosh v}F(z,\tau)\exp(-i\varphi)dv\right]$$

We are looking for asymptotically stable solution of Equations (3.1)-(3.4) near $\eta = \eta_o, \xi = 0$, and $\theta = 0$, that is why the most interesting is behavior of the Equations (3.1)-(3.4) near that point. Assuming $\eta = \eta_o + \Delta\eta$, where $|\Delta\eta| \ll \eta_o$ we expand the right-hand sides of Equations (3.1)-(3.4) near $\eta = \eta_o, \xi = 0, \theta = 0$ and neglecting the higher order terms we have

$$\frac{d\Delta\eta}{dz} = \varepsilon[(C_1 + C_2\Delta\eta)\sin 2\delta_1 + C_3 + C_4\Delta\eta + \varepsilon\Gamma_\eta(z)], \tag{3.5}$$

$$\frac{d\xi}{dz} = -\varepsilon[D_1\xi\sin 2\delta_1 + D_2\theta\cos 2\delta_1 + D_3\xi - \varepsilon\Gamma_\xi(z)], \tag{3.6}$$

$$\frac{d\theta}{dz} = \xi - \varepsilon[K_1\xi\cos 2\delta_1 - K_2\theta\sin 2\delta_1 - \varepsilon\Gamma_\theta(z)], \tag{3.7}$$

$$\frac{d\delta_1}{dz} = 2|q_2|^2 + \eta_o\Delta\eta + \varepsilon[(Q_1 + Q_2\Delta\eta)\cos 2\delta_1 + \varepsilon\Gamma_\delta(z)], \tag{3.8}$$

where we denoted

$$C_1 = 2\eta_o(p_o + 2p_1a), \quad C_2 = 2(p_o + 2p_1a - \pi p_1 b\Omega/\eta_o),$$
$$C_3 = 2\eta_o(g - I\eta_o^2/3), \quad C_4 = 2(g - I\eta_o^2),$$
$$D_1 = 2[p_o + 2p_1(a + \pi b\Omega/(2\eta_o))], \quad D_2 = 2p_1a\Omega^2, \quad D_3 = 4I\eta_o^2/3$$
$$K_1 = \frac{\pi^2}{\eta_o^2}\left(\frac{p_o}{3} - 2p_1b'\right), \quad K_2 = 2\pi p_1 b\Omega/\eta_o, \tag{3.9}$$
$$Q_1 = p_o + 2p_1h, Q_2 = -2p_1x_oh'/\eta_o$$
$$a = \Phi(x_o), \quad b = Y(x_o), \quad b' = Y'(x_o), \quad h = H(x_o), \quad h' = H'(x_o)$$
$$x_o = \frac{\pi\Omega}{2\eta_o}.$$

4. Equation for Soliton Time Position

Differentiating Equation (3.7) with respect to z, combining it with Equations (3.6)-(3.8) and picking up only linear terms, including the Langevin forces, we get the equation for θ

$$\frac{d^2\theta}{dz^2} + \varepsilon D(z)\frac{d\theta}{dz} + \varepsilon U(z)\theta = \varepsilon^2\Gamma(z), \tag{4.1}$$

where

$$D(z) = (D_1 - K_2 + 2K_1\mu_1)\sin 2\delta_1 - \varepsilon K_1^2\mu\sin 4\delta_1 + D_3, \quad \mu = 2\overline{|q_2|^2},$$
$$\mu_1 = \mu + \varepsilon Q_1\cos 2\delta_1$$
$$U(z) = [D_2(1 + \varepsilon K_1\cos 2\delta_1) - 2K_2\mu_1]\cos 2\delta_1 - \varepsilon K_2\sin 2\delta_1[(D_1 + 2K_1\mu)\sin 2\delta_1 + D_3]$$

$$\Gamma(z) = \Gamma_\xi(z) + \Gamma_\theta'(z),$$

(we neglected the ε^3-order terms in left-hand side of Equation (4.1)). If $\Gamma_\theta(z)$ is a broken function we, first, approximate it by a smooth function and then, in finite result (see Equation (5.8)), transfer to the limit in the points of breaks.

5. Solution of Equation for Soliton Parameters

The set of equations (3.5)-(3.8) relates to the class of equations with slowly varying parameters and fast varying phase. The general method of solution of such equations was developed by V.M.Volosov, described in monograph of Yu.A.Mitropolksy [8]. The basic idea of that method is finding out a transformation which reduces Equations (3.5)-(3.8) to averaged equations where slowly varying parameters are separated from fast varying phase. It is worth to notice that this method is asymptotic one in a sense that the authors proved rigorously the convergence of the approximate solution to exact solution. First, let us reduce Equations (3.5)-(3.8) to the standard form more convenient for using. We introduce new variables Φ and Ψ instead of θ and ξ:

$$\theta = \Phi\left(1 - \varepsilon\sigma\cos 2\delta_1\right), \tag{5.1}$$

$$\frac{d\theta}{dz} = \varepsilon\Psi + 2\varepsilon\sigma\mu\Phi\sin 2\delta_1, \tag{5.2}$$

where

$$\sigma = -\frac{D_2 - 2K_2\mu}{4\mu^2}.$$

Differentiating Equation (5.1) with respect to z and comparing it with Equation (5.2) we obtain taking into account the terms of up to ε -order

$$\frac{d\Phi}{dz} = \varepsilon\Psi. \tag{5.3}$$

Further, differentiating Equation (5.2) with respect to z and comparing it with Equation (4.1) we have equation for Ψ:

$$\frac{d\Psi}{dz} = -\varepsilon\Psi\left[2\sigma\mu\sin 2\delta_1 + \left(D_1 - K_2 + 2K_1\mu\right)\sin 2\delta_1 + D_3\right] - \varepsilon\Phi 4\sigma\mu Q_1\cos^2 2\delta_1 -$$

$$\varepsilon\Phi\left\{2\sigma\mu\left[\left(D_1 - K_2 + 2K_1\mu\right)\sin^2 2\delta_1 + D_3\sin 2\delta_1\right] + \left(D_2K_1 - 2K_2Q_1\right)\cos^2 2\delta_1\right\} +$$

$$\varepsilon\Phi\Big\{K_2\sin 2\delta_1\big[(D_1+2K_1\mu)\sin 2\delta_1+D_3\big]+\gamma(D_2-2K_2\mu)\cos^2 2\delta_1\Big\}+\varepsilon\Gamma(z)$$

$$(5.4)$$

In derivation of Equations (5.3) and (5.4) we neglected in right-hand sides the terms of ε^2-order and higher as well as the square terms $\Phi\Delta\eta$ and $\Phi\xi$, as we assumed above linear approximation . To get the full set of equations we should add Equations (5.3) and (5.4) to Equations (3.5) and (3.8).

5.1. SOLUTION OF AVERAGED EQUATIONS

In the first order approximation the averaged equations are produced simply with the help of averaging of Equations (5.3),(5.4) with respect to δ_1. The averaging yields

$$\frac{d\overline{\Phi}}{dz}=\varepsilon\overline{\Psi},\qquad(5.5)$$

$$\frac{d\overline{\Psi}}{dz}=\varepsilon\big[-D_3\overline{\Psi}-V\overline{\Phi}+\Gamma(z)\big],\qquad(5.6)$$

where

$$V=\frac{1}{8\mu^2}\big[D_2^2-2D_2(D_1+K_2+2Q_1)\mu\big].$$

Differentiating Equation (5.5) and comparing with Equation (5.6) we get the equation for the averaged soliton time position $\overline{\Phi}$:

$$\frac{d^2\overline{\Phi}}{dz^2}+\varepsilon D_3\frac{d\overline{\Phi}}{dz}+\varepsilon^2 V\overline{\Phi}=\varepsilon^2\Gamma(z).\qquad(5.7)$$

Equation (5.7) represents the basic result of the present paper. It resembles the equation of stimulated oscillations with damping. Integration of Equation (5.7) yields

$$\overline{\Phi}=\exp(-\varepsilon D_3 z/2)\big[\mathcal{C}_1\exp(\varepsilon\lambda z/2)+\mathcal{C}_2\exp(-\varepsilon\lambda z/2)\big]+$$

$$+\varepsilon\frac{2}{\lambda}\int_0^z\Gamma(z')\exp\big[\varepsilon D_3(z'-z)/2\big]\sinh\big[\varepsilon\lambda(z-z')/2\big]dz',\quad(5.8)$$

where from, using Equation (2.6), one can find out the solution for mean-square time coordinate of the soliton

$$\langle\overline{\Phi}^2\rangle=\exp(-\varepsilon D_3 z)\big[\mathcal{C}_1\exp(\varepsilon\lambda z/2)+\mathcal{C}_2\exp(-\varepsilon\lambda z/2)\big]+$$

$$+\varepsilon\frac{2}{3}\frac{Q_F\eta_o}{\lambda^2 D_3}\left\{\frac{\lambda^2}{4V^2}+\left[1-\frac{D_3}{2V}\cosh(\lambda z+\gamma)\right]\exp(-\varepsilon D_3 z)\right\},(5.9)$$

where $\mathcal{C}_{1,2}$ are constants defined by initial conditions,

$$\lambda^2 = D_3^2 - 4V, \quad \exp(-2\gamma) = \frac{D_3-\lambda}{D_3+\lambda}.$$

When $\lambda^2<0$ one can put in right-hand side of Equation (5.9) a formal change $\lambda \to i\lambda$. If we suppose $D_3>0$ and $V>0$ then one can easy obtain from Equation (5.9) that, with z increasing, the mean-square time coordinate of the soliton tends to the constant

$$\left\langle\overline{\Phi^2}\right\rangle = \varepsilon\frac{Q_F\eta_o}{6D_3 V}. \tag{5.10}$$

That is the desirable locking of the soliton time coordinate occurs. Using solution (5.8) one can find out any higher order momenta as well.

The averaged equation for soliton amplitude is produced in the same manner from Equation (5.5):

$$\frac{d\overline{\Delta\eta}}{dz} = \varepsilon\left(C_3 + C_4\overline{\Delta\eta}\right), \tag{5.11}$$

If $\eta_o^2 = 3g/I$ then $C_3 = 0, C_4 < 0$ and $\overline{\Delta\eta}=0$ is stable solution of Equation (5.11).

6. Discussion of the Results

Thus we found that under condition $V>0$ the soliton time position at zero point (and hence at the points $\theta_n = 2\pi n/\Omega, \ n = 0,\pm 1,\pm 2,...$) is asymptotically stable with respect to noise perturbations. Its mean-square deviation remains constant defined by Equation (5.10). In limit case $V=0$ (when controlling wave $q_2=0$ or wave vector mismatch $\delta k = \infty$) it follows from Equation (5.9) that the mean-square time coordinate of soliton grows linearly in z:

$$\left\langle\overline{\Phi^2}\right\rangle = const + \varepsilon^2\frac{Q_F\eta_o}{3D_3^2}z, \tag{6.1}$$

and we get well known result for the case considered in [2,3]. Let us determine a condition imposed on the beating frequency Ω and controlling field intensity μ when $V > 0$. Using definition of V in Equation (5.6) and expressions for parameters D_1, D_2 and K_2 from Equations (3.9) one can obtain that $V > 0$ when

$$R = \frac{1}{x_o^3}\left(\frac{p_o}{p_1}\sinh x_o + 3x_o - x_o^2 ctgh(x_o)\right) < \frac{\eta_o^2}{\pi^2\mu}, \qquad (6.2)$$

where

$$x_o = \pi\Omega/2\eta_o.$$

Inequality (6.2) determines a boundary condition imposed on the frequency Ω at given μ when the soliton position locking does take place. There can exist two cases:

1. $R > 0$. There is the upper boundary of controlling wave intensity μ:

$$\mu = 2|q_2|^2 < \frac{\eta_o^2}{\pi^2 R(x_o)}$$

2. $R < 0$. There is no any limitation for controlling wave intensity μ to be $V > 0$.

It is seen from Equation (6.2) that for large enough q_2 the condition of asymptotic stability of the soliton time position (6.2) may be violated. It does not contradict to the result obtained in our previous work [5] for asymptotic stability of the soliton for the case of strong controlling fields, much more than the soliton amplitude. Actually, in present case the violation is caused by the fact that nonlinear wave vector mismatch Δk increases with q_2 increasing, that leads to a break-down of the soliton locking regime. Whereas in our previous work [5] the exact phase matching condition was assumed. In our further works we are going to discuss in details the question of the threshold break-down for the soliton locking effect. In opposite case of small q_2, as follows from definitions of V in Equation (5.6) and definitions of D_1, D_2 from Equations (3.9) and (A.8)

$$V \approx \frac{D_2^2}{8\mu^2} = V_0, \qquad (6.3)$$

where V_0 is constant independent of q_2 (D_1, D_2 are proportional to $\mu = |q_2|$). Hence the soliton mean-square time coordinate (5.10) does not depend on controlling field amplitude q_2 in wide enough region. The lower boundary of this region is determined by the framework of approximation of averaging theory used in Section 5.

In other limit case (without amplification and filtering of the soliton) when $g = I = 0$ and $D_3 = 0$ we have from Equation (5.7)

$$\left\langle \overline{\Phi}^2 \right\rangle = \left[\mathcal{C}_1 \sin(\varepsilon|\lambda|z/2) + \mathcal{C}_2 \cos(\varepsilon|\lambda|z/2) \right]^2 + \varepsilon \frac{\eta_o Q_F}{12 V^{3/2}} \left[\varepsilon|\lambda|z - \sin(\varepsilon|\lambda|z) \right]$$

$$. (6.4)$$

Thus in both cases without filtering-amplification but in the presence of controlling fields (Equation 6.4), and with filtering-amplification but without controlling fields (Equation 6.1) the mean-square soliton coordinate increases in z. That is, for these cases there is no asymptotic stability for the soliton time position. It is curious that when both these factors (filtering-amplification and controlling fields) work together we get the asymptotic stability for the soliton time coordinate. The reason here is the same that of for harmonic oscillator with damping (see Equation (5.7) with zero right-hand side). If damping is zero (filtering term $D_3 = 0$) and $V > 0$ we have oscillating solution with arbitrary amplitude. If there is a damping ($D_3 > 0$) but oscillator frequency is zero ($V = 0$) then any constant coordinate is solution of the equation. In both cases there is no whatever fixed coordinate which is asymptotically stable solution. However when both those factors ($V > 0$ and $D_3 > 0$) work together there appears the asymptotically stable solution at zero point.

7. Conclusion

Thus we found the effect of cooling and self-ordering of solitons in a sense that we suppressed the entropy growth of the soliton in the presence of noise (associated with growth of mean-square soliton time coordinate) provided that the solitons occupy predetermined, ordered, asymptotically stable positions. The mechanism of the effect is based on injecting a monochromatic controlling wave (at the frequency higher then the soliton frequency) co-propagating with the soliton train. Owing to four-wave parametric interaction the second controlling wave is generated itself. If we use a narrow-band comb-like filtering (with frequency interval between adjacent peaks $= 2\pi/T_o$, where T_o is the average time interval between adjacent solitons in the soliton train) for the second controlling wave, then we get nonlinear parametric phased grating in time with period T_o. We found out analytical solution for the soliton position near its rest points in the presence of noise using the method of averaging theory, developed in classical works of N.N.Bogolubov etc. It was shown that the mean-square soliton time coordinate evolves to the fixed value determined by the medium parameters and the noise correlation coefficient. Thus the locking of the soliton time position at the ordered rest points nT_o ($n = 0, \pm 1, \pm 2, \ldots$) occurs.

Appendix

Neglecting in left-hand side of Equation (2.1) the term $q_1\left(|q_1|^2 + 2|q_s|^2 + 2|q_2|^2\right)$ in comparison with $q_s^2 q_2^* \, exp(-i\delta kz)$ we get in retarded time framework $\tau = t - z/v$, z

$$i\frac{\partial q_1}{\partial z} + \frac{1}{2}d_1\frac{\partial^2 q_1}{\partial \tau^2} = -r_1 q_s^2 q_2^* \, exp(-i\delta kz) + ig_1 q_1 + \int_{-\infty}^{\tau}\chi_1(\tau - t')q_1(t')dt'$$

(A.1)

Using the Fourier transform

$$q_1 = \int_{-\infty}^{\infty} c_\Delta \, exp(i\Delta\tau)d\Delta$$

we get from Equation (A.1)

$$i\frac{dc_\Delta}{dz} = -r_1 S_\Delta \, exp(-i\delta kz) + \left(ig_1 - iG_1(\Delta) + \frac{1}{2}d_1\Delta^2\right)c_\Delta, \quad \text{(A.2)}$$

where

$$S_\Delta = \frac{1}{2\pi}\int_{-\infty}^{\infty} q_s^2 q_2^* \, exp(-i\Delta\tau)d\tau, \quad G_1(\Delta) = \frac{i}{2\pi}\int_0^{\infty}\chi_1(x)\,exp(-i\Delta x)dx.$$

Let us focus on the source term S_Δ. The soliton sequence field

$$q_s(z,\tau) = \sum_{n=1}^{N}\alpha_n a_n(\tau - nT_0), \tag{A.3}$$

where

$$a_n(\tau) = \eta_n \, sec \, h\left[\eta_n(\tau - \theta_n)\right]exp \, i\left[\xi_n(\tau - \theta_n) + \delta_n\right],$$

$$\frac{d\delta_n}{dz} = \frac{1}{2}\left(\eta_n^2 + \xi_n^2\right), \quad \frac{d\theta_n}{dz} = \xi_n, \quad \frac{d\delta_n}{d\tau} = \frac{d\theta_n}{d\tau} = 0, \quad \alpha_n = 1 \text{ or } 0.$$

Integrating Equation (A.3) and neglecting slow dependence of S_Δ on z in comparison with $exp(-i\delta kz)$ we have

$$c_\Delta = ir_1 \frac{S_\Delta(z)exp(-\delta kz) - S_\Delta(0)exp\left[g_1 - G_1(\Delta) - id_1\Delta^2/2\right]z}{i\left(d_1\Delta^2/2 - \delta k\right) - g_1 + G_1(\Delta)}.$$

(A.4)

Integrating we have at the resonant frequencies $\Delta_m = 2\pi m/T_o$

$$S_{\Delta_m} = \frac{1}{2\pi} \sum_{l=1}^{N} q_2^* P_l(\Delta_m - 2\xi_l) \, exp \, i[2\delta_l - \Delta_m(\theta_l + lT_o)] =$$

$$\frac{1}{2\pi} N \langle q_2^* P(\Delta_m - 2\xi) \, exp \, i(2\delta - \Delta_m \theta) \rangle,$$

(A.5)

where

$$P_l(\Delta) = \eta_l^2 \int_{-\infty}^{\infty} sec \, h^2(\eta_l t) \, exp(-i\Delta t) dt = \pi\Delta \, cos \, ech(\pi\Delta/2\eta_l),$$

and angle brackets denote statistical average on the ensemble of N solitons in the soliton train. As was shown in section 3, statistical properties of the soliton parameters η, ξ, θ and δ are determined by the Langevin forces $\Gamma_\eta, \Gamma_\xi, \Gamma_\theta$ and Γ_δ. As follows from their definitions these forces are independent as far as they may be interpreted as linear combinations of first four weight coefficients of the expansion of noise term $F(z, \tau)$ into set of orthogonal functions. Hence η, ξ, θ and δ are statistically independent and we get from Equation (A.5)

$$S_{\Delta_m} = \frac{1}{2\pi} N q_2^* \left\langle P(\Delta_m) + \frac{\partial P}{\partial \Delta_m}(-2\xi) + \frac{1}{2}\frac{\partial^2 P}{\partial \Delta_m^2}(4\xi^2) + ... \right\rangle \langle e^{-i\Delta_m\theta} \rangle \langle e^{i2\delta} \rangle$$

(A.6)

Assuming odd momenta of ξ and θ to be zero and noticing that $P(\Delta)$ and its even derivatives are symmetrical functions on Δ one can obtain that $S_{\Delta_m} = S_{-\Delta_m}$ and as follows from Equation (A.4) c_{Δ_m} is symmetrical function too:

$$c_{\Delta_m} = c_{-\Delta_m}.$$

(A.7)

Using the results of averaging theory in second order approximation (see Section 5) one can see that $\langle e^{-i\Delta_m\theta} \rangle \to const_1$ whereas $\langle e^{i2\delta} \rangle \to const_2 \cdot e^{i\eta_o^2 z}$ with z increasing where $\eta_o = \langle \eta \rangle$ is average soliton amplitude. Thus restoring q_1 we can write the phased grating term in left-hand side of Equation (2.3)

$$2q_1 q_2 \, exp(i\delta kz) = \mathcal{P} exp(i\eta_o^2 z),$$

where

$$\mathcal{P} = p_o + 2\sum_m p_m \, cos(m\Omega\tau),$$

(A.8)

$$p_m = \nu q_2 c_{\Delta_m} \exp(i\delta k z) = -\nu \frac{r_1 s_{\Delta_m}}{\delta k} e^{-i\eta_o^2 z} + O\left(e^{i\delta k z}\right),$$

$$s_{\Delta_m} = \frac{1}{2\pi} N |q_2|^2 \left\langle P(\Delta_m - 2\xi) \right\rangle \left\langle \exp(-i\Delta_m \theta) \right\rangle \left\langle \exp(i2\delta) \right\rangle,$$

where ν is characteristic width of comb filter, $O\left(e^{i\delta k z}\right)$ is fast oscillating term in z. In deriving Equation (A.8) we took into account that typically δk is much larger then both dispersion term $\sim d_1$, gain factor g_1 and η_o^2; at the resonant frequency of the filter we took unit transmission: $G(\Delta_m) = 0$. Using Equation (A.8) we obtain finely the equation for soliton wave

$$i\frac{\partial q_s}{\partial z} + \left(\frac{1}{2} - iI\right)\frac{\partial^2 q_s}{\partial \tau^2} + q_s|q_s|^2 + 2q_s|q_2|^2 = -q_s^* \mathcal{P} \exp(i\eta_o^2 z) + ig_s q_s + \vec{F}(z,\tau)$$

$$(A.9)$$

Let us notice that in considered communication line g_s and I are periodical functions of z with period being equal to inter-repeater distance z_a. Besides, controlling field amplitude $|q_2|$ is also periodical function of z with the same period because a new controlling wave is injected after each repeater. As $z_a < 1$ one can average Equation (A9). One can show that in the first order of averaging we get the equation for the averaged soliton wave $\overline{q_s}$ of the same form that of Equation (A9) with averaged constants $\overline{I},\ \overline{g_s},\ \overline{|q_2|^2}$ and $\overline{s_{\Delta_m}}$.

It is worth to note that in reality the width of peaks of comb filter ν is always finite. In that case it is easy to see from Fourier transform that not all soliton train will contribute to the field q_1 but only part of them which are present in the time interval of length of ν^{-1}- order. It is this part of solitons that determines the statistical ensemble of N solitons used in Equations (A.5) and (A.6). It follows from that a requirement for coherence time of monochromatic controlling wave q_2:

$$t_{coh.} \gg 1/\nu.$$

References

1.   Gordon, J.P. and Haus, H.A.: Random walk of coherently amplified solitons in optical fibers, *Opt.Lett.* **11** (1986), 665-667.
2.   Mecozzi, A., Moores, J.D., Haus, H,A., and Lai, Y.: Soliton transmission control, *Opt.Lett.* **16** (1991), 1841-1843.
3.   Kodama, Y. and Hasegawa, A.: Generation of asymptotically stable optical solitons and suppression of Gordon-Haus effect, *Opt.Lett.* **17** (1992), 31-33.
4.   Kubota, H. and Nakazawa, M.: Soliton transmission control in time and frequencydomains, *IEEE J.Quant.El.* **29** (1993), 2189-2197.

5. Grigoryan, V.S., Hasegawa, A., and Maruta. A.: Control of solitons by light waves, *Opt.Lett.* **20** (1995), 857-859.

6. Kodama, Y. and Hasegawa, A. (1992) Theoretical foundation of optical-soliton concept in fibers, in E.Wolf (ed.), *Progress in optics, volume XXX*, Elsevier Science Publishers B.V., Amsterdam - London - N.Y.- Tokyo. pp.205-259.

7. Karpman, V.I. and Maslov, E.M.: Perturbation theory for solitons, *ZhETF* **73** (1977), 537-559 (*Sov.Phys.JETP* **46** (1977), 281-299).

8. Mitropolsky, Yu.A.: Averaging method in nonlinear mechanics, *Int. J. Nonlinear mechanics* **2** (1967), 69-96.
    Mitropolsky, Yu.A.: *Averaging method in nonlinear mechanics*, Naukova Dumka, Academy of sciences of Ukrainian SSR, Kiev, 1971.

NONLINEAR THEORY OF SOLITON INSTABILITIES

Y. S. KIVSHAR

Optical Sciences Centre, The Australian National University
ACT 0200 Canberra, Australia

D. E. PELINOVSKY

Department of Mathematics, Monash University
Clayton Victoria 3168, Australia

AND

V. V. AFANASJEV, A. V. BURYAK

Optical Sciences Centre, The Australian National University
ACT 0200 Canberra, Australia

Abstract. We describe the development of solitons instabilities using a multi-scale asymptotic expansion technique valid near the instability threshold. This approach allows us to describe analytically both linear and nonlinear regime of the soliton evolution. Examples considered include: (i) bright solitons supported by non-Kerr nonlinearities, (ii) parametric bright solitons in $\chi^{(2)}$ optical materials, and (iii) dark solitons in non-Kerr materials.

1. Introduction

As is well known, propagation of nonlinear pulses in optical fibers and self-guided beams in nonlinear optical waveguides is governed by the nonlinear Schrödinger (NLS) equation [1]-[3]. However, practical materials often display the physical effects, such as *nonlinearity saturation*, which can be only understood by analyzing the generalized models of the nonlinear refractive index (see, e.g., [4]-[6] and references therein). In dimensionless units, these models are described by the generalized NLS (GNLS) equation,

$$i\frac{\partial \Psi}{\partial t} + \frac{\partial^2 \Psi}{\partial x^2} + f(|\Psi|^2)\Psi = 0, \tag{1}$$

A. Hasegawa (ed.), Physics and Applications of Optical Solitons in Fibres '95, 75–87.

where Ψ is a slowly varying envelope of the electric field and, for example, in the case of *spatial solitons*, the variables t and x stand for the longitudinal and transverse coordinates, respectively. The function $f(|\Psi|^2)$ characterizes a nonlinear correction to the material refractive index.

The model (1) has been intensively investigated in the context of the existence and stability of bright optical solitons which are stationary solutions of Equation (1) of the form $\Psi_s(x,t) = \Phi(x;\omega)e^{i\omega t}$, where the real function $\Phi(x;\omega)$ vanishes for $|x| \to \infty$, and ω is the nonlinearity-induced shift of the propagation constant which, for simplicity, we call below 'frequency'. For the generalized nonlinearities, such solitons may become *unstable*, and a standard approach is to analyze the soliton instability using *the linear stability analysis*. For the case of bright solitons of the GNLS equation (1), the soliton stability is given by the Vakhitov-Kolokolov criterion [7] (see also [8]-[10]). However, the linear stability analysis *does not allow to understand* the subsequent evolution of unstable solitons when the linearized equations, describing *exponentially growing* perturbations on the soliton profile, become not valid, and numerics is used for every particular problem.

The main objective of this chapter is to present a brief review of the analytical asymptotic models which describe not only linear instabilities but also the nonlinear long-term evolution of the unstable solitons. Our approach is based on a nontrivial modification of the soliton perturbation theory [11] near the instability threshold. Assuming application of our results to the problems of nonlinear optics, here we consider the instabilities of (i) bright solitons in a non-Kerr medium, (ii) bright solitons supported by resonant parametric interactions in a $\chi^{(2)}$ optical material, and (iii) dark solitons in a non-Kerr defocusing waveguide. However, the basic ideas and method itself can be easily extended to cover solitons of other nonintegrable models, e.g., those describing wave interactions and coupling between different wavelengths or polarizations.

2. Bright Solitons in a Non-Kerr Medium

The standard soliton perturbation theory [11] is usually applied to analyse the soliton dynamics under the action of external perturbations. Here we deal with *a qualitatively different physical problem* when an unstable bright soliton evolves under the action of its 'own' perturbations. As a result of the development of the instability, the soliton frequency ω varies in time, $\omega = \omega(t)$. The linear stability of bright solitons is determined by the slope of the derivative $N_s'(\omega) \equiv dN_s/d\omega$ [7, 9], where $N_s(\omega)$ is the soliton energy

$$N_s(\omega) = \frac{1}{2}\int_{-\infty}^{+\infty} \Phi^2(x;\omega)dx.$$

Near the instability threshold, when the derivative $N'_s(\omega)$ vanishes, the growth rate is small, and we can assume that the instability-induced evolution of the perturbed soliton is, first, slow in t and, second, almost adiabatic (i.e. self-similar). Therefore, we can develop an asymptotic theory representing the solution to the original model (1) as $\Psi = \phi(x; \omega; T) \exp[i\omega_0 t + i\epsilon \int_0^T \Omega(T')dT']$, where $\omega = \omega_0 + \epsilon^2\Omega(T)$, $T = \epsilon t$, and $\epsilon \ll 1$. Here the constant value ω_0 is chosen near the critical value ω_c where the derivative $N'_s(\omega)$ vanishes. Then, using the asymptotic multi-scale expansion for $\phi(x; \omega; T)$, $\phi = \Phi(x; \omega) + \epsilon^2\phi_2(x; \omega; T) + O(\epsilon^3)$, we obtain the following equation for the soliton propagation constant Ω (details can be found in [12]),

$$M_s(\omega_c)\frac{d^2\Omega}{dT^2} + \frac{1}{\epsilon^2}\frac{dN_s}{d\omega}\bigg|_{\omega=\omega_0}\Omega + \frac{1}{2}\frac{d^2N_s}{d\omega^2}\bigg|_{\omega=\omega_c}\Omega^2 = 0, \qquad (2)$$

where $M_s(\omega)$ is calculated through the stationary soliton solution,

$$M_s = \int_{-\infty}^{+\infty}\left[\frac{1}{\Phi(x; \omega)}\int_0^x \Phi(x'; \omega)\frac{\partial\Phi(x'; \omega)}{\partial\omega}dx'\right]^2 dx > 0.$$

Thus, the remarkable result which follows from this asymptotic analysis is the following. In the GNLS equation (1) the dynamics of solitons near the instability threshold can be described by Equation (2) which is equivalent to the equation for motion of an effective *inertial and conservative* particle of the mass $M_s(\omega_c)$ and coordinate Ω moving under the action of the potential force which is proportional to the difference $N_0 - N_s(\omega)$, where $N_0 = N_s(\omega_0)$.

First two terms in Equation (2) give the result of the linear stability analysis [7, 9], i.e. the soliton of Equation (1) is linearly unstable provided $N'_s(\omega_0) < 0$. Nonlinear term in Equation (2) allows us to consider not only linear but also long-term (nonlinear) dynamics of unstable solitons and, moreover, to identify all scenarios of the soliton instability dynamics.

As an example, we consider the GNLS equation with two power-law nonlinearities, $f(I) = -I^{p/2} + \beta I^p$, which possesses an explicit soliton solution for any $p > 0$. For $1 < p < 2$ the function $N_s(\omega)$ always has a minimum, and the derivative $N''_s(\omega_c)$ is positive [see Figure 1, left]. It is convenient to analyse the dynamical system (2) on the phase plane $(\Omega, \dot{\Omega})$ or, equivalently, on the plane $(\omega, \dot{\omega})$. When the value N_0 exceeds the extremum value of $N_s(\omega_c)$, the corresponding phase plane is presented in Figure 1 (right).

The standard initial-value problem to the GNLS model (1) corresponds to the initial condition for (2) lying on the axis $\dot{\omega}(0) = 0$. For such initial conditions we reveal *three different regimes* of the soliton dynamics. They are depicted by the curves $1, 2$ and 3 in Figure 1.

If the amplitude of the initially perturbed soliton is taken to be smaller than the amplitude of the (unstable) stationary solution [i.e., $\omega(0) < \omega_0$,

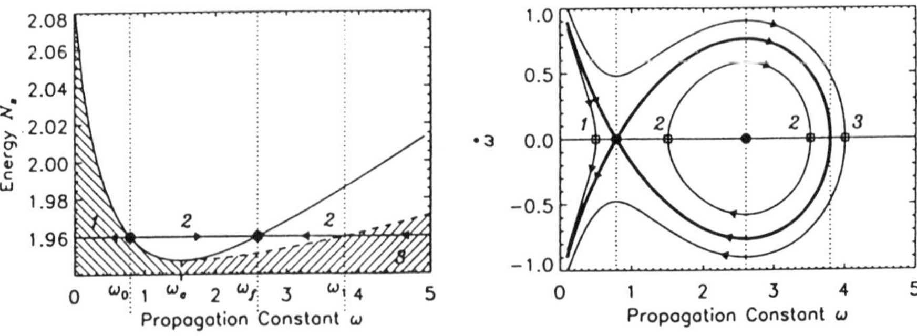

Figure 1. Left: Characteristic dependence of the soliton energy $N_s(\omega)$ in the model (1) with $f(I) = -I^{p/2} + \beta I^p$ for $1 < p < 2$. Minimum point ω_c corresponds to the instability threshold, ω_0 and ω_f, to the unstable and stable solitons at $N_s(\omega_0) = N_0 = 1.96$, respectively. Right: Phase plane $(\omega, \dot{\omega})$ of the asymptotic model describing different regimes of the soliton dynamics: soliton decay (curves 1 and 3) and amplitude oscillations (curve 2).

curves 1 in Figure 1], the instability leads to a decrease of ω (proportional to the soliton amplitude) and this process results in the soliton spreading into a small-amplitude wave packet which finally decays into linear waves.

On the other hand, if the initially perturbed unstable soliton has the amplitude slightly bigger than that of the stationary soliton solution [curves 2 in Figure 1], the exponentially growing instability 'pushes' the soliton into the stability region where there exists *a stable stationary* soliton solution with $\omega = \omega_f$, corresponding to the same value of the energy invariant N_0 [see Figure 1, left]. However, due to the inertial nature of the soliton evolution, the transition from the unstable to stable states is accompanied by the long-lived periodic oscillations of the soliton amplitude. These oscillations can be explained by the existence of a nontrivial *internal mode* of a bright soliton in the GNLS model (1). Similar oscillations have been recently observed numerically for the solitons of the so-called threshold nonlinearity [13].

We note that, according to our analytical theory, the periodic oscillations of the soliton amplitude must disappear for larger deviations from the stable equilibrium state [$\omega(0) > \omega_1$, curves 3 in Figure 1]. In spite of the fact that larger deviations of the soliton amplitude do not remove the soliton from the stability region, the evolution of perturbations is predicted to lead again to decreasing of the soliton amplitude. As soon as the soli-

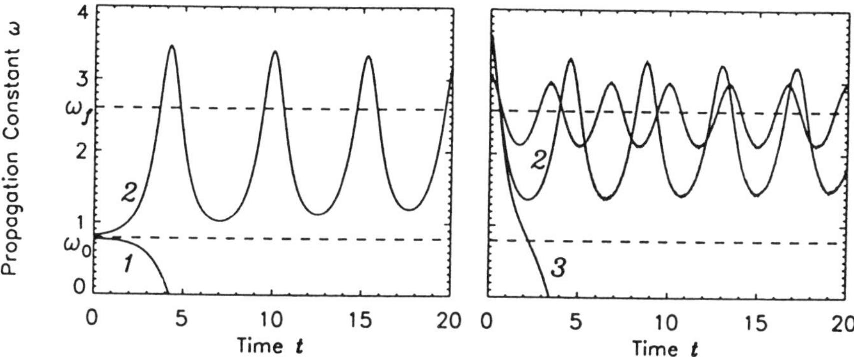

Figure 2.   Results of numerical simulations of the soliton instabilities in the GNLS equation with $f(I) = -I^{p/2} + \beta I^p$ at $p = 1.35$. Left: Initial conditions are selected near the unstable soliton with $\omega = \omega_0$. Right: Initial conditions are selected near the stable soliton with $\omega = \omega_f$.

ton enters the unstable region, it finally spreads out due to dispersion (or diffraction).

All predicted regimes of the soliton instability have been found in numerical simulations. Figure 2 shows several types of the soliton evolution for the soliton frequency $\omega(t)$ determined as the first-order derivative of the soliton phase calculated numerically at the pulse maximum. The results are in an excellent agreement with the predictions of the analytical model (2) [cf. curves in Figure 2 with the trajectories on the phase plane shown in Figure 1 (right)].

3. Parametric Solitons Due to $\chi^{(2)}$ Nonlinearity

It is well-known that the NLS equation becomes not valid near resonances with the higher harmonics excited due to a nonlinear (generally *quadratic*) response of a medium. When dispersion or diffraction become important, nontrivial effects can be observed for two interacting waves due to parametric interactions. In application to nonlinear optics, this means that the nonlinearity-induced phase shift and self-trapping of light beams can be achieved in the so-called $\chi^{(2)}$ materials due to cascaded nonlinearities. It has been also shown that the cascaded nonlinearities can support different types of two-wave (spatial or temporal) parametric solitons (see, e.g., [14, 15]) which have been recently observed experimentally in nonlinear planar waveguides [16].

Resonant interaction between the fundamental (w) and second (v) harmonics in a diffractive (or dispersive) $\chi^{(2)}$ medium can be described by the coupled equations for the dimensionless variables [15],

$$i\frac{\partial w}{\partial z} + \frac{\partial^2 w}{\partial x^2} - w + w^* v = 0,$$

$$i\sigma\frac{\partial v}{\partial z} + \frac{\partial^2 v}{\partial x^2} - \alpha v + \frac{1}{2}w^2 = 0, \tag{3}$$

where z is the propagation distance, x stands for the transverse coordinate (spatial solitons) or retarded time (temporal solitons). The parameter σ describes either the ratio of the wave vectors (spatial solitons) or the ratio of the group-velocity dispersions (temporal solitons) of two interacting waves. The parameter α can be presented as $\alpha = 2\sigma - \Delta$, where Δ is proportional to the wave vector mismatch $\Delta k = k_2 - 2k_1$ between the harmonics and it can also include the spatial walk-off effect.

The diffraction (dispersion) effects described by the second-order derivatives in Equations (3) are crucial for the existence of soliton solutions. Stationary soliton solutions of Equations (3) are presented by real functions $w = w_0(x;\alpha)$ and $v = v_{0;\alpha}(x)$ which are independent on z. For the particular value $\alpha = 1$ these solutions were first found by Karamzin and Sukhorukov [14]. Recent numerical analysis [15] revealed the existence of the soliton solutions of the system (3) for any positive value of α and, moreover, it was also shown that these solitons can be generated from certain classes of localized initial conditions.

To analyze the stability of the soliton solutions w_0 and v_0, we should linearize Equations (3) on the soliton background as the following,

$$w(x,z) = w_0(x) + [W_r(x) + iW_i(x)]e^{\lambda z}, v(x,z) = v_0(x) + [V_r(x) + iV_i(x)]e^{\lambda z}$$

and investigate the corresponding linear eigenvalue problem for the corrections (W_r, W_i) and (V_r, V_i). Similar to Section 2, this problem can be solved by the asymptotic method for *nonzero but small* λ. Near the instability threshold, such solutions are expected to exist only for special values of the parameters α near the critical curve $\alpha = \alpha_c(\sigma)$. The instability threshold, as well as the general dependence $\lambda(\alpha;\sigma)$, can be found from the corresponding solvability conditions to the linear problem. This analysis has been recently presented in Reference [17], and it was shown that the instability threshold is given by the generalized Vakhitov–Kolokolov criterion, $\partial\tilde{Q}/\partial\omega = 0$, where $\omega(\alpha;\sigma) = (2\sigma - \alpha)^{-1}$ is the renormalized soliton propagation constant and $\tilde{Q} = (2\sigma - \alpha)^{-3/2}Q$,

$$Q(\alpha;\sigma) = \frac{1}{2}\int_{-\infty}^{+\infty} \left[w_0^2(x;\alpha) + 2\sigma v_0^2(x;\alpha)\right] dx$$

is the renormalized energy (Menley-Rowe) invariant of Equations (3). In all these formulas the parameter σ is considered as an arbitrary parameter. Using this criterion and the numerical results on the stationary soliton solutions, the instability threshold curve $\alpha = \alpha_c(\sigma)$ has been calculated in Reference [17]. Moreover, it has been shown that the two-wave parametric solitons are unstable for $\alpha < \alpha_c$.

The important physical question is the development of linear instability in the subsequent dynamics of the two-wave solitons. In order to describe this analytically, we should take into account *nonlinear effects* similar to Section 2. We select $\alpha \equiv \alpha_0$ close to α_c, so that the small parameter ϵ characterizes the deviation $(\alpha_0 - \alpha_c) \sim O(\epsilon^2)$. Then, it follows from the linear theory that the growth rate has the order $O(\epsilon)$ and, therefore, the unstable linear perturbations grow on the 'slow' scale $Z = \epsilon z$. This allows us to introduce the slowly varying complex phase $S = S(Z)$ and look for the perturbed solutions of Equations (3) in the form of asymptotic series $w = [w_0(x; \alpha) + \epsilon^2 \Omega W_r(x; \alpha) + O(\epsilon^3)]e^{i\epsilon S}$, $v = [v_0(x; \alpha) + \epsilon^2 \Omega V_r(x; \alpha) + O(\epsilon^3)]e^{2i\epsilon S}$, where the functions W_r and V_r are the solutions of the linearized problem, and $\Omega = (2\sigma - \alpha)^{-1} dS/dZ$ describes a correction to the soliton propagation constant. Using the asymptotic multi-scale technique we find the nonlinear equation for the function Ω,

$$\tilde{M}_s(\alpha_c) \frac{d^2\Omega}{dZ^2} + \frac{1}{\epsilon^2} \frac{\partial \tilde{Q}}{\partial \omega}\bigg|_{\alpha=\alpha_0} \Omega + \frac{1}{2} \frac{\partial^2 \tilde{Q}}{\partial \omega^2}\bigg|_{\alpha=\alpha_c} \Omega^2 = 0, \qquad (4)$$

where the renormalized soliton mass is defined as $\tilde{M}_s = (2\sigma - \alpha)^{-1/2} M_s(\alpha; \sigma)$,

$$M_s = \int_{-\infty}^{+\infty} \left[w_0^2 \left(\frac{W_i}{w_0}\right)_x^2 + v_0^2 \left(\frac{V_i}{v_0}\right)_x^2 + \frac{1}{2v_0} (w_0 V_i - 2v_0 W_i)^2 \right] dx > 0,$$

where the index x stand for the corresponding partial derivatives, and the functions W_i and V_i are the solutions of the linearized problem. We mention that the derivative $\partial^2 \tilde{Q}/\partial \omega^2|_{\alpha=\alpha_c}$ is always positive for the two-wave parametric solitons.

It turns out that the asymptotic theory applied to the two-wave $\chi^{(2)}$ solitons gives in the renormalized form essentially the same equation of motion of the equivalent particle as Equation (2) for the (one-component) bright solitons in non-Kerr optical materials. As we discussed in Section 2, this equation describes two main scenarios of the instability of bright solitons, either long-lived periodic amplitude oscillations or soliton decay.

Indeed, according to Equation (4), the exponential growth of linear perturbations with $\Omega(0) > 0$ is stabilized by nonlinearity leading to oscillations around a novel stable equilibrium state. This equilibrium state corresponds to a stationary soliton which can be also described by Equations (3) but for

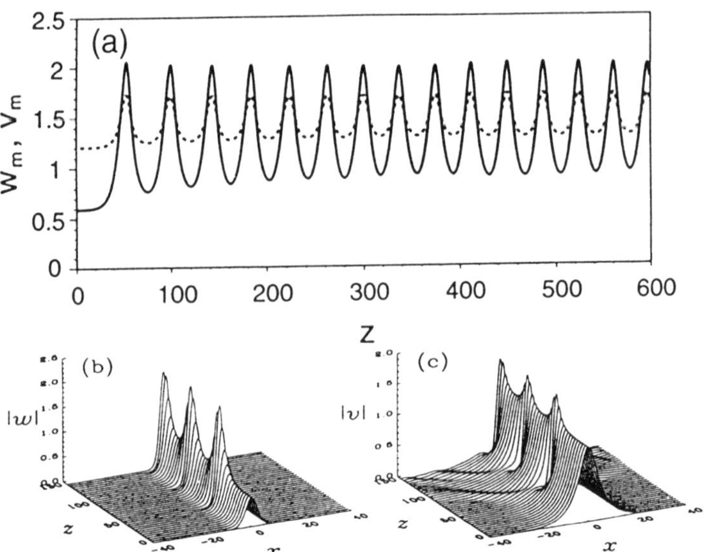

Figure 3. Periodic oscillations of the two-wave soliton of the model (3) after the development of instability ($\sigma = 2$, $\alpha = 0.05$). (a) Evolution of the harmonics amplitudes $w_m \equiv |w(0,z)|$ (solid) and $v_m \equiv |v(0,z)|$ (dashed). (b),(c) Propagation of the soliton components w and v, respectively.

a renormalized parameter α lying *inside* the stability region. Therefore, for a slightly increased amplitude of an unstable soliton our analytical model (4) predicts *in-phase* pulsations of the fundamental and second harmonics around a novel stable soliton, and this exactly corresponds to the evolution observed numerically and presented in Figure 3. For $\Omega(0) < 0$, according to Equation (4), such a stabilization is not possible and, as a result, a slightly decreased amplitude of an unstable soliton gradually decreases further.

4. Dark Solitons in a Non-Kerr Medium

The asymptotic analysis presented in Sections 2 and 3 for one- and two-component bright solitons can be also developed for dark solitons. Here we consider the dark solitons described by the GNLS equation (1) and derive an asymptotic equation governing their nonlinear dynamics near the instability threshold.

Dark soliton of Equation (1) can be presented in the form,

$$\Psi_s = \Phi_s(\xi; v, q)e^{if(q)t}, \quad \Phi_s = \Phi e^{i\theta},$$

where $\xi = x - vt$, $d\theta/d\xi = \frac{v}{2}\left(1 - q/\Phi^2\right)$, and the real function $\Phi(\xi; v, q)$ tends to the *nonzero boundary conditions* at infinity, $\Phi^2 \to q$ as $|\xi| \to \infty$.

Here we consider only the case of "classical" dark solitons for which $\Phi^2 < q$ for any ξ. Thus, the dark soliton depends on two parameters, the velocity v and the background intensity q. For $v \to c$, where $c \equiv \sqrt{-2qf'(q)}$ is the phase velocity of linear waves propagating on the constant background $|\Psi|^2 = q$, the dark soliton has small amplitude, while for $v \to 0$ the soliton amplitude (or 'darkness') reaches its maximum (but always $I_{\min} > 0$).

In spite of the fact that the background intensity is determined by the boundary conditions, its *local variations* are still possible, e.g. due to radiation emitted by the perturbed dark soliton during its nonstationary evolution. Therefore, the dynamics of unstable dark solitons is much more complicated than that of bright solitons. However, we show below that even in this case, but near the instability threshold, the effective analytical model for the nonlinear dynamics of the dark solitons can be derived with the help of the asymptotic method.

As has been recently shown in References [18, 19], the renormalized field momentum defined as

$$\tilde{P}_s(v,q) = 2\mathrm{Im} \int_{-\infty}^{+\infty} \frac{\partial \Psi_s^*}{\partial \xi} \Psi_s \left(1 - \frac{q}{|\Psi_s|^2}\right) d\xi = -v \int_{-\infty}^{+\infty} \frac{(\Phi^2 - q)^2}{\Phi^2} d\xi \quad (5)$$

plays a key role in the formulation of the variational principle for dark solitons, and also it defines the criterion for their linear instability (the latter is given by the slope of the derivative $d\tilde{P}_s/dv$, see [19]). Thus, the linear stability analysis for dark solitons is based on the renormalized momentum (5).

Following the approach outlined in Section 2, we present the perturbed dark soliton of the GNLS equation (1) in the form

$$\Psi = \phi(\xi; v, q; X, T) \exp[if(q)t + \epsilon R(X, T)],$$

where $\xi = x - v_0 t - \int_0^T V(T')dT'$, $v = v_0 + \epsilon V(T)$, $X = \epsilon x$, $T = \epsilon t$, and $\epsilon \ll 1$. The constant value v_0 stands for the velocity of the (unstable) unperturbed dark soliton which is selected near the critical value v_c where the derivative $d\tilde{P}_s/dv$ vanishes. Then, we expand ϕ in the asymptotic series,

$$\phi = \Phi_0(\xi; v, q) + \epsilon^2 \Phi_2(\xi; v, q; X, T) + O(\epsilon^3)$$

and reduce the GNLS equation (1) to the sequence of linear inhomogeneous equations, similar to the case of bright solitons. However, the important difference between the asymptotic procedure for bright and dark solitons is the existence of the linear correction term to the dark soliton which does not vanish at infinity, $\frac{1}{\epsilon^2}(|\phi|^2 - 1) \to u^{\pm}$ as $\xi \to \infty$. This nonlocalized part of Φ_2 in the region of the soliton $[\xi \sim O(1)]$ indicates the generation of the radiation field escaping the perturbed dark soliton. This radiation field can be found as a solution to the GNLS equation (1) in the outer region $X \sim O(1)$ where $|\Psi|^2 = q + \epsilon^2 U_r^{\pm}(X,T) + O(\epsilon^3)$. Analysis shows that, in the leading order, the radiation propagates to the right and to the left with the limiting phase velocity c, $U_r^{\pm} = U_r^{\pm}(X \mp cT)$. At the more slowly time scale, the evolution of the radiation field obeys the Korteweg–de Vries (KdV) equations [20],

$$\mp 4c\frac{\partial U_r^{\pm}}{\partial \tau} + \alpha U_r^{\pm}\frac{\partial U_r^{\pm}}{\partial X} + \frac{\partial^3 U_r^{\pm}}{\partial X^3} = 0, \tag{6}$$

where $\tau = \epsilon^3 t$ and $\alpha = 4\left[3f'(q) + qf''(q)\right]$. Then, the matching conditions produce the profile of the radiation field generated by the perturbed dark soliton, $U_r^{\pm} = u^{\pm}$ as $\xi \to \pm\infty$, $X \to 0$, and $T = X/v_0$.

After such a nontrivial procedure, in the third-order approximation, we can derive the following differential equation for $V(T)$

$$M_s(v_c)\frac{dV}{dT} + \frac{1}{\epsilon}\frac{\partial \tilde{P}_s}{\partial v}\bigg|_{v=v_0} V + \frac{1}{2}\frac{\partial^2 \tilde{P}_s}{\partial v^2}\bigg|_{v=v_c} V^2 = 0, \tag{7}$$

where the positive coefficient $M_s(v,q)$ has the form,

$$M_s(v,q) = \frac{q}{c}\left(\frac{\partial S_s}{\partial v}\right)^2 + \frac{2c}{q}\left(\frac{\partial N_s}{\partial v}\right)^2 > 0$$

Here $S_s(v,q) = v\int_{-\infty}^{+\infty}\left(1 - \frac{q}{\Phi^2}\right)d\xi$ and $N_s(v,q) = \frac{1}{2}\int_{-\infty}^{+\infty}\left(\Phi^2 - q\right)d\xi$ are proportional to the total phase shift across the dark soliton and its complementary power, respectively. Thus, due to the radiation the dark solitons near the instability threshold can be described by an equation for motion of an effective *inertial and dissipative* particle of mass M_s and velocity V under the action of the frictional force which is proportional to the difference $\tilde{P}_0 - \tilde{P}_s(v)$, where $\tilde{P}_0 = \tilde{P}_s(v_0)$.

As follows from Equation (7), the dark soliton becomes unstable for $\partial \tilde{P}_s/\partial v|_{v=v_0} < 0$ and the instability dynamics essentially depends on the sign of the perturbation. Usually (see, i.e. [19]) the instability is observed for small velocities so that $\partial^2 \tilde{P}_s/\partial v^2|_{v=v_c} > 0$. Therefore, if we decrease the intensity of the unstable dark soliton, the instability results in a monotonic transition from unstable to stable soliton realized at the same value of the renormalized momentum $\tilde{P}_0$. This scenario of the soliton instability is described asymptotically by the following solution of Equation (7),

$$V = \Delta V\frac{e^{\lambda T}}{1 + e^{\lambda T}}, \tag{8}$$

where λ and ΔV are defined through the coefficients of Equation (7). For the other sign of the perturbation, $\Delta V < 0$, the instability cannot be described by bounded solutions. This implies that an increase of the intensity of the dark soliton can not be suppressed by weak nonlinear effects described by Equation (7), and it can lead to an essential transformation of the unstable dark soliton. As a result of this transformation, the intensity of the dark soliton falls to zero at the minimum point with the subsequent evolution depending on the global nonlinear properties of the particular case of the GNLS model. This process of the soliton decay can be called *"collapse of dark solitons"*.

Radiation field excited due to the transition of the dark soliton from unstable to stable states can be also found, $U_r^{\pm} = (\zeta_{\pm}\Delta V \lambda/4)\mathrm{sech}^2(kX)$,

$$\zeta_{\pm} = \mp\frac{1}{c(c\mp v)}\left(\frac{q}{2}\frac{\partial S_s}{\partial v} \pm c\frac{\partial N_s}{\partial v}\right)$$

and $k = \lambda/(2v)$. It can be shown that for small v the amplitude ζ_{+} is positive while ζ_{-} is always negative. Since the radiation field developes into a soliton of the KdV equation (6) only for negative amplitude the secondary dark soliton is formed in the wave U_r^{-} which propogates in the opposite direction with respect to the original perturbed soliton. On the other hand, the radiation in a co-propagating wave U_r^{+} decays into quasi-linear dispersive waves.

To confirm the analytical theory of the dark soliton instability we consider the GNLS equation with the nonlinearity in the form, $f(I) = I - \beta I^2$. Figure 4 displays two scenarios of the instability development, namely, the decay of the unstable dark soliton into two stable solitons of smaller amplitudes which propagate in the opposite directions [Figure 4, left] and the collapse of a dark soliton and formation of two kink-type fronts [Figure 4, right]. Thus, the basic predictions of the asymptotic analytical theory are in direct agreement with the results of numerical simulations.

5. Conclusions

We have presented the general approach for deriving the analytical asymptotic models describing the dynamics of solitons in nonintegrable systems near the instability threshold. Our theory can describe *nonlinear regimes* of the soliton instabilities, and it gives the analytical description of the longterm evolution of unstable solitons. We have demonstrated this method considering three important examples which find their applications in the problems of nonlinear optics, namely, bright and dark solitons supported by generalized (non-Kerr) nonlinearities, and two-wave parametric solitons in an optical medium with a quadratic responce near the second-harmonic

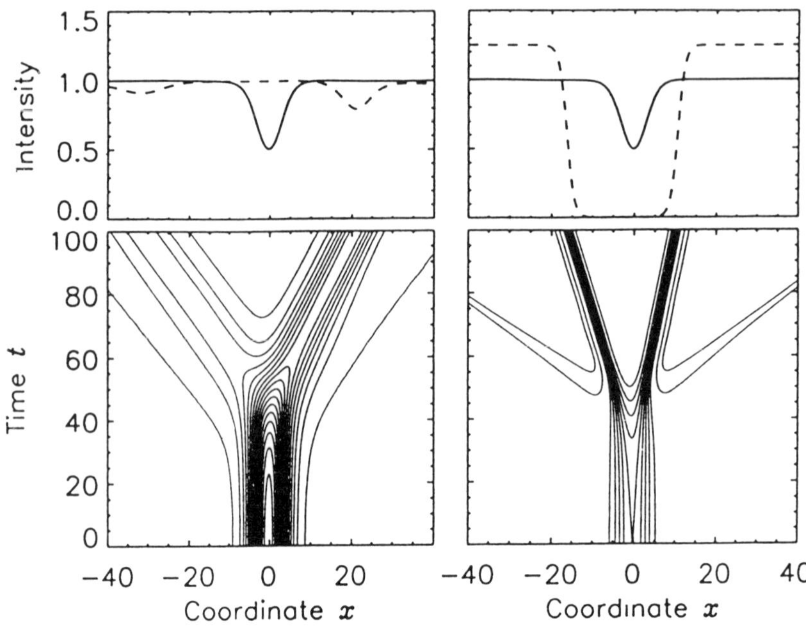

Figure 4. Evolution of the unstable dark soliton in the model (1) with $f(I) = I - \beta I^2$. Left: Transformation to a stable dark soliton via decay of an unstable soliton into two stable ones. Right: Collapse of the unstable dark soliton into two kink-type fronts. Top: initial (solid) and final (dashed) states; bottom: the corresponding contour plots.

resonance. However, we expect that the approach itself and the most of the results described here for the nonlinear regimes of the soliton instabilities will be also valid for other nonintegrable models of various physical context.

References

1. Hasegawa, A. and Tappert, F.: *Appl. Phys. Lett.* **23** (1973), 142.
2. Chiao, R. Y., Garmire, E. and Townes, C. H.: *Phys. Rev. Lett.* **13** (1964), 479.
3. Zakharov, V. E. and Shabat, A. B.: *Zh. Eksp. Teor. Fiz.* **61** (1972), 118 [English translation: *Sov. Phys. JETP* **34** (1972), 62].
4. Zakharov, V. E. and Synakh, V. S.: *Zh. Eksp. Teor. Fiz.* **68** (1975), 940 [English translation: *Sov. Phys. JETP* **41** (1975), 465].
5. Kaplan, A. E.: *Phys. Rev. Lett.* **55** (1985), 1291.
6. Mulder, L. J. and Enns, R. H.: *IEEE J. Quantum Electron.* **25** (1989), 2205.
7. Vakhitov, M. G. and Kolokolov, A. A.: *Radiophys. Quantum Electron* **16** (1973), 783 [*Izv. Vyssh. Uch. Zav. Radiofizika* **16** (1973), 1020].
8. Kuznetsov, E. A., Rubenchik, A. M. and Zakharov, V. E.: *Phys. Rep.* **142** (1986), 103 (1986).
9. Weinstein, M. I.: *Comm. Pure. Appl. Math.* **39** (1986), 51.
10. Mitchell, D. J. and Snyder, A. W.: *J. Opt. Soc. Am. B* **10** (1993), 1572.
11. Kivshar, Yu. S. and Malomed, B. A.: *Rev. Mod. Phys.* **61** (1989), 763.

12. Pelinovsky, D. E., Afanasjev, V. V. and Kivshar, Yu. S.: *Phys. Rev. E* **52** (1995).
13. Snyder, A. W., Hewlett, S. and Mitchel, D. J.: *Phys. Rev. E* **51** (1995), 6297.
14. Karamzin, Yu. N. and Sukhorukov, A. P.: *JETP Lett.* **20** (1974), 339.
15. Buryak, A. V. and Kivshar, Yu. S.: *Opt. Lett.* **19** (1994), 1612; *Phys. Lett. A* **197** (1995), 407.
16. Baek, Y., Schiek, R. and Stegeman, G. I.: *Nonlinear Guided Waves and Their Applications*, OSA Technical Digest Series, **Vol. 6** NThB1 (1995), 24.
17. Pelinovsky, D. E., Buryak, A. V. and Kivshar, Yu. S.: *Phys. Rev. Lett.* **73** (1995), 591.
18. Kivshar, Yu. S. and Krolikowski, W.: *Opt. Commun.* **114** (1995), 353.
19. Kivshar, Yu. S. and Krolikowski, W.: *Opt. Lett.* **20** (1995), 1527.
20. Pelinovsky, D. E., Stepanyants, Yu. A. and Kivshar, Yu. S.: *Phys. Rev. E* **51** (1995), 5016.

STOCHASTIC INSTABILITY OF CHIRPED OPTICAL SOLITONS IN MEDIA WITH PERIODIC AMPLIFICATION

F. KH. ABDULLAEV

Theoretical Division, Physical-Technical Institute
of Uzbek Academy of Sciences
700084, Tashkent-84, G. Mavlyanov Str. 2-B, Uzbekistan

Abstract. We consider the propagation of chirped optical pulses in media with periodic amplification. The process is described by means of a variational approach. The nonlinear resonances and chaotic oscillations of the pulse width are studied. The stochastic instability with respect to the evolution of the width and the stochastic decay of pulses under periodic amplification are predicted. The decay length is calculated and compared with numerical simulations. The application for long-haul optical communications lines with amplifiers is discussed.

1. Introduction

As can be shown by different methods, the evolution of an initially chirped pulse has some interesting properties. Namely, at propagation of initially chirped pulses like

$$u(x, t) = N \operatorname{sech}(\tau/a) \exp(i\phi_0 \tau^2), \tag{1}$$

oscillations of width and amplitude are observed [1]. Such inherently frequency chirped pulses can be obtained for example from pulsed injection lasers. A good description of the oscillations is given by a variational approach [2, 3]. The inhomogeneity of the medium or periodic gain can lead to changes in dynamics of soliton and that can provide the control of soliton parameters. As an example the periodic variation of dispersion may be considered which lead to nonlinear resonances in the oscillating of the soliton width [4]. In the numerical modelling performed in this work stochastic oscillations and decay of the soliton was also observed. The process of soliton

A. Hasegawa (ed.), Physics and Applications of Optical Solitons in Fibres '95, 89–98.

decay under stochastic variation of the dispersion has been investigated in [5]. In this work we calculated also the distortion length of solitons. Another natural example of perodic variation of the medium parameters is a chirped pulse propagation in optical fibers with periodic gain. The soliton instability for this problem has been investigated in [6]. Recently it was suggested to use a periodic variation of the fiber core for the control of soliton parameters [7] (see also [8]). It should be noted, that the problem of spatial soliton propagation in periodically modulated media, namely in media with periodic modulation of the refractive index along z direction (the transverse direction is x) can be reduced to the considered problem.

Thus, the periodic variation of the gain or a diameter of core in optical fibers can induce nonlinear oscillations in the width and amplitude of chirped pulses and in particular a stochastic instability of chirped soliton. The purpose of the present work is to investigate the nonlinear dynamics of the parameters of the chirped soliton at propagation in media with a variable nonlinearity and in particular the nonlinear resonances and dynamical chaos of the soliton width and amplitude oscillations. We obtain the stochasticity criterion and estimate the boundary frequency of stochastic motion. The chaotic decay length of the soliton is calculated as well. The considered problem is important for the investigation of dynamical stochasticity of nonlinear waves in inhomogeneous media. This work presents an example of the application of the variational approach to the description of the dynamical stochasticity in the dynamics of chirped solitons. In general the investigation of the dynamical stochasticity of separate solitons deals with the randomization of evolution of soliton parameters (see for example [9, 10]). The distortion of a soliton due to dynamical stochastization which leads to diffusive growth of the soliton width is investigated here. The analogy with the depinning of particle from oscillating in time potential of the Kepler problem is used for the solution.

The paper is organized as follows: In section 2 the model is described and the equations for the parameters of the chirped soliton are obtained by means of a variational approach. It is shown, that the problem is equivalent to the perturbed Kepler problem. In section 3 we review the unperturbed Kepler problem. Finally, the nonlinear resonances and chaos in the soliton width and amplitude oscillations are studied in section 4.

2. Description of the Model

Let us consider the optical pulses propagation in the optical fibers with a periodically varying parameter of nonlinearity. Introducing standard variables we have the modified nonlinear Schrödinger equation (NLSE) for the dimensionless envelope of the electric field [6, 8]

$$iu_x + \frac{1}{2}u_{\tau\tau} + f(x)\mid u\mid^2 u = 0, \tag{2}$$

where x, τ are the coordinate along the direction of propagation and time is given in a moving reference frame respectively. The function $f(x)$ describes the periodic modulation of nonlinearity of the media. In the case of periodic gain $f(x)$ may be represented as

$$f(x) = \sum_{-\infty}^{\infty} c_n e^{-k_n x}, \quad c_n = \frac{1 - e^{-2\Gamma l}}{2\Gamma l}\frac{1}{1 - ik_n/\Gamma}, \quad k_n = 2\pi n/l. \tag{3}$$

Here Γ is the damping constant, l is the distance between amplifiers. For $\Gamma l \ll 1$ we have $f(x) = 1 + \frac{2\Gamma l}{\pi}\sin k_1 x$. It should be noted, that in practice we deal with the small scale modulations and strong perturbations [11]. Here we will consider the small perturbation case like in [6]. Taking into consideration applications to many problems of this model and possibility of analytical treatments we restricted our analysis by this simplified case. For the propagation of spatial solitons in inhomogeneous media it is necessary to change variables $x \to z, \tau \to x$ and assume that the periodical modulation of nonlinear part of refraction index along axis z takes place. Below we will consider the evolution of chirped pulses in the form (1). As we have mentioned before, the variational analysis of unperturbed NLSE showed that the soliton width exhibits oscillating behavior at propagation along the fiber. According to the Inverse Scattering Transform method the amplitude is slowly damped due to radiation effect and the pulse evolves into the asymptotic soliton. The period of oscillations depends on the deviation of initial conditions from the solitonic solution. For the analysis we take a trial function in the form of chirped soliton

$$u(x,t) = A(x)\mathrm{sech}[\frac{\tau}{a(x)}]\exp\left[i\phi_0(x)\tau^2\right], \tag{4}$$

where $A(x), a(x), \phi_0(x)$ describe the amplitude, width and soliton frequency chirp respectively. To investigate the evolution in x of these parameters, following to the idea of the variational approach [3], we must calculate the average Lagrangian $< L >$

$$< L >= \int_{-\infty}^{\infty} d\tau L, \quad L = i(uu_x^* - u^*u_x) + \mid u_\tau \mid^2 - f(x)\mid u \mid^4. \tag{5}$$

The equations for the soliton parameters A, a, ϕ_0 are obtained by using of variation of $< L >$

$$\delta \int_0^x < L > dx = 0. \tag{6}$$

From (4)-(6) we obtain the system of equations

$$\phi_0 = \frac{1}{2}(\ln a)_x, \tag{7}$$

$$\frac{d}{dx}(a \mid A \mid^2) = 0, \quad N^2 = a \mid A \mid^2, \tag{8}$$

$$a_{xx} = \frac{4}{\pi^2 a^3} - \frac{4N^2 f(x)}{\pi^2 a^2}, \tag{9}$$

$$\frac{d}{dx}\arg A = -\frac{1}{3a^2} + \frac{5}{6}\frac{N^2}{a}f(x). \tag{10}$$

As can be seen from (9) the soliton width evolution is described by the motion of unit mass effective particle in the nonstationary effective anharmonic potential. When $f(x)$ is the periodic function we deal with a periodically perturbed potential of the Kepler problem. Consequently, the investigation of soliton width oscillations under the periodic gain is reduced to the study of particle dynamics in the periodically perturbed potential of the Kepler problem. Before proceeding to the solution we will first discuss the unperturbed Kepler problem.

3. Description of the Unperturbed Kepler Problem

We give here the information necessary for the further analysis the unperturbed Kepler problem [12]. The corresponding potential energy is expressed by

$$U = \frac{2}{\pi^2 a^2} - \frac{4N^2}{\pi^2 a}. \tag{11}$$

The minimum of this potential is achieved at $a_c = 1/N^2$ and is equal to $U_0 = -2N^4/\pi^2$. From (11) the frequency ω_0 of small oscillations of the particle near the bottom of potential can be found. The result is $\omega_0 = 2N^4/\pi$. It is the frequency of chirped soliton width oscillations during its propagation in the homogeneous fiber. The character of oscillations of soliton width and amplitude is defined by the quantity $E_0 = 2/\pi^2 a_0^2 - 4N^2/\pi^2 a_0 + 2\phi_0^2$. When $E_0 < 0$ i.e. $1 + \phi_0^2 \pi^2 a_0^2 < 2N^2 a_0$, we have oscillatory regime, but the case $E_0 > 0$, i.e. if $1 + \phi_0^2 \pi^2 a_0^2 > 2N^2 a_0$ corresponds to the unlimited motion when progressively soliton distortion takes place($a \to \infty$). Below we will investigate the case when the first condition ($E_0 < 0$) is valid. From (9) the total energy of the effective particle is

$$E = \frac{a_x^2}{2} + \frac{2}{\pi^2 a^2} - \frac{4N^2}{\pi^2 a}. \tag{12}$$

For the oscillatory regime the action variable is

$$J = \oint a_x dx = \frac{4\sqrt{2}N^2}{\pi\sqrt{-E}} - 4. \tag{13}$$

The total energy is expressed in terms of the action J by relation

$$H = E = -\frac{32N^4}{\pi^2}\frac{1}{(J+4)^2}. \tag{14}$$

From this expression the frequency of soliton width oscillations can be derived

$$\omega(J) = \frac{dH}{dJ} = 2\frac{32N^4}{\pi^2}\frac{1}{(J+4)^3} = \frac{\pi}{2\sqrt{2}N^2}(-E)^{3/2}. \tag{15}$$

There is also need in the knowledge of a spectral properities of trajectories of the Kepler motion [13]. In the parametric form the solution for trajectory is:

$$a = b(1 - e_0\cos\xi), \quad \omega x = \xi - e_0\sin\xi,$$
$$e_0 = (1 - \frac{\pi^2\,|\,E\,|}{2N^4}), \quad b = \frac{2N^2}{\pi^2\,|\,E\,|}. \tag{16}$$

Here e_0 is the excentricity. From (16) it follows that $a_{min} = b(1-e_0)$, $a_{max} = b(1+e_0)$ in agreement with [3]. The Fourier component for $a(x)$ and $1/a(x)$ can be expressed from (16)

$$a = \sum_m a_m e^{-im\omega x}, \quad a_m = \frac{2b}{m}J'_m(e_0 m). \tag{17}$$

Here (') denotes the differentiatiin of the Bessel function with respect to its argument. For the function $1/a(x)$ we have

$$\frac{1}{a(x)} = \sum_m c_m e^{-im\omega x}, \quad c_m = \frac{2}{b}J_m(e_0 m). \tag{18}$$

For further analysis it is also important to know the number of harmonics N_0 in the unperturbed motion spectrum.

$$N_0 = (1 - e_0^2)^{-3/2} = \frac{2^{3/2}N^6}{\pi^3\,|\,E\,|^{3/2}}. \tag{19}$$

We have near the separatrix, when $e_0 \to 1$, $N_0 \to \infty$. This circumstance is of a major importance when the investigation of the stochasticity of motion is concerned.

4. Nonlinear Resonances and Chaos in the Soliton Width and Amplitude Oscillations

At propagation of chirped soliton in optical fibers the interplay between two periodic processes - the soliton width oscillations and periodic modulation of medium nonlinearity occurs. Here the nonlinear resonances and dynamical chaos phenomena are possible. This problem has a close analogy with the problem of hydrogen atom ionization in a periodic field [14]. Below for the purpose of analysis the theory of nonlinear resonances [15, 16] will be applied. It is conveinient to use the action-angle variables for solution of the problem. The total Hamiltonian in the action-angle variables J, θ has the form

$$H = H_0 + V = H_0(J) + V(J, \theta; x). \tag{20}$$

The unperturbed Hamiltonian H_0 is

$$H_0 = -\frac{32N^4}{\pi^2}\frac{1}{(J+4)^2}. \tag{21}$$

The interaction term of the Hamiltonian V is

$$V(J, \theta; x) = -\frac{4N^2 f(x)}{\pi^2 a(J, \theta)}. \tag{22}$$

From (20)-(22) we have in the variables J, θ the equations of motion as

$$\frac{dJ}{dx} = -f(x)\frac{\partial V}{\partial \theta},$$

$$\frac{d\theta}{dx} = \omega(J) + f(x)\frac{\partial V}{\partial J}. \tag{23}$$

The influence of perodic force $f(x)$ initiated nonlinear resonances in oscillations of the soliton parameters. The regions of nonlinear resonances can be obtained using the expansion of $V(J, \theta; x)$ in Fourier series. Using (19) we found from (22)

$$V = -\frac{4N^2}{\pi^2}\sum_k\sum_m \frac{2}{a_0} J_m(e_0 m) f_k e^{-im\theta - ik\Omega x} + c.c. \tag{24}$$

Below for simplicity we consider the case when there exists only one harmonic term in $f(x)$, i.e. when $k = 1$. The resonances appeared if the condition

$$m\omega(J) - \Omega = 0, \tag{25}$$

is satisfied, i.e.when $\omega = \Omega/m$. The oscillations of soliton width a and amplitude resonates with the periodcal modulations of fiber parameters.

This effect can be used for the control of soliton parameters. Another interesting phenomenon occurring here is the overlaping of resonances. This phenomenon occurs when the distance between resonances $\delta\omega$ become comparable with the width of resonances $\Delta\omega$. The distance between resonances can be estimated as follows

$$\delta\omega = \mid \omega_{m+1} - \omega_m \mid \approx \frac{\Omega}{m^2} = \frac{\omega^2}{\Omega}. \tag{26}$$

The width of m-th resonance is

$$\Delta\omega = \frac{\omega'\omega \mid V_m \mid}{\Delta\omega\Omega}, \omega' = \frac{d\omega}{dJ}, \tag{27}$$

where

$$V_m \approx -\frac{8N^2}{\pi^2 b} f_0 J_m(e_0 m). \tag{28}$$

The stochasticity criterion can be derived from the requirement that the chaotic motion manifests itself when the resonances are overlaped, i.e. if the condition (we use (26),(27))

$$K = \frac{\Delta\omega^2}{\delta\omega^2} = \frac{\omega' \mid V_m \mid \Omega}{\omega^3} \geq 1 \tag{29}$$

is valid. Near the separatrix a stochastic layer is appeared. So, in the region $1 + \phi_0^2\pi^2 a_0^2 > 2N^2 a_0$, where according to the standard analysis soliton must exist, due to stochasticity the loss of solitonic properies of the pulse takes place. The main reason consists in the diffusion over resonances and transition from the oscillatory regime to the unlimited motion, where $E > 0$. The diffusion time (in our case diffusion length) can be estimated as $L = 2\pi/\bar{\omega}$. It is convenient to consider two limiting cases, when the explicit estimates for stochasticity criterion can be obtained:
1. Let $1 \ll m \ll N_0$ and $1 - e_0 \ll 1$. Then for the Bessel function J_m the estimate is valid:

$$J_m(e_0 m) \approx \frac{1}{2\pi}\left(\frac{4}{3}\right)^{1/6}\frac{1}{m^{1/3}}. \tag{30}$$

Substituting (26)-(28),(30) into (29), we have the following criterion for stochasticity of oscillations of the soliton's width

$$K = \frac{3^{5/6} f_0 \Omega^{2/3}(J + 4)^4}{2^{17/3} b N^{14/3}} \geq 1. \tag{31}$$

The boundary of stochastic layer in frequency is obtained by equalizing the parameter of stochasticity $K = 1$. From (31) and (15) we obtain

$$\bar{\omega} = \omega_0\left[\frac{3^{5/6} 2^{1/3}\pi^{1/3} f_0 \Omega^{2/3}}{b N^{14/3}}\right]^{3/4}, \tag{32}$$

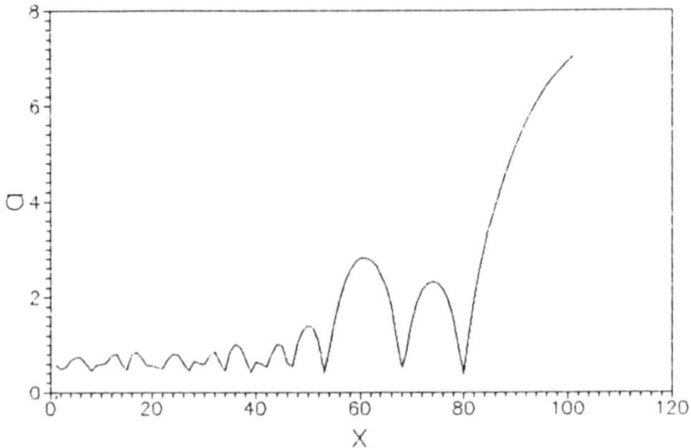

Figure 1. The soliton width versus distances of propagation from Equation (9)

where ω_0 is defined by (15). The length of stochastic decay of soliton is order of

$$L = \frac{2\pi}{\bar{\omega}} \sim \frac{1}{f_0^{3/4}(\Omega N)^{1/2}}. \tag{33}$$

As can be seen from (31),(32) when the initial conditions are inside the stochastic layer, the oscillations of soliton width become chaotic. The diffusion over resonances will occur and in consequence the soliton width tends to infinity i.e. soliton decays. In Figure 1 the soliton width versus x from numerical solution of (9) for value of parameters $a_0 = 1, N = 1.3, \Omega = 1, f_0 = 0.16$ is plotted.

It is clearly seen the chaotic oscillations of soliton width. Let us estimate the critical value of perturbation f_c when the soliton is decaying. We have from (31)

$$f_c \geq \frac{\pi^{11/3} b E^2}{2^5 3^{5/6} \Omega^{2/3}}. \tag{34}$$

In Figure 2 the results of numerical modelling of (9) for the parameters as in Figure 1 are presented. The stochastic oscillations and decay of solitons are observed.

2. Let $m \gg N_0$. Then the next asymptotic behavior for $J_m(e_0 m)$ is valid

$$J_m(e_0 m) \approx (\frac{3}{3\pi})^{1/2} \sqrt{\frac{1}{m}} N_0^{1/6} \exp\left(-\frac{m}{3N_0}\right). \tag{35}$$

Using (28),(29),(35) we can found

$$K = \frac{3^{1/2} f_0 \Omega^{3/2} (J+4)^7}{2^{27/2} N^8} \exp\left(-\frac{2\pi^2 \Omega}{3N^4}\right) \geq 1. \tag{36}$$

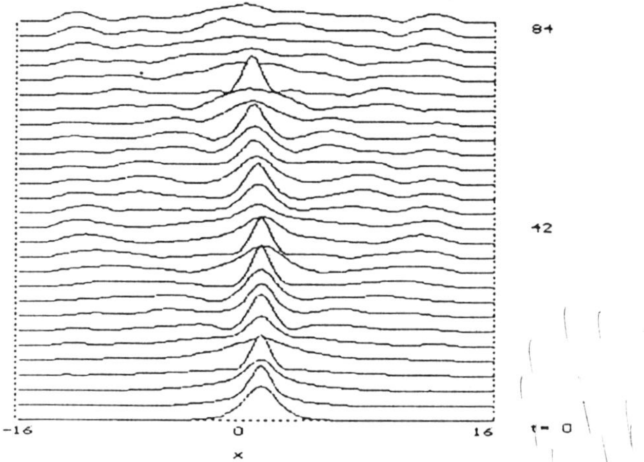

Figure 2.   Space-time plot of $| u |$ from numerical solution of Equation (2)

The boundary value of stochastic layer in frequency is

$$\bar{\omega} = \omega_0 \frac{3^{9/4} 2^{3/14} f_0^{3/7} \Omega^{9/14}}{N^{24/7}} \exp\left(-\frac{2\pi^2 \Omega}{7N^4}\right). \tag{37}$$

The stochastic decay length is estimated as

$$L = \frac{2\pi}{\bar{\omega}} \sim \Omega^{-9/14} \exp\left(\frac{2\pi^2 \Omega}{7N^4}\right). \tag{38}$$

It is seen that in this case the decay length is exponentially large.

5. Conclusion

In conclusion we have investigated the propagation of chirped solitons in a Kerr-like nonlinear media with periodically varying nonlinearity. In the variables of standard NLSE it corresponds to the temporal variation of the nonlinearity. The existence of nonlinear resonances in chirped soliton amplitude and width oscillations due to periodic modulations of the medium nonlinearity is shown. When the nonlinear resonances are overlaped a new kind of instability of chirped pulses called dynamical stochasticity present itself. This instability leads to the chaotic oscillations of the amplitude and width with the subsequent decay of the chirped soliton. We have obtained stochasticity criterion, when the stochastic decay of a soliton can occur. The critical value of modulation depth for the soliton distortion is estimated. The stochastic decay length is calculated as well. We note that the mechanism of instability is universal and independent from the character of periodicity of modulation. The obtained results can be applied to the analogous problems in the theory of spatial solitons in modulated media, Langmuir solitons in nonstationary plasma.

6. Acknowledgments

Author is grateful to JSPS and Deutsche FG for support of this work, and to A.A.Abdumalikov, B.Baizakov for collaboration. The useful discussions with S.Darmanyan, F.Lederer and I.Uzunov are appreciated.

References

1. Satsuma, J. and Yajima, N.: Initial value problems of one-dimensional self modulation of nonlinear waves in dispersive media, *Prog. Theor. Phys. Suppl.* **55** (1974), 284-306.
2. Anderson, D.: Variational approach to nonlinear pulse propagation in optical fibers, *Phys.Rev.* **A27** (1983), 3135-3145.
3. Anderson, D., Lisak, M. and Reichel, T.: Asymptotic propagation properties of pulses in a soliton-based optical-fiber communication system, *J.Opt.Soc.Am.* *B* **5** (1988), 207-210.
4. Malomed, B. A., Parker, D. F. and Smith, N. F.: Resonant shape oscillations and decay of a soliton in a periodically inhomogeneous nonlinear optical fiber, *Phys.Rev.* **E48** (1993),1418-1425.
5. Abdullaev, F. Kh., Baizakov, B. B. and Umarov, B. A.: Propagation of optical pulses in media with random dispersion, *JTP Letters* **20** (1994), 23-27.
6. Gordon, J. P.: Dispersive perturbations of solitons of the nonlinear Schrödinger equation, *J.Opt.Soc.Am.B* **9** (1992), 91-97.
7. Bauer, R. G. and Melnikov, L. A.: Multi-soliton fission and quasi-periodicity in a fiber with a periodically modulated core diameter, *Opt.Commun.* **115** (1995), 190-198.
8. Abdullaev, F. Kh.: Propagation of optical soliton in fiber with randomly varying parameters *JTP Letters* **9** (1983), 305-307.
9. Abdullaev, F. Kh.: Dynamical chaos of solitons and nonlinear periodic waves, *Phys.Rep.* **179** (1989), 1-78.
10. Scharf, R. and Bishop, A. R.: Soliton chaos in the nonlinear Schrödinger equation with spatially periodic perturbations, *Phys.Rev.* *A* **46** (1992), 2973-2976.
11. Kodama, Y., Maruta, A. and Hasegawa, A.: Long distance communications with solitons, *Quantum Opt.* **6** (1994), 463-516.
12. Goldstein, H.: *Classical Mechanics*, Wiley, New York, (1965).
13. Landau, L. D. and Lifshitz, E. M.: *Field Theory*, Pergamon, New York, (1975).
14. Delone, N. B., Krainov, V. P. and Shepelyansky, D. L.: *Sov. Phys.-Uspekhi* **140** (1983), 355-370.
15. Chirikov, B. V.: Universal instability of many dimensional oscillator systems, *Phys. Rep.* **52** (1979), 263-343.
16. Zaslavsky, G. M. and Sagdeev, R. Z.: *Introduction to Nonlinear Physics*, Nauka Publishers, Moscow (In Russian), (1988).

FEMTOSECOND PULSE GENERATION
FROM SEMICONDUCTOR LASERS
USING SOLITON EFFECT COMPRESSION
AND ITS APPLICATION TO ULTRAFAST
ELECTRO-OPTIC SAMPLING MEASUREMENTS

H. F. LIU
Photonics Research Laboratory
Department of Electrical and Electronic Engineering
The University of Melbourne
Parkville, Vic.3052, AUSTRALIA
T. KAMIYA
Department of Electronic Engineering
University of Tokyo
7-3-1 Hongo, Bunkyo-ku, Tokyo 113, JAPAN

Abstract. An experimental investigation of femtosecond pulse generation from semiconductor lasers using the soliton-effect compression technique is presented. By making use of the fiber birefringence advantageously with a wave plate, the inherent pedestal associated with this technique is eliminated, which results in a train of 185 fs pedestal-free pulses, the shortest pulses ever reported for semiconductor lasers. 400 fs clean pulses are also generated by an alternative approach of adiabatic compression in a dispersion decreasing fiber. The generated femtosecond pulses are implemented in a non-invasive ultrafast electro-optic sampling system, by which a high speed metal-semiconductor-metal photodetector is characterized with a temporal resolution of 2 ps.

1. Introduction

The generation of ultrashort optical pulses using semiconductor lasers is of great interest and practical importance for the development of high speed optical communications including soliton-based systems [1,2] owing to their distinct

99

A. Hasegawa (ed.), Physics and Applications of Optical Solitons in Fibres '95, 99–114.
© *1996 Kluwer Academic Publishers. Printed in the Netherlands.*

advantages over other laser systems, such as small and compact in size, high efficiency, low cost, ease of monolithic integration and their ability of direct current modulation up to gigahertz rates. In addition to the applications in telecommunications, such semiconductor laser-based pulse sources are also being used in ultrafast data processing and switching, electro-optic sampling, optical sampling, millimeter-wave signal generation and many other areas of ultrafast laser technology [3]-[7]. Over the past decades, there have been many advances in ultrashort pulse generation from semiconductor lasers to improve the performance of such pulses in terms of pulse width, peak power, repetition rates, spectral purity, wavelength tunability and temporal stability.

Two commonly used techniques for generating short pulses from semiconductor lasers are gain-switching and mode-locking. Gain-switching can be easily achieved by exciting the first peak of relaxation oscillation while biasing the laser just below the threshold. This method is simple and robust and can provide pulse trains with variable repetition rates up to gigahertz ranges. The inherent frequency chirp associated with gain-switched pulses can be compensated for by compressing the pulses with a dispersion element. By doing so, pulses down to a few picoseconds have been generated, which is essentially limited by the chirped bandwidth of the laser diodes [8,9]. Mode-locking, on the other hand, is realized by modulating the laser at an intercavity mode spacing frequency in a monolithic or an external extended cavity configuration, which can provide much shorter pulses at a predetermined repetition rate with very pure spectral properties. If the whole gain bandwidth of the semiconductor laser materials can be fully utilized, it should be possible to generate very short pulses ($\approx$ 50 fs). Experimentally, most of the mode-locked pulses achieved so far are still in the picosecond regime, although femtosecond pulses have been reported for devices with some special structures, or by combining with chirp compensation techniques [10,11].

To further reduce the pulse width, the use of nonlinear effects in optical fibers is most promising. Among others, the soliton-effect pulse compression technique that utilizes the self-phase modulation (SPM) in optical fibers has been shown to be very effective for achieving femtosecond pulses [12]. By combining with a grating pair compressor, pulses as short as 18 fs have been realized in solid-state laser systems [13]. Since new frequency components are generated through the SPM, which increases the optical bandwidth of the pulses, the finite gain bandwidth of semiconductor laser materials will essentially no longer impose limitations on the final resultant pulses. Conceptual simplicity and ease of implementation in practical systems are the key advantages of this technique. However, an inherent drawback is that the compressed pulses are always accompanied by a broad low intensity pedestal that carries a large proportion of the pulse energy, which obviously is undesirable. Hence, an important issue in implementing this compression technique is to remove the pedestal.

In this paper, we present a series of experimental investigation on ultrashort pulse generation from semiconductor lasers using the soliton-effect compression technique, and demonstrate the implementation of thus generated short pulses in non-invasive ultrafast electro-optic sampling measurements. The paper is organized as follows. A brief description of the soliton effect compression in optical fibers is given in Section 2. Section 3 highlights the experimental investigation of the soliton effect compression of semiconductor laser pulses. In particular, we demonstrate that an appropriate length of the soliton-effect compression fiber, when used in conjunction with a wave plate and a polarizer, can act simultaneously as a compressor and an intensity discriminator. By this novel technique, a train of 185 fs pedestal-free pulses is generated, which are the shortest pulses ever reported for semiconductor lasers. Section 4 encompasses the implementation of femtosecond pulses generated by the soliton effect compression in an electro-optic sampling system. Using such a sampling system, a high speed photodetector is characterized. Conclusions of this paper are summarized in Section 5.

2. Soliton Effect Pulse Compression

Optical solitons are the results of interaction between the group velocity dispersion (GVD) and SPM in fibers. Soliton theory [14,15] shows that if a pulse has a $\mathrm{sech}^2(t)$ profile and the pulse parameters are such that the two effects exactly balance each other, the optical pulse travels along the fiber as a fundamental soliton with neither temporal nor spectral change. For this to occur, the pulse peak power needs to satisfy

$$P_N = 3.11N^2 \frac{D\lambda^2}{2\pi c\gamma\tau^2} \tag{1}$$

with $N = 1$, where D is the fiber dispersion parameter, c is the light velocity, τ is the pulse width, λ is the signal wavelength and γ is the fiber nonlinearity coefficient. If the peak power is greater than that required to excite fundamental solitons, higher order solitons will be formed, whose order, N, is determined by the pulse peak power.

An interesting property of higher order ($N > 1$) solitons is their shapes change periodically during the propagation, where they always experience an initial narrowing phase before recovering their original shape at integral multiples of the soliton period. The soliton effect compression is realized by exploiting this property. It is this unique property that makes the pulse compression possible. With an appropriate choice of fiber length, the input pulses can be compressed by a factor that depends on the soliton order N. As the compression phenomenon is due to the pulse narrowing of solitons, this technique is referred to as the soliton effect compression.

The soliton-effect compression can, therefore, be realized by simply amplifying transform-limited pulses to form higher order solitons, and then launching them into a fiber, whose length is determined by the distance that solitons attain a shape with the narrowest width during their periodic shape change. Numerical studies have shown that the compression factors increase with the increase in soliton orders. As attractive as it may appear, this technique is not without its problems. First, the compressed pulses are always accompanied inherently by an undesirable pedestal component. Secondly, as the widths of the compressed pulses are very narrow, higher order fiber nonlinearities will largely influence the compression process. These issues will be addressed at Section 3.

3. Femtosecond Pulse Generation from Semiconductor Lasers Using the Soliton Effect Compression Technique

Although the usefulness of the soliton effect compression in achieving ultrashort pulses was demonstrated using solid-state lasers in early 1980's, the required high power restrained its application to semiconductor laser pulses. However, the situation changed dramatically with the development in optical amplifiers in late 1980's. The first experimental demonstration was reported in 1990 by Liu, et al., using a 1.3 μm gain-switched distributed feedback (DFB) laser in conjunction with two semiconductor laser amplifiers. By exciting $N = 3$ solitons, a train of 1.26 ps pulses were obtained [16,17]. A similar experiment at 1.55 μm was subsequently reported using multi-stage high gain fiber amplifiers to produce 250 fs pulses [18]. In this section, our recent investigation on femtosecond pulse generation using the soliton effect is described. We will discuss the pulse compression, pulse reshaping and effects of higher order fiber nonlinearities. Pulse compression experiments using an alternative approach will also be presented.

3.1 SOLITON-EFFECT COMPRESSION OF SEMICONDUCTOR LASER PULSES

Figure 1 illustrates schematically the experimental set-up used in the soliton effect compression experiments. A 1.55 µm InGaAsP DFB laser diode was gain-switched by pumping with intense short electrical pulses from a comb generator operating at 100 MHz, which yielded a train of 27 ps pulses with red-shifted frequency chirp from the laser. These pulses were subsequently launched into a 1 km-long dispersion-shifted fiber (DSF) of normal dispersion (D = - 26.6 ps/km/nm @ 1.55 µm) to remove the undesired chirp. After the chirp compensation, the pulses were compressed down to 3.6 ps, with a time-bandwidth product of 0.6 [9]. Hence, the pulses are nearly transform-limited and can be used as the soliton sources for the compression experiments.

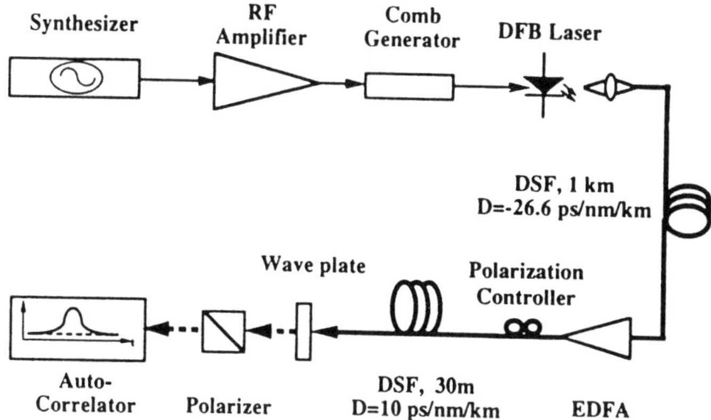

Figure 1. The experimental set-up used for the soliton effect compression.

From the average power measurements, we estimated the peak power of these linearly compressed pulses was about 120 mW. To perform the soliton effect compression, these pulses were then amplified by a high gain Erbium-dope fiber amplifier and coupled into another DSF (D = 10 ps/km/nm). We aimed for the compression of eighth order solitons, which according to Equation (1), require the pulse peak power be about 50 W with an optimum fiber length of 30 m. The amplifier gain was adjusted through the pump currents, and when the peak power reached the required level to form an eighth order soliton, significant pulse narrowing was observed. The autocorrelation trace and the time-averaged spectrum of the compressed pulses is shown in Figure 2(a) and Figure 2(b), respectively. As can be seen clearly the optical spectrum has been broadened considerably due to the SPM. If we assume the central spike has a sech^2 shape, the inferred pulse width has a full width at half maximum of 200 fs. As expected, the central spike is accompanied by a big pedestal component, which carries approximately 60% of the total pulse energy.

In a separate experiment, dependence of the pulse output on laser DC bias was investigated, where the soliton order was fixed at N = 4. Autocorrelation traces of the compressed pulses at different bias levels are shown in Figure 3, which indicated that the pedestal can be suppressed to some extent by adjusting the bias currents of the laser [19]. It is interesting to note that the pedestal suppression was achieved when the compressed pulse width was also at minimum, which was less than 500 fs.

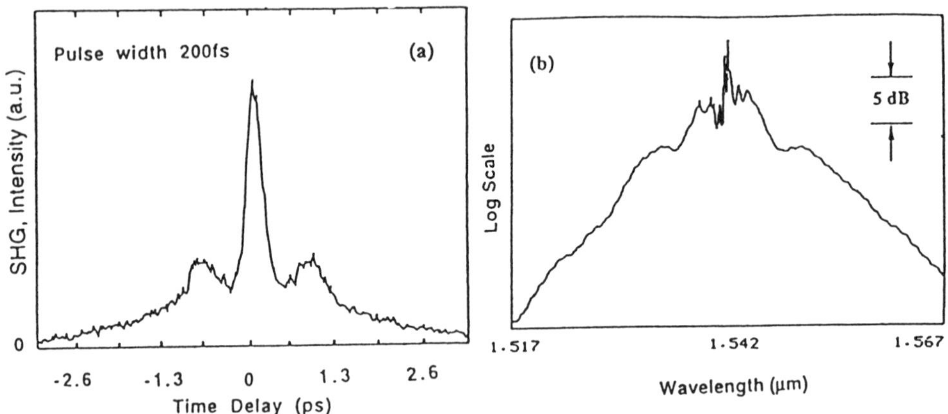

Figure 2. Autocorrelation trace and time-averaged spectrum of the compressed pulses

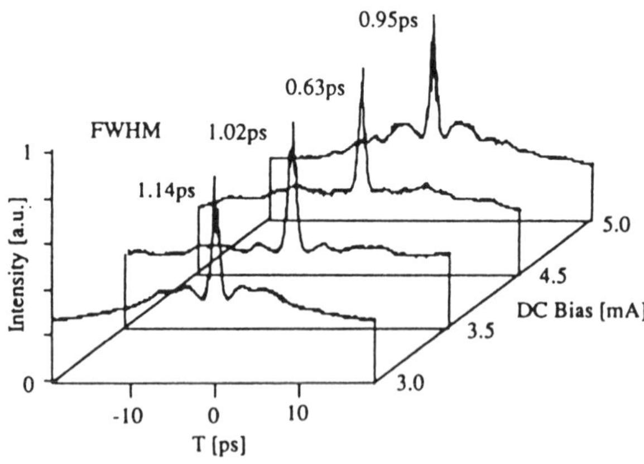

Figure 3. DC bias dependence of the soliton effect compressed pulses from a gain-switched laser diode.

3.2 PULSE RESHAPING

As has been noticed from the Figure 2(a), the compressed pulses are accompanied by an undesirable broad pedestal that carries a significant amount of pulse energy. This inherent drawback is due to the fact that compression is only effective at the central part of the pulse where the chirp is linear. This pedestal will not only degrade the pulse quality but also interfere with adjacent pulses to produce further pulse distortions. Previously, a wide band optical filter had been used to achieve nearly pedestal-free 250 fs pulses [18]. Recently, nonlinear optical loop mirrors (NOLM) have been proposed as an alternative for pulse reshaping, and 250 fs pedestal-free pulses have been reported [20]. Another technique is to use an intensity discriminator [21] in conjunction with the soliton-effect compression to separate the low intensity pedestal from the high intensity central spike. With this technique, Mollenauer, et al, have realized 260 fs nearly pedestal-free pulses from a color center laser [12].

In this work, we demonstrate that by making use of the fiber birefringence advantageously in conjunction with a wave plate, a polarizer and a polarization controller, the soliton effect compression can be realized with simultaneous suppression of the low intensity pedestal [22,23].

The pedestal suppression is realized as a result of the intensity dependent polarization state of light in fibers. This can be understood as follows by referring to Figure 4, which qualitatively illustrates the intensity discrimination process. As a higher order soliton pulse propagates inside a fiber, the narrowing process caused by the soliton-effect first comes into effect, which increases the pulse peak power. At the optimum fiber length, the compressed pulse consists of an extremely high intensity peak and a low intensity pedestal. In a birefringent fiber, we can consider the total electric field consists of two orthogonal components, aligned along the two different optic axes. Due to the intensity-dependent nonlinear birefringence, the intense central peak and the weak pedestal will experience different phase shifts, $\Delta\phi_1$ and $\Delta\phi_2$ for the central spike and the pedestal. This in turn will result in different polarization states at the fiber output for the high intensity peak and the low intensity pedestal. At appropriate power levels, the phase shift difference between them can be such that they have completely different polarization states. By rotating the wave plate to compensate for the phase shift of the intense peak, it will change to a linear polarization state while leaving the pedestal components in a different polarization states.

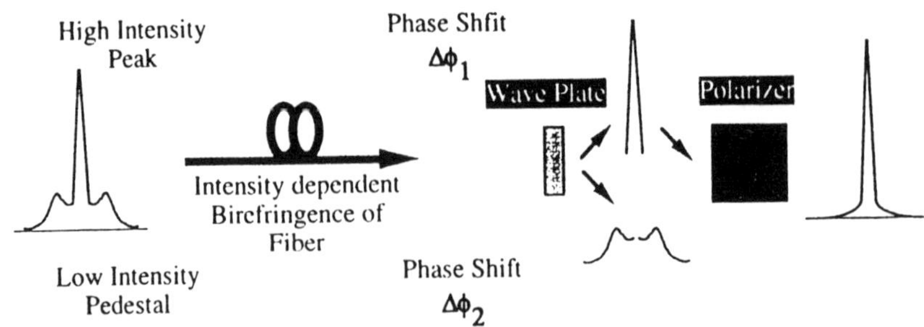

Figure 4 Schematic diagram showing the pedestal removal in a birefringent fiber.

The subsequent use of the polarizer will then reject the pedestal components and ensure that light in only one direction is produced at the output. Hence, with the help of the wave plate and the polarizer, intensity discrimination against the pedestal is realized. As the polarization state of the input light and its launching angle with respect to the optic axes of the fiber are also very crucial, a polarization controller is used at the fiber input to control the input polarization state and the launching angle into the fiber.

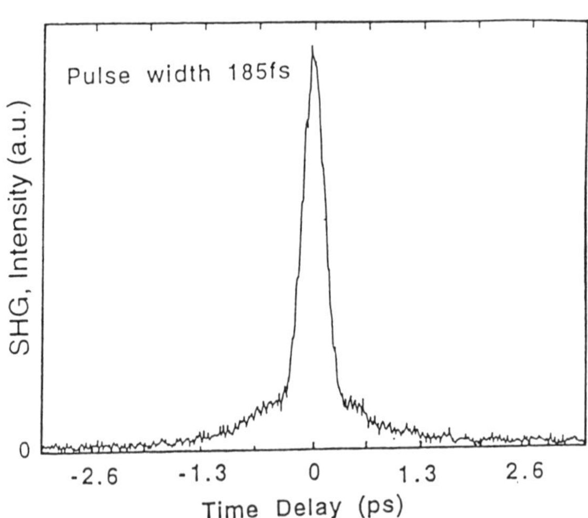

Figure 5. Autocorrelation trace of the same pulses as in Figure 2 after the pulse shaping.

We applied this technique to eighth order solitons and successfully suppressed the pedestal. Figure 5 shows the autocorrelation trace of the same pulses in Figure 2 after passing through the wave plate and the polarizer. As can be clearly seen, the inherently associated pedestal component has been removed and the pulse width has further been reduced to 185 fs. The pedestal suppression has been confirmed by an autocorrelation measurement with a longer time scale. These represent the shortest pedestal-free pulses ever generated from semiconductor lasers. Since a significant amount of pulse energy resided in the pedestal, a considerable decrease in the average power was observed, which was estimated to be half of that of the original pulses from the autocorrelator measurements. This, however, is still more than adequate for most applications. This pedestal-suppression technique was also applied to solitons of different orders. It was found that at higher soliton orders, although the pulse width could be even shorter, pedestal suppression was not as effective as that for eighth order solitons. This can be attributed to the fact that the optimum fiber length for higher order solitons was too short to cause appropriate phase shifts between the peak and the pedestal components. Therefore the discrimination between the peak and pedestal was not very effective resulting in pulses that were less clean. However, with further optimization in fiber parameters such as nonlinear birefringence and dispersion, we believe even shorter pedestal-free pulses can be generated.

The main advantage of this technique, in comparison with the NOLM [20], is the realization of higher order compression at the same input power level. As the input power is spilt almost equally into two arms in a NOLM, the soliton order will be effectively lower for the same input power. Using exactly the same input pulses (3.6 ps at the same power level) in conjunction with a NOLM, the shortest pedestal-free pulses we obtained were about 400 fs. Moreover, careful adjustments of the balance between two arms were found to be essential to achieve complete discrimination. On the other hand, our approach of combining the soliton-effect pulse compression and intensity discrimination is more straightforward and can be optimized by using fibers of appropriate birefringence and dispersion.

3.3 INFLUENCE OF HIGHER ORDER FIBER NONLINEARITIES

The soliton effect compression described in Section 2 was based on an ideal fiber model, which only includes SPM and GVD. However, as the compressed pulses become shorter and shorter, other higher order fiber nonlinear effects such as stimulated Raman scattering (SRS) [24] and third-order dispersion [25] will come into play. These effects will result in either a secondary pulse or an increased pedestal component. Although they can be removed, how to suppress their effects to achieve clean pulses with higher peak power is an important issue. To understand the influence of these effects, a modified nonlinear Schroedinger equation incorporating extra terms for SRS and third order dispersion as shown below has been used for numerical simulations.

$$\frac{\partial A}{\partial z} = i\,\gamma(|A|^2\,A - T_R\,A\,\frac{\partial |A|^2}{\partial T}) - \frac{i}{2}\beta_2\frac{\partial A}{\partial T^2} + \frac{1}{6}\beta_3\frac{\partial^3 A}{\partial T^3} - \frac{\alpha}{2}A \quad (2)$$

where A is the pulse amplitude, z and T are the spatial and temporal variables, T_R is the delayed noninear response that models SRS, γ is the fiber nonlinearity coefficient, β_2 represents the GVD, β_3 is the third order dispersion and α is the fiber loss. Figure 6 shows the calculated optimum compression ratio as a function of soliton order N for 1 ps input pulses in a DSF, where curve (a) is the results for the ideal case, curve (b) and (c) show the results with only SRS and β_3, respectively, and curve (d) shows the combined results of SRS and β_3. It can be seen clearly that these effects reduces the degree of achievable compression considerably, particularly at higher orders. Since SRS depends on the soliton power levels, which is proportional to the fiber dispersion parameter, it is preferable to use DSFs to facilitate the soliton-effect pulse compression at lower power levels. On the other hand, β_3 will have a more significant effect in such fibers. Hence, optimizing the fiber dispersion parameter to minimize the influence of these two effects is highly desired [26].

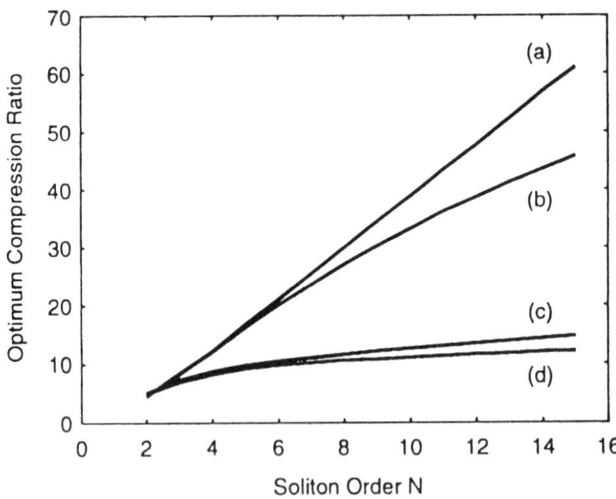

Figure 6. Calculated optimum compression ratio as a function of soliton order for the ideal compression case (a), with only SRS (b), with only β_3(c), and with both SRS and β_3(d)

One promising solution is to use dispersion flattened fibers to perform the soliton effect compression. If the GVD in such fibers is kept at a low value, the detrimental effects of both SRS and β_3 will be largely suppressed.

3.4 ADIABATIC SOLITON COMPRESSION

An alternative method to compress short pulses is to use the adiabatic soliton compression technique[27], which can provide much cleaner fundamental solitons. The basic idea behind this is to provide slow amplification over a certain length of fiber so that the solitons being launched can be adiabatically compressed while preserving their pulse areas. Mathematically, this can be expressed by $gz_0 \ll 1$, where g is the average gain coefficient and z_0 is the soliton period. Such slow amplification of solitons has been realized through both distributed fiber Raman amplification and effective amplification in dispersion decreasing fibers.

Since the first observation using a color-center laser [28], there have been many experimental demonstration in other laser systems. The first attempt of applying this technique to semiconductor laser pulses was made by Iwatsuki, et al., in 1991, where 5.8 ps pulses were compressed to 3.6 ps by distributed Raman amplification over a long DSF [29]. Using a dispersion decreasing fiber, Schell, et al., realized compression of 3 ps pulses to 540 fs [30]. With 0.8 ps pulses achieved by this technique, Suzuki, et al., demonstrated 160 Gb/s data stream generation through time-domain multiplexing [31].

Recently, we have generated even shorter pulses using the same technique. This was achieved by compressing 15 ps gain-switched pulses from a 1.55 μm DFB laser diode first to 4 ps by the usual linear chirp compensation in a 1 km-long DSF, and then by the adiabatic soliton compression to 400 fs in a 2 km-long dispersion decreasing fiber. In addition to being able to generate clean pulses, not critically depending on the fiber length is another key advantage over the soliton effect compression technique.

4. Applications of Short Pulses to Ultrafast Electro-Optic Sampling Measurements

In recent years, there has been tremendous progress in the development of high speed optoelectronic devices, where accurate performance characterization over the device operating bandwidth is always required to gain better understandings of the undergoing physical processes. The conventional methods for characterizing high speed devices includes time-domain techniques, based on sampling-scopes, and frequency-domain techniques, based on network analyzers. However, they are bandwidth limited to 50 GHz, and 100 GHz, respectively. The recent development of electro-optic (EO) sampling techniques has greatly alleviated the limitations imposed on the conventional methods and the measurement bandwidth has been increased to the THz regime.

Most of the laser sources used in such measurement systems are high power solid state lasers, which are usually operated at fixed repetition rates, making the synchronization with the circuits difficult. Recently, we have developed a compact desk-top EO-sampling system using gain-switched semiconductor laser sources, the repetition rate of which are capable of varying over a wide frequency range. By introducing a novel probe beam polarization modulation scheme and a balanced detection scheme, the major drawback of inferior sensitivity associated with the weak semiconductor laser pulses has been overcome [32,33]. The use of femtosecond pulses obtained by the soliton effect compression has improved the temporal resolution of the system to a comparable level to that of solid-state laser-based systems.

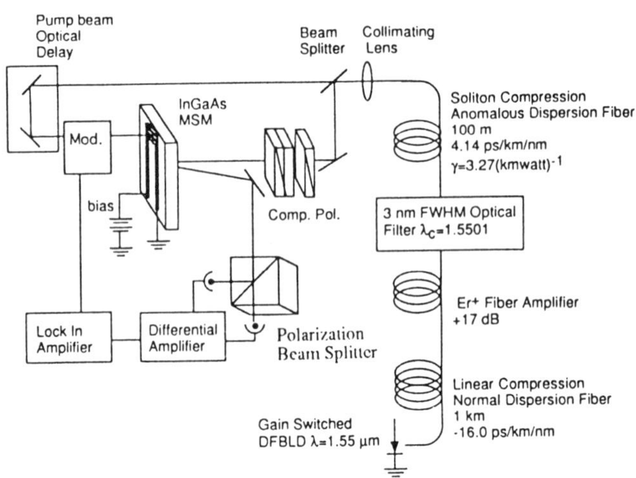

Figure 7. Experimental layout of the semiconductor laser-based electro-optic sampling system using femtosecond pulses generated by the soliton effect compression.

The system performance was examined by evaluating the pulse response of an InGaAs-InP metal-semiconductor-metal photodiode (MSM-PD). Figure 7 shows the semiconductor laser-based EO-sampling system set-up for such a measurement. Femtosecond optical output from the soliton effect compressor is divided into two beams, one pumping the MSM-PD under test and the other synchronously sampling the response. The pump beam was focused onto the MSM-PD after passing through a variable delay line and a modulator, and the sampling beam was focused on the transmission line via a polarizer and a compensator. The ultrafast transients in the MSM-PD induce modulation in the electric field seen by the sampling pulses, and modifies its polarization states. The reflected signals were then, after a polarization beam splitter, fed into two slow detectors for balanced detection. A lock-in amplifier is used to measure the DC output that is proportional to the amplitude of the sampled electrical signals while varying the delay between the two optical beams.

The MSM-PD was monolithically integrated with a 50 Ω coplanar stripline to minimize the waveform deterioration due to the bonding wire parasitic. The typical response measured by the EO-sampling system is shown in Figure 8, which has a rise time of 6.5 ps. Also shown is the calculated system response by deconvoluting the input pulse autocorrelation with an assumed detector-transmission line impulse response, $V(t)$.

$$V(t) = V_0(1 - e^{-t/\tau_1})e^{-t/\tau_2} \tag{3}$$

Using a rise time constant of $\tau_1 = 2.5$ ps and a fall time constant $\tau_2 = 110$ ps, the calculated curve matched well with the measured trace with a 10-90% rise time of 5.8 ps. This time constant is mainly due to the propagation time of the electric field penetration through the 50 µm substrate, and the RC time constant of the detector-transmission line.

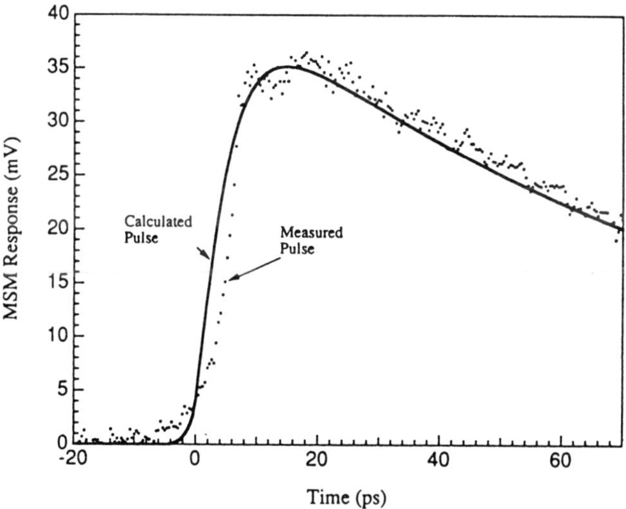

Figure 8. Measured response of the MSM-PD, which has a 10-90% rise time of 6.5 ps.

The temporal resolution of this EO-sampling system was limited by the timing jitter of the gain-switched laser diode, which has a typical value of 1-2 ps. Nevertheless, this is the best temporal resolution reported for a semiconductor laser-based EO sampling system [33]. By incorporating the injection seeding technique to reduce the timing jitter of such gain-switched laser diodes [34], we believe the system resolution can be further improved. Because of its compactness, this system offers a practical solution for non-invasive characterization of high speed optoelectronic devices.

5. Summary

We have investigated the generation of femtosecond pulses from semiconductor lasers using the soliton-effect compression technique. By making use of the fiber birefringence advantageously with a wave plate, we have demonstrated simultaneous pulse compression and intensity discrimination, which has resulted in a train of 185 fs pedestal-free pulses, the shortest pulses ever reported for semiconductor lasers. The total compression factor is more than 150. 400 fs pulses have been generated by an alternative approach of soliton adiabatic compression in a dispersion decreasing fiber. The application of thus generated femtosecond pulses in a non-invasive ultrafast electro-optic sampling system has been demonstrated. Using such a sampling system, a high speed InGaAs metal-semiconductor-metal photodetector has been characterized with a temporal resolution of 2 ps.

6. Acknowledgments

We would like to thank our colleagues K. A. Ahmed and K. C. Chan of the University of Melbourne and R. T. Sahara, H. Takeshita, K. Miwa and M. Tsuchiya of the University of Tokyo for their contributions to this work.

References

1. Lau, K.Y. : Short pulse and high frequency signal generation in semiconductor lasers, *J. Lightwave Technol.*, **7** (1989), 400-419.
2. Nakazawa, M.: Soliton transmission in telecommunication networks, *IEEE Commun. Mag.*, **32** (1994), 34-41.
3. Islam, M.N., Soccolich, C.E. and Gordon, J.P.: Ultrafast digital soliton logic gates, *Opt.Quantum.Electron.*, **24** (1992), 1216-1235.

4. Nagatsuma, T., Yaita, M., Shinagawa, M., Kato, K., Kozen, A., Iwatsuki, K. and Suzuki, K.: Electro-optic characterisation of ultrafast photodetectors using adiabatically compressed soliton pulses, Electron.Lett., **30** (1994), 815-817.

5. Kanada, T. and Franzen, D.L.: Optical waveform measurement by optical sampling with a mode-locked laser diode, *Electron.Lett.*, **22** (1986), 61-62,

6. Ahmed, Z., Liu, H.F., Novak, D., Pelusi, M.D., Ogawa, Y. and Kim, D.Y.: Low phase noise mm-wave signal generation using a passively mode-locked monolithic DBR laser injection locked by an optical DSBSC signal, *Electron.Lett.*, **31** (1995), 1254-1255.

7. Vasil'ev, P.: *Ultrafast Diode Lasers*, Artech House, Inc., Boston, London (1995).

8. Liu, H.F., Ogawa, Y. and Oshiba, S.: Generation of an extremely short single mode pulse (≈ 2 ps) by fiber compression of a gain-switched pulse from a 1.3 μm distributed-feedback laser diode, *Appl.Phys.Lett.*, **59** (1991), 1284-1286.

9. Ahmed, K.A., Liu, H.F., Onodera, N., Lee, P., Tucker, R.S. and Ogawa, Y.: Nearly transform-limited pulses (3.6 ps) generation from a gain-switched 1.55 μm distributed feedback laser by using fiber compression technique. *Electron.Lett.*, **29** (1993), 54-56.

10. Arahira, S., Oshiba, S., Matsui, Y., Kunii, T. and Ogawa, Y.: 500 GHz optical short pulse generation from a monolithic passively mode-locked distributed Bragg reflector laser diode, *Appl.Phys.Lett.*, **64** (1994), 1917-1919.

11. Azouz, A., Stelmakh, N., Langlois, P., Lourtioz, J-M. and Gavrilovic, P.: Nonlinear chirp compensation in high power broad-spectrum pulse from single-stripe mode-locked laser diode, *IEEE J.Selected Topics in Quantum Electron.*, **1** (1995), 577-582.

12. Mollenauer, L. F., Stolen, R. H., Gordon, J. P. and Tomlinson, W. J.: Extreme picosecond narrowing by means of soliton effect in single-mode optical fibers, *Opt. Lett.* (1983), 289-291.

13. Gouveia-Neto, A.S., Gomes, A.S.L. and Taylor, J.R.: Pulse of four optical cycles from an optimized optical fiber/grating pair/soliton pulse compressor at 1.32 μm, *J.Mod.Opt.*, **35** (1988), 7-10.

14. Agrawal, G. P.: *Nonlinear Fiber Optics*, 2nd ed., Academic Press, Boston, Massachusetts (1995).

15. Hasegawa, A.: *Optical Solitons in Fibers*, Springer-Verlag, Berlin (1989).

16. Liu, H. F., Ogawa, Y., Oshiba, S. and Nonaka, T.: Picosecond pulse generation from a 1.3 μm distributed feedback laser diode using soliton-effect compression, *IEEE J. Quantum Electron.*, **27** (1991), 1655-1660.

17. Liu, H.F., Ogawa, Y., Kunii, T. and Oshiba, S.: Generation of an extremely short (~1.26 ps) single mode pulse using soliton effect compression. 12th IEEE International Semiconductor Laser Conference, Post deadline paper PD-8, Davos, Switzerland (1990).

18. Nakazawa, M., Suzuki, K. and Yamada, E.: Femtosecond optical pulse generation using a distributed-feedback laser diode, *Electron. Lett.*, **26** (1990), 2038-2040.

19. Ong, J. T., Takahashi, R., Tsuchiya, M., Wong, S. H., Sahara, R. T., Ogawa, Y. and Kamiya, T.: Subpicosecond soliton compressionof gain-switched diode laser pulses using an erbium-doped fiber amplifier, *IEEE J. Quantum Electron.*, **29** (1993), 1701-1707.

20. Chusseau, L. and Delevaque, E.: 250-fs optical pulse generation by simultaneous soliton compression and shaping in a nonlinear optical loop mirror including a weak attenuation, *Opt. Lett.*, **19** (1994), 734-736.

21. Stolen, R. H., Botineau, J. and Ashkin, A.: Intensity discrimination of optical pulses with birefringent fibers, *Opt. Lett.*, **7** (1982), 512- 514.

22. Ahmed, K.A., Cheng, H.H.Y. and Liu, H.F.: Generation of 185 fs pedestal-free pulse using a 1.55 μm distributed feedback semiconductor laser, *Electron.Lett.*, **31** (1995), 195-196.

23. Ahmed, K.A., Chan, K.C. and Liu, H.F.: Femtosecond pulse generation from semiconductor lasers using the soliton-effect compression technique, *IEEE J.Selected Topics in Quantum Electron.*, **1** (1995), 592-600.

24. Chan, K.C. and Liu, H.F.: Effects of Raman scattering and frequency chirping on soliton-effect pulse compression, *Opt. Lett.*, **18 (1993)**, 1150-1152.

25. Chan, K.C. and Liu, H.F.: Effects of third order dispersion on soliton-effect pulse compression, *Opt. Lett.*, **19** (1994), 49-51.
26. Chan, K.C. and Liu, H.F.: Effects of fiber characterisitcs on soliton effect pulse compression. To be published at *IEEE J. Quantum Electron.*, **31** (1995).
27. Mamyshev, P.V.: Generation and compression of femtosecond solitons in optical fibers, in *Optical Solitons - Theory and Experiment,* Taylor, J.R. Ed., Cambridge University Press, Cambridge (1992).
28. Smith, K. and Mollenauer, L.F. : Experimental observation of adiabatic comprssion and expansion of soliton pulses over long fiber paths, *Opt.Lett.*, **14** (1989), 751-753.
29. Iwatsuki, K., Suzuki, K. and Nishi, S.: Adiabatic soliton compression of gain-switched DFB-LD pulse by distributed fiber Raman amplification, *IEEE Photon.Tech.Lett.*, **3** (1991), 1074-1076.
30. Schell, M., Bimberg, D., Bogatyrjov, V.A., Dianov, E.M., Kurkov, A.S., Semenov, V.A. and Sysoliatin, A.A.: 540 fs light pulses at 1.5 μm with variable repetition rate using a tunable twin guide laser and soliton compression in a dispersion decreasing fiber, *IEEE Photon.Tech.Lett.,* **6** (1994), 1191-1193.
31. Suzuki, K., Iwatsuki, K., Nishi, S. and Saruwatari, M.: 160 Gb/s sub-picosecond transform-limited pulse data stream ultilizing adiabatic soliton compressin and optical time-division multiplexer, *IEEE Photon.Tech.Lett.*, **6** (1994), 352-354.
32. Takahashi, R. and Kamiya, T.: Noise suppression by a novel probe beam polarization modulation in electro-optic sampling, *Appl.Phys.Lett.*, **58** (1991), 2200-2222.
33. Sahara, R. T., Takeshita, H., Miwa, K., Tsuchiya, M. and Kamiya, T.: Electrooptic sampling evaluation of 1.5 μm metal-semiconductor-metal photodiodes by soliton compressed semiconductor laser pulses, *IEEE J. Quantum Electron.*, **31** (1995), 120-125.
34. Seo, D.S., Liu, H.F., Kim, D.Y. and Sampson, D.D.: Injection power and wavelength dependence of an external-seeded gain-switched Fabry-Perot laser. *Appl.Phys.Lett.,* **67** (1995), 1503-1505.

ULTRAHIGH BIT RATE GENERATION AND APPLICATIONS

S. V. CHERNIKOV, M. J. GUY AND J. R. TAYLOR
Femtosecond Optics Group
Physics Department, Imperial College
Prince Consort Road, London SW7 2BZ, England

Abstract. Various techniques are described for the generation of high repetition rate, high quality soliton pulse trains in the range 10 GHz to 1 THz. Emphasis is placed on the technique of nonlinear conversion of an optical beat signal in dispersion modified fibre. In addition, self-Raman scattering has been demonstrated as a powerful technique for ultrashort pulse generation and excellent mark space ratio operation. Techniques based on phase modulated sideband extraction and direct modulation using electro-absorption modulators operating in the 10 GHz - 60 GHz regime are also described.

1. Introduction

Currently, considerable research interest is being directed towards soliton transmission and optical switching schemes capable of operating with single channel rates in the region of 100 Gbit/s and above. Although split contact, passively mode locked semiconductor lasers can operate effectively at these repetition rates, as a laboratory test source they are somewhat inflexible to selectable variations in repetition rate. Here we describe our research programme on the development of alternative, much more widely applicable sources of high bit rate soliton pulse trains based upon the non-linear optical beat signal conversion technique. Selectable repetition rates have been demonstrated from 40 GHz - 1 THz. These pulse trains have been widely applied to investigations of dispersion compensation using chirped Bragg gratings and in studies of loss compensation over communication scale lengths of dispersion decreasing fibre. We have also used the pulse

115

A. Hasegawa (ed.), Physics and Applications of Optical Solitons in Fibres '95, 115–128.
© 1996 *Kluwer Academic Publishers. Printed in the Netherlands.*

trains in ultrafast switching studies in semiconductor amplifiers and in high
bit rate transmission studies in distributed amplifiers. For repetition rates
below 40 GHz, the lengths of dispersion decreasing optical fibre required by
the optical beat signal conversion technique become excessively long and
other more direct techniques are more attractive and applicable. We report
here on one such method which is based upon direct modulation employing
an electro-absorption modulator and chirped Bragg grating for chirp correction. This provides a source of transform limited pulses at widely selectable
repetition rates, at present up to 20 GHz. Using non-linear processing, the
pulses can be temporally compressed and from the spectral continuum so
generated spectral selection permits a source of transform limited pulses for
WDM applications. The high quality of the pulses from the base source,
in terms of jitter and inter pulse noise make it widely suitable for OTDM
operation up to 100 GHz, as is described below.

2. The Non-linear Beat Conversion Technique

The crucial concept of this technique was based on an idea [1], for the
compensation of the temporal broadening experienced by solitons through
propagation in optical fibres exhibiting loss. On experiencing loss through
propagation over a distance which is long compared to the characteristic
soliton period propagation, solitons experience an exponential broadening
of the pulsewidth with propagation distance. Tajima suggested that if the
dispersion of the fibre was constructed such that it exponentially decreased
with increasing fibre length, then the broadening associated with the propagation loss could be compensated and the soliton would propagate with
no temporal distortion. The required dispersion profile can be achieved
through the construction of a fibre with a tapered core, since the group
velocity dispersion can be precisely controlled via the size of the core diameter. Such a tapered fibre is equivalent to providing adabatic amplification,
the action of which can be used to compress an input soliton structure
with no energy loss to a dispersive wave. Dianov et al. were the first to
propose the application of such a fibre to convert an optical beat signal
into a train of fundamental solitons [2]. Mamyshev et al. theoretically expanded and experimentally demonstrated the technique for the first time
[3]. Clearly, the dispersion decreasing fibre used must be carefully selected
to allow the compression required at the chosen repetition rate. The upper
repetition rate is essentially limited by the input power requirement while
the lower repetition rate is limited by the fibre length. The lower repetition
rates by definition requiring longer characteristic soliton lengths, while the
maximum length is experimentally limited by the control of the minimum
group velocity dispersion. Note that at 20 GHz for a 5 psec soliton, the

characteristic soliton period for a group delay dispersion of 1 ps/nm/km is approximately 700 km. The compression or output pulsewidth is determined by the total effective amplification. On the input side, it is essential to carefully control the input power, i.e. the input amplitude should be less than the amplitude of a fundamental soliton with a fwhm duration of $T/2$, where T is the repetition time and the amplification experienced over the length $(T/2)^2$ should be small. The first realisation of the technique using diode laser inputs was by Chernikov et al. [4].

3. Dispersion Decreasing Fibre

The first demonstration of pulse compression and and ultrahigh bit rate generation employing diode laser input was by Chernikov et al. [4]. The beat signal was derived from two identical single frequency DFB lasers operating around 1530 nm. Through varying their relative temperatures, their wavelength separation could be varied by up to 2 nm. The signal was amplified to non-linear power levels employing a 20 m length of Er doped fibre, pumped by the second harmonic of a cw pumped mode-locked Nd: YAG laser. This amplified signal was injected into a tapered fibre, manufactured at the General Physics Institute Moscow [5] which was approximately 2.2 km long with a loss of 0.6 dB/km and an optimised configuration giving a dispersion varying from 6.5 ps/nm/km at input to 1.1 ps/nm/km at the output at 1530 nm, limiting the possible compression.

The linewidth of the diode lasers was ~ 25 MHz, however at the operational average power levels, stimulated Brillouin scattering was in evidence, clearly limiting the efficiency of the conversion technique. In order to reduce the SBS, it was necessary to modulate the current to the diodes at a frequency ~ 100 MHz. It can be seen that two problems therefore exist with this technique. At the required power levels in the dispersion decreasing fibre, SBS is a limiting process, and its reduction through modulation of the source linewidth introduces phase noise on the generated signal. An additional problem associated with passive techniques is the difficult in obtaining clock synchronisation. This can be accomplished through the implementation of a high bandwidth phase-locked loop, however, the latency of the compression fibre can introduce additional complications to such a scheme.

The initial experiments did however, demonstrate, through varying the input power and wavelength separation, the possibility of generating pedestal free, picosecond, fundamental soliton outputs selectable in the range 80 - 130 GHz, with the pulse durations varying from 3.0 - 1.5 ps and mark space ratios of 4 - 7. Using the dispersion decreasing fibre, the technique was further refined and demonstrated at 70 Gbit/s for a completely cw pumped system [6].

4. Step-Like Dispersion Profiled Fibre

The production of tapered, dispersion decreasing fibres requires precise control of the changing fibre core diameter with length and fibres need to be selectively produced for specific repetition rates. A split step theoretical modelling of soliton pulse transmission can be applied to the construction of a multisegmented fibre of different and decreasing dispersion sections, optimally selected such that the overall profile presented approximates a fibre with an exponentially decreasing dispersion and the application of such a step-like dispersion profiled fibre for soliton generation and compression was predicted by Mamyshev et al. [3].

The experimental verification was undertaken using a dispersion stepped profiled fibre in an experimental configuration which was a simple refinement on the scheme used with the dispersion decreasing fibre [7]. The beat signal source was similar to that described above and based on a pair of temperature tuned and stabilised single frequency DFB lasers. The combined signal was amplified in a low noise Yb-Er doped fibre amplifier to an average power level of up to 800 mW and then launched into the step-like dispersion profiled fibre (SDPF).

The SDPF was constructed from six segments of dispersion shifted fibre of different dispersions at the operational wavelength of 1553 nm. The first section was normally dispersive, 1 ps/nm/km and 500 m long. It was introduced to spectrally enhance the beat signal through four wave mixing prior to soliton formation and compression in the following SDPF. An exponentially decreasing dispersion profile was approximated in the SDPF through the use of five, optimal length, fusion-spliced, anomalously dispersive sections of dispersion shifted fibre 6, 1.9, 1.3, 0.9 and 0.5 ps/nm/km respectively at 1553 nm, designed for operation around 100 GHz. For adiabatic conversion of the beat signal into a train of solitons without background, the relative variations of the dispersion from segment to segment should not exceed a ratio of 1.5. In order to compensate for the large variation between the second and third and between the fifth and sixth segments, it was necessary to introduce a step loss of -2.0 dB and -0.8 dB respectively. This could be undertaken since the variations in dispersion can be analogously associated with a variation in the intensity of a soliton in the dispersion profiled fibre and as such allows considerable leeway in the actual design of a SDPF for this application. A cw pulse train of 670 fs solitons at 104 GHz was obtained for an average input power of 400 mW. Due to the large ratio of the input to output dispersions, a relatively large compression ratio was achieved and a mark space ratio of 1:14 was measured. It was possible to tune the repetition rate in a 10 GHz range while still maintaining the high quality of pulses and mark space ratio.

5. Comb-Like Dispersion Profiled Fibre

The step like dispersion profiled fibre offers system design flexibility through the use of optimised lengths of differing dispersion fibres. Additional optimisation and flexibility can be achieved through the use of a comb-like dispersion profiled fibre, where the effects of dispersion and nonlinearity can be separately treated and optimised. Optical solitons are stable to perturbations in two extremes. In one, the adiabatic regime, the perturbation to the soliton takes place over a length scale which is long compared to the characteristic soliton length. In the other, the so called average soliton regime [8], [9], stability is exhibited provided the length scale of any perturbing effect, be it gain or dispersion etc., is short compared to the soliton length. In the comb-like dispersion profiled fibre both situations are addressed. The dispersion and non-linear contributions essentially of each step are separately addressed. Experimentally this is achieved by alternating short segments of high anomalous dispersion together with longer lengths of very low dispersion at the operational wavelength. In this way the dispersion profile of the fibre resembles that of a comb. In the low dispersion section, the effects of nonlinearity (self phase modulation) dominate, while in the shorter highly anomalously dispersive section it may be neglected but pulse compression and chirp correction is significant. Within the average soliton model however, an average dispersion of the two sections may be considered, while overall, an adiabatic pulse compression takes place due to the designed decreasing function of this dual section averaged dispersion. This approach has the advantage in that both nonlinearity and dispersion can be separately considered and optimised. The technique was first considered both theoretically and experimentally [7], [10]. For stable compression, the non-linear phase shift over each segment of the composite fibre should not exceed $\pi/2$.

The comb-like dispersion profiled fibre (CDPF) employed was constructed from twenty segments of standard telecommunications fibre (STF) and dispersion shifted fibre (DSF), spliced together and alternating DSF-STF-DSF-STF-DSF... etc. The measured splice loss was ~ 0.1 dB, which needed to be accounted for in the overall design and projected compression. Also requiring consideration was the average fibre propagation loss of ~ 0.25 dB/km. The lengths of the DSF segments are determined by the input power level, and provided that the length of the STF sections are considerably less than the DSF segments, then the length of the STF sections is determined by the beat frequency. This is in contrast to the dispersion decreasing fibre, where the input power and fibre length depend on the square of the repetition frequency. Therefore, the CDPF allows considerable flexibility in the system design and permits significant expansion of the upper and lower limits of the practical frequency range. In the experimental configuration, the length of the DSF sections was in the range 500 - 600 m, while the STF sections varied from 220 - 230 m to permit the monotonically decreasing average dispersion profile required. The non-linear phase shift in each section was $< \pi/2$. Like the step-like dispersion profiled fibre, there is also an added advantage in the use of various lengths of dispersion shifted fibre of differing dispersion-zero wavelengths. The dispersion shifting is undertaken by varying the Germanium content. For example in the CDPF described above, five different DSF's were used with Germanium concentrations varying between $8-14\%$. However, the frequency shift due to Brillouin scattering is dependent on the Germanium concentration, being measured as 89 MHz/wt% [11]. Therefore, the cascading of various relatively short lengths of fibres of differing Germanium content, inhibits the buid-up of stimulated Brillouin scattering, since a Germanium doping difference of $\sim 0.5\%$ should be sufficient to prevent overlap of the SBS gain spectra. At the input side of the CDPF, two 1.1 km lengths of DSF with different dispersion zero, were included to reduce the problem of SBS.

Conceptually, the other parts of the experimental configuration employing the CDPF were similar to those described above. However, the actual configuration was a totally integrated fibre device. The dual frequency source was based on coupled cavity, single frequency, fibre lasers incorporating intra-core fibre Bragg gratings as the cavity reflectors [12]. The coupled cavity fibre laser was 1.65 cm in total length and provided a dual frequency output, without mode hops, centred around 1545 nm with a selected frequency separation of 59.1 GHz. The long term measured average linewidth was ~ 16 kHz, which was effectively instrument limited, and the stability of the dual linewidth was also instrumentation limited, measured as 3 MHz. Pump radiation not required in the pumping of the dual frequency source was transmitted and used to pump a following Yb:Er fibre preamplifier, the output of which was further amplified in a Yb:Er power amplifier, giving an average power in the beat signal of up to 200 mW. This was converted in the integrated CDPF into a train of high quality fundamental solitons which exhibited no detectable pedestal or intra-pulse noise. The measured pulse duration was 2.2 ps, corresponding to a mark-space ratio of 1:7.7 and indicated the suitability of the technique for high repetition rate pulse train generation in an integrated all fibre configuration. Through the use of additional segments it is also possible to increase the mark-space ratio.

6. Application of Raman Self-Scattering

For ultrashort pulse solitons propagating in a fibre with a low net dispersion, it is possible for the high frequency components of the spectrally broad ultrashort pulse soliton to provide Raman gain for the low frequency component with a relatively high efficiency. This so called soliton self-frequency shifting leads to an asymmetric spectrum and a continuous evolution in the long wavelength region of the spectrum as the pulse propagates [13], [14]. Generally this self-pumped low frequency shift continues until the power in the pulse is insufficient to maintain the soliton condition, since most usually the dispersion (and associated soliton power required) increases with wavelength. The addition of amplification, excess loss or dispersion modification can, however, change this and arrest early self termination.

On the simplest level, it can be seen that once, for example, ultrashort pulse solitons have been formed or are forming within a dispersion decreasing fibre or one of its modifications, then the self frequency shift should play a significant role. Chernikov et al. first reported this technique in combination with a dispersion decreasing fibre, to generate low repetition rate pulsed trains containing 250 fs soliton pulses at 114 GHz [15]. We have developed this technique and in association with high power Yb:Er amplifier technology have demonstrated a truely cw 1 THz soliton pulse source. The experimental configuration utilised two DFB diode lasers to generate a 125 GHz beat signal at 1555 nm which was pre and power amplified in diode pumped Yb:Er fibre amplifiers to an average power level of 1.0 W. This was launched into a 1.6 km long dispersion decreasing fibre with an input and output group delay disperson of 10 and 0.5 ps/nm/km respectively. Continuous trains of 230 fs soliton pulse were obtained, with an exceptionally high mark space ratio of 1:35. The Raman shifted, 11 nm broad spectral band was spectrally distinct from the remaining carrier which was around 1555 nm.

In order to generate a pulse train at a repetition rate of 1 THz, a polarisation division multiplexing technique was employed, using bulk optics. The 125 GHz pulse train was made to be incident on a calcite crystal such that the input polarisation was at 45 degrees to the optic axis. For a crystal of 7.68 mm thick, the e- and o- rays temporally separate by 4 ps. Thus the input 125 GHz signal can be converted to 250 GHz. Each of the pulses so generated can be further sub-divided by a second calcite crystal of half the thickness of the original and with its optic axis rotated by 45 degrees to that of the first crystal. By cascading three crystals, with the third having a thickness of 1.92 mm and placing these between crossed polarisers, an efficient eight times multiplexer can be constructed. Figure 1 shows an autocorrelation trace of a 250 GHz and a 1 THz multiplexed pulse train.

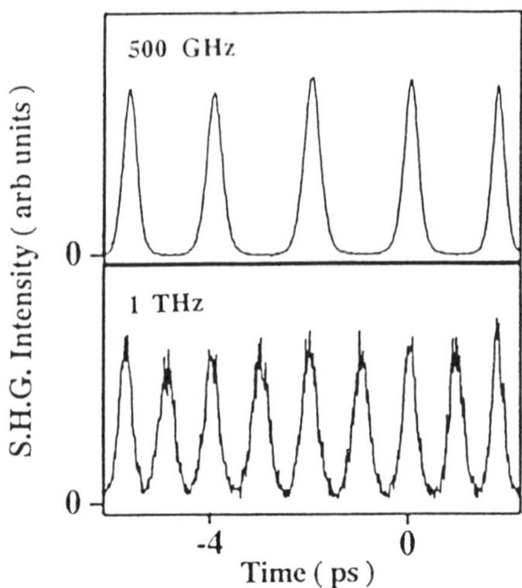

Figure 1.   Autocorrelation traces of 250 GHz and 1 THz pulse trains, generated using polarisation multiplexing of a base 125 GHz pulse train derived from a Raman self frequency shifted optical beat signal source.

It should be noted that the average power in these is in excess of 100 mW, and currently developed diode pumped Yb:Er amplifiers should readily allow average power levels well in excess of 1.0 W.

7. Modulation Sideband Extraction Techniques

The problems of ease of clock synchronisation and phase noise in the above techniques, although solvable, detract from the simplicity of the source. One method to overcome this is based on sideband extraction of a frequency modulated single frequency source [16]. We have investigated this using a single frequency DFB laser source operating around 1562 nm. This was coupled into a fibre pigtailed lithium niobate phase modulator, which was driven at 10 GHz, which gave rise to efficient sideband generation of up to eight side band orders. Employing an in-line Fabry-Perot with a free spectral range of 50 GHz, although this can be constructed for any sideband extraction frequency separation, two stable, phase-locked single frequencies were extracted. The Fabry-Perot was stabilised and electronically locked on the transmisson peaks using an electronic feedback technique. Once the two sidebands were filtered, the beat signal conversion technique was readily applied to generate a high repetion rate soliton pulse train (at 50 GHz in this case) with a synchronous 10 GHz electronic clock signal.

8. Electro-Absorption Modulator Applications

For lower repetition rates, the beat signal conversion technique requires impractical lengths of dispersion decreasing fibre or its analogues. Below 40 GHz, direct modulation techniques provide the simplest and potentially most applicable method for short pulse generation and control. In the past few years, electro-absorption modulators have attracted considerable interest demonstrating modulation depths greater than 30 dB for only a few volts applied and they have been shown to be capable of generating short pulses at up to 30 GHz. One problem with such devices is that in general the generated pulses are not transform limited and some form of chirp compensation is required following the modulator. Moodie et al. have produced pulses as short as 6.3 ps at 10 GHz, employing a dispersion compensating fibre [17]. With a duty ratio of 6.3%, time division multiplexing readily allowed the production of a data rates of 40 Gbit/s. Some problems can be introduced through the application of a dispersion correcting fibre, such as phase and polarisation drift if the temperature and birefringence are not actively controlled. We have recently developed an alternative geometry, replacing the fibre with a chirped fibre Bragg grating and demonstrating a versatile source of ultrashort pulses as short as 5.8 ps at 10 GHz directly out of the grating corrector, and with non-linear compression have generated soliton pulses as short as 200 fs. At lower repetition rates, employing a double modulation Vernier technique 6.0 ps pulses have been generated at a demonstration frequency of 1 GHz. Following this, nonlinear conversion allowed the production of stable 1.2 ps pulses, highly applicable to time division multiplexing.

Figure 2 shows a schematic of the experimental arrangement. A DFB laser operating at 1.5625 μm was coupled into an InGaAsP multiple quantum well electro-absorption (EA) modulator. This was supplied with a reverse bias of 7.5 V and driven at 10 GHz (although this particular device can readily operate up to 20 GHz) by a sinusoidal peak to peak voltage signal of 8 V into 50 Ω. The pulses generated were amplified in a diode pumped Yb:Er preamplifier and chirp compensated by a thermally chirped fibre Bragg grating of 1.75 nm bandwidth incorporated in a low loss transmission filter based on a polarisation dependent polarisation coupler [18]. This permitted the production of transform limited pulses of duration ~ 6 ps [19]. These were amplified to an average power level of ~ 200 mW in a second diode pumped Yb:Er fibre amplifier, and following transmission through a 1 km length of standard telecommunications fibre, were com-

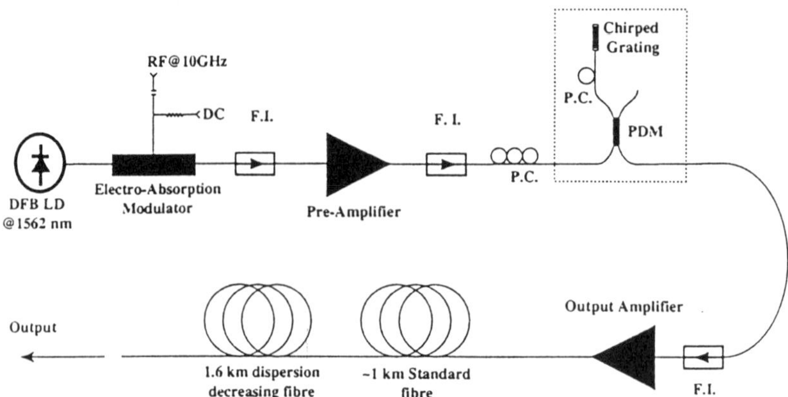

Figure 2.   Experimental configuration of soliton pulse generation based upon direct modulation of a cw source using an electro-absorption modulator, chirp compensation and non-linear conversion.

pressed in a 1.6 km length of dispersion decreasing fibre, which had a dispersion at 1.55 μm varying from 10 ps/nm/km at input to 0.5 ps/nm/km at the output [20]. Figure 3 shows a representative second harmonic generation autocorrelation of the pulses and their associated spectrum. It can be seen that self-Raman shifting dominated the process and the carrier frequency is distinct from the broader soliton spectrum. One clear advantage of this is that the carrier can be readily removed simply by spectral filtration, for example with an in-line Bragg grating. The time bandwidth product of the 187 fs pulses was measured to be 0.32. On a logarithmic intensity trace, a relatively small $< 0.5\%$ pedestal was shown to be present on the pulses.

The dispersion decreasing fibre employed for compression is optimised for a particular input pulse duration and does not permit significant deviation from the optimum criteria. Hence large variation and wide selectivity of the output pulse duration is not possible. However, spectral filtration within the spectrum of the generated femtosecond solitons does permit the production of transform limited, wavelength tunable pulses and by varying the bandwidth of the spectral filter, pulse duration selectable picosecond and femtosecond high repetition rate soliton pulses have been obtained. We have already demonstrated using a periodic polarisation filter the generation of seven, 5 ps, 10 GHz channels separated by 3.7 nm [21]. It should be possible to provide up to 50 x 10 GHz channels from this source, suitable for OTDM and/or WDM applications.

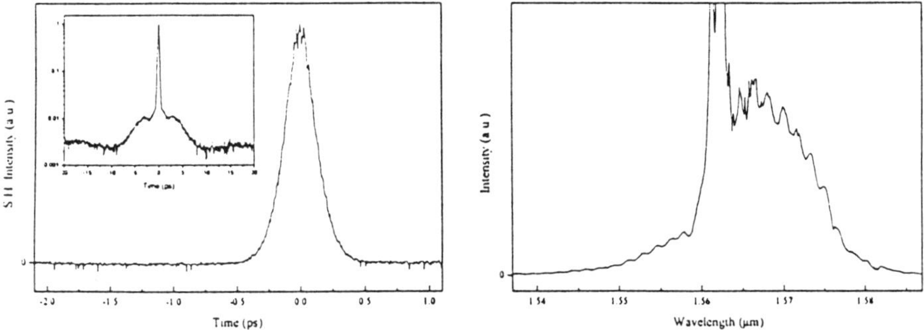

Figure 3.   Autocorrelation trace and corresponding spectrum of femtosecond pulses generated using the non-linear conversion of dispersion compensated pulses from an electro-absorption modulator source of pulses.

Currently considerable efforts being directed towards the development of transmission and switching schemes operating with single channel data rates around 100 Gbit/s, generally constructed through optical time domain multiplexing. Consequently we have investigated a means of generating high quality pulses of ~ 1 ps duration with low background levels, at base rates of 1 - 2.5 GHz, suitable for OTDM application at 100 Gbit/s. To achieve this we have employed double modulation technique, whereby the source DFB semiconductor laser was gain switched at the base frequency required (1 GHz in this example) in tandem with the electro-absorption modulator which was driven at a high frequency multiple, which in the case reported here was 10 GHz. Figure 4 shows the experimental configuration. The combination of the gain switched DFB and electro-absorption modulator generated a 1 GHz source of chirped 15 ps pulses. These were dispersion compensated in a ~ 7 mm long, 2 nm bandwidth chirped Bragg grating filter, which compressed the pulses to 6 ps with a 0.32 time-bandwidth product, indicative of a transform limited output. This signal was amplified in a diode pumped Yb:Er fibre amplifier to an average power of 50 mW. A 2 km long section of optimised dispersion decreasing fibre was used to compress the pulses to 1.2 ps, with a spectral width of 2.1 nm ($\Delta\nu\Delta\tau \sim 0.32$). Figure 5 shows an autocorrelation of the 1.2 ps pulses together with the corresponding spectrum. It can be seen that a contribution from the carrier frequency is present. However, integrating the signals inferred that at maximum 0.4% of the total signal was in the carrier, in good agreement with

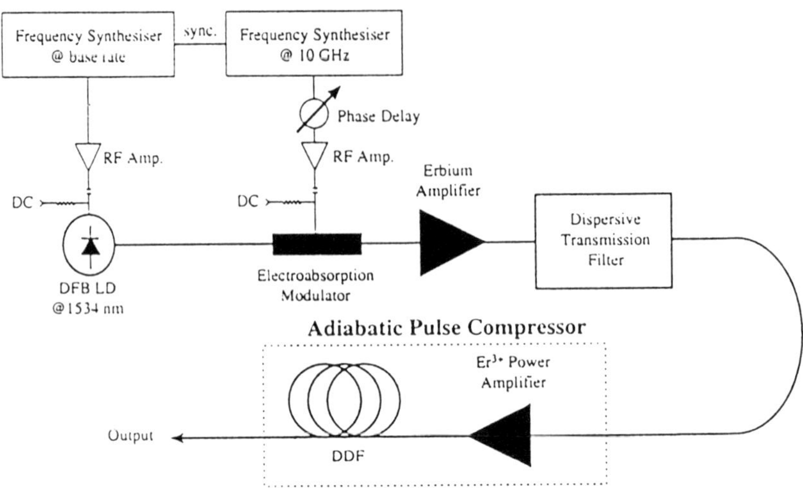

Figure 4.   Experimental configuration employed to generate picosecond pulses at low repetition rates (1 - 2.5 GHz) suitable for OTDM applications.

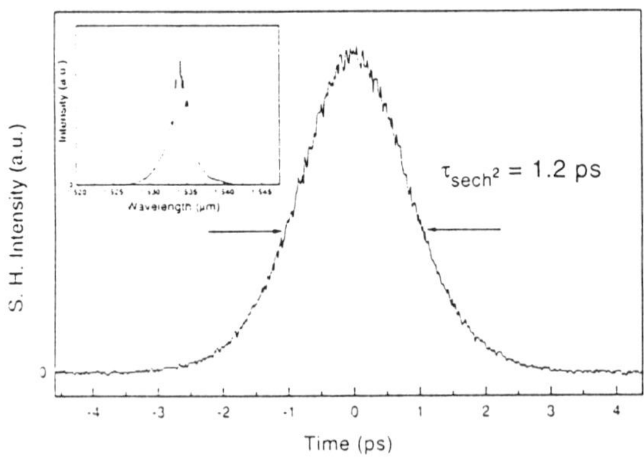

Figure 5.   Autocorrelation trace and (inset) corresponding spectrum of 1.2 ps pulses at 1 GHz derived employing the double modulation technique illustrated in Figure 4.

estimates based upon the maximum attenuation of 24 dB of the modulator. This adiabatically compressed source, with such a low pedestal component should be widely applicable to OTDM.

9. Acknowledgement

We are grateful to the SERC and the EPSRC for their financial support of the research programme reported here. We would also like to express our sincere thanks to IRE Polus and in particular Dr. V. P. Gapontsev for his continued support and provision of high power diode pumped Yb:Er fibre amplifiers, without which much of this research could not have been undertaken. We also gratefully acknowledge the contribution of our collaborators, Dr. D. G. Moodie and Dr. R. Kashyap of British Telecom Laboratories in particular for provision of the electro-absorption modulators and fibre Bragg gratings respectively. We also thank Dr. A. F. Evans (Corning Inc.) for the provision of long lengths of dispersion decreasing fibres.

References

1. Tajima, K.: Compensation of soliton broadening in nonlinear optical fibre with loss, *Opt. Lett.* **12** (1987), 54-56

2. Dianov, E. M., Mamyshev, P. V., Prokhorov, A. M. and Chernikov, S. V.: Generation of a train of fundamental solitons at a high repetition rate in optical fibres, *Opt. Lett.* **14** (1989), 1008-1010

3. Mamyshev, P. V., Chernikov, S. V. and Dianov, E. M.: Generation of fundamental soliton trains for high-bit-rate optical fibre communication lines, *IEEE J. Quantum Elect.* **27** (1991), 2347-2355

4. Chernikov, S. V., Taylor, J. R., Mamyshev, P. V. and Dianov, E. M.: Generation of soliton pulse train in optical fibre using two cw singlemode diode lasers, *Electron. Lett.*, **28**, (1992) 931-932

5. Bogatyrev, V. A., Bubnov, M. M., Dianov, E. M., Kurkov, A. S., Mamyshev, P. V., Prokhorov, A. M., Miroshnichenko, S. I., Semenov, V. A., Semenov, S. L., Sysoliatin, A. A. and Chernikov, S. V.: Single mode fibre with chromatic dispersion varying along the length, *J. Lightwave Tech.* **9** (1991), 561-566

6. Chernikov, S. V., Richardson, D. J., Laming, R. I., Dianov, E. M. and Payne, D. N., 70Gbit/s fibre based source of fundamental solitons at 1550 nm. *Electron. Lett.*, **28**, (1992) 1210-1212

7. Chernikov, S. V., Taylor, J. R. and Kashyap, R., Comb-like dispersion profiled fibre for soliton pulse train generation. *Opt. Lett.*, **19**, (1994) 539-541

8. Mollenauer, L. F., Evangelides, S. G. and Haus, H. A., Long-distance soliton propagation using lumped amplifiers and dispersion shifted fibre. *J. Lightwave Tech.*, **9**, (1991) 194-197

9. Hasegawa, A. and Kodama, Y., Guiding-centre soliton in optical fibres. *Opt. Lett.*, **15**, (1990) 1443-1445

10. Chernikov, S. V., Taylor, J. R. and Kashyap, R., Integrated all optical fibre source of multigigahertz soliton pulse train. *Electron. Lett.*, **29**, (1993) 1788-1789

11. Tkach, R. W., Chraplyvy, A. R. and Derosier, R. M.: Spontaneous Brillouin scattering for single mode optical fibre characterisation, *Electron. Lett.* **22** (1986), 1011-1013

12. Chernikov, S. V., Taylor, J. R. and Kashyap, R.: Coupled-cavity erbium fiber lasers incorporating fiber grating reflectors, *Opt. Lett.* **18** (1993), 2023-2025

13. Dianov, E. M., Karasik, A. Ya., Mamyshev, P. V., Prokhorov, A. M., Serkin, V. N., Stel'makh, M. F. and Fomichev, A. A.: Stimulated Raman conversion of multisoliton pulses in quartz optical fibres, *JETP Lett.* **41** (1985), 294-297

14. Mitschke, F. M. and Mollenauer, L. F.: Discovery of the soliton self-frequency shift,

Opt. Lett. **11** (1986), 659-661

15. Chernikov, S. V., Dianov, E. M., Richardson, D. J., Laming, R. I. and Payne, D. N.: 114 Gbit/s soliton train generation through Raman self scattering of a dual frequency beat signal in dispersion decreasing optical fibre, *Appl. Phys. Lett.* **63** (1993), 293-295

16. Swanson, E. A., Chinn, S. R., Hall, K., Rauschenbach, K. A., Bondurant, R. S. and Miller, J. W.: 100 GHz soliton pulse train generation using soliton compression of two phase sidebands from a single DFB laser, *IEEE Photonics Tech. Lett.* **6** (1994), 1194-1196

17. Moodie, D. G., Ellis, A. D. and Ford, C. W.: Generation of 6.3 ps optical pulses at 10 GHz repetition rate using a packaged electroabsorption modulator and dispersion compensating fibre, *Electron. Lett.* **30** (1994), 1700-1701

18. Guy, M. J., Chernikov, S. V., Taylor, J. R. and Kashyap, R.: Low-loss fibre Bragg grating transmission filter based on a fibre polarisation splitter, *Electron. Lett.* **30** (1994), 1512-1513

19. Guy, M. J., Chernikov, S. V., Taylor, J. R., Moodie, D. G. and Kashyap, R.: Generation of transform-limited optical pulses at 10 GHz using an electroabsorption modulator and a chirped fibre Bragg grating, *Electron. Lett.* **31** (1995), 671-672

20. Guy, M. J., Chernikov, S. V., Taylor, J. R., Moodie, D. G. and Kashyap, R.: 200 fs soliton pulse generation at 10 GHz through non-linear compression of transform-limited pulses from an electro-absorption modulator, *Electron. Lett.* **31** (1995), 740-741

21. Chernikov, S. V., Guy, M. J., Taylor, J. R., Moodie, D. G. and Kashyap, R.: A 10 GHz source capable of providing up to 50 channels by OTDM and/or WDM, *Conference on Lasers and Electro-Optics CLEO'95* **CPD14**, 1-2

MONOLITHIC MODE-LOCKED SEMICONDUCTOR LASERS FOR HIGH SPEED OPTICAL COMMUNICATION SYSTEMS

S. ARAHIRA, Y. OGAWA AND Y. MATSUI
Oki Electric Industry Co., Ltd.
550-5 Higashiasakawa, Hachioji-shi
Tokyo 193, Japan

Abstract. Recent progress in the development of ultrafast semiconducor laser diodes generating optical short pulses at very high repetition rates plays an important role in increasing bit rates in large-capacity optical communication systems. Of the several methods used to obtain optical short pulses, the mode-locking technique is very attractive in optical soliton transmission, because it can offer transform-limited pulses directly without using optical fibers and bandpass filters. In particular, passive mode-locking based on an intracavity saturable absorber has an advantage over active mode-locking in order to increase pulse repetition rates, because it requires no electrical modulation and hence no electrical limitation is imposed on the resultant pulse widths and repetition rates. In the first half of this chapter, we review recent topics in ultrafast pulse generation by passive mode-locking in monolithic semiconducor lasers diodes. All-optical clock extraction technique from ultrafast optical data signals is another attractive application of mode-locked semiconducor laser diodes (LDs) for large capacity optical communication systems, and is also reviewed in this chapter.

1. Fabrication of Monolithic Mode-locked Semiconductor Laser Diodes

The key component necessary for successful achievement of passive mode-locking is a suitable saturable absorber. Theoretical studies have suggested some necessary conditions for the stable achievement of passive mode-locking [1]-[3]. Because the net loss must be seen before and after the passage of the pulse, the recovery time of the absorption must be shorter

A. Hasegawa (ed.), Physics and Applications of Optical Solitons in Fibres '95, 129–144.
© 1996 *Kluwer Academic Publishers. Printed in the Netherlands.*

than that of the gain. As for LDs, the gain recovery time is dominated by the spontaneous carrier lifetime and typically ranges in nanoseconds.

Damage sites or traps created in the semiconductors by ion implantation shorten the carrier lifetimes to several picoseconds due to the decrease of nonradiative lifetimes. Degrated laser facet by implantation through one of the mirrors has been used as an intracavity saturable absorber in monolithic passively mode-locked LDs [4,5].
However, saturable absorbers obtained through this process are unstable and unreliable for practical system application.

A more robust and reliable technique for achieving passive mode-locking in the LDs is to incorporate separately biased saturable absorber section in a multi-section laser [6]-[10]. A waveguide saturable absorber is formed by reverse biasing the short segment of the laser. Reverse-biased laser segment behaves as an waveguide photo-detector. The incoming optical pulse saturates the band-to-band absorption and bleaches the short segment. After the passage of the optical pulse, the electrical field applied to the absorber sweeps the photocarriers out of the active region and returns the absorber to a high attenuation state. The absorption recovery time through this process is strongly dependent on the reverse voltage applied to the absorber and ranges several picoseconds [11]. This value is much shorter than the gain recovery time of the forward-biased gain segment incorporated in the same cavity, verifying this method

The saturation energies of the gain and the absorption media are also important parameters in achieving passive mode-locking. For successful passive mode-locking, the saturable absorber must be more easily "bleached" than the gain material [2]. This means that the saturation energy of the absorber must be smaller than that of the gain material. A small ratio of saturation energy between the gain and absorber media is an especially important feature in increasing the passive mode-locking frequency up to millimeter-wave and higher frequencies, because it sets an upper limitation of the achievable pulse repetition rates for the laser structure used [3].

The use of semiconductor multiple quantum well (MQW) structure under different bias conditions for gain and absorber media looks very promising for obtaining a small ratio of the saturation energy [12]. Figure 1 shows an example of the saturation characteristics of the semiconductor quantum well as a function of carrier density. The reduction in differential gain with increased carrier density is particularly large in semiconductor quantum well lasers compared with bulk lasers, because of the step-like density-of-state function in quantum wells. Saturation energy is inversely proportional to differential gain. As shown in Figure 1, if the absorber is biased at a lower carrier density than the gain section, the saturation energy of the absorber is smaller than that of the gain section. This makes it possible to utilize the same medium under the different bias conditions to provide both saturable absorption and the gain functions; monolithic fabrication of mode-locked LDs.

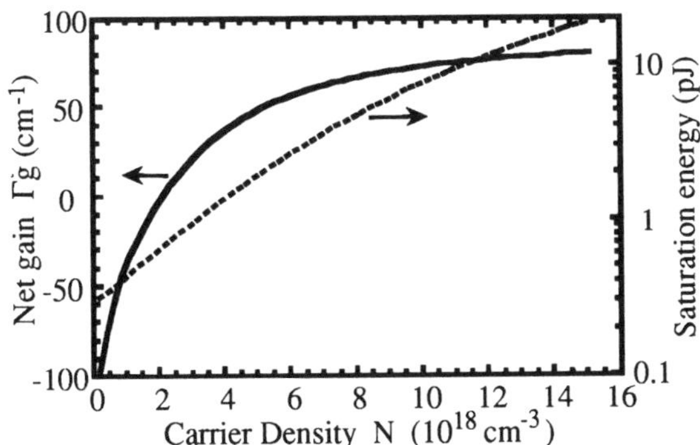

Figure 1. Saturation characteristics of the semiconductor quantum wells as a function of carrier density.

Figure 2 shows a schematic of the monolithic passively mode-locked LD we have developed; a multi-electrodes <u>D</u>istributed-<u>B</u>ragg-<u>R</u>eflector (DBR)-LD [10]. The LD was fabricated using five-step metal-organic vapor phase epitaxy (MO-VPE) and ordinary chemical etching.

The LD has four sections: a 75 μm saturable absorber section, a 750 μm gain section, a 150 μm phase control section, and a 90 μm intracavity Bragg reflector. The isolation resistance between the sections exceeded 1kΩ. Both the gain and the absorber sections consisted of three InGaAs strained quantum wells (7.5 nm thick) separated by InGaAsP quaternary barriers with a bandgap wavelength of 1.3 μm (13 nm thick), and the upper and the lower InGaAsP graded-index confining layers. A 1.3 μm band-gap wavelength InGaAsP was used for the waveguides of the phase control and the DBR sections. The LD has a buried heterostructure. The mesa was approximately 1 μm wide. The coupling efficiency of the butt-joint between the gain section and the phase control section exceeded 90%.

The absorber facet was coated for high-reflectivity (~95%) and output power was obtained from the DBR facet. The high-reflectivity coating to the absorber facet led to the self colliding-pulse operation at the absorber [13]. The colliding pulse effect reduces the effective saturation energy of the absorber to approximately one-third of its intrinsic value. This causes reduction of the unsaturated gain required for the achievement of passive mode-locking.

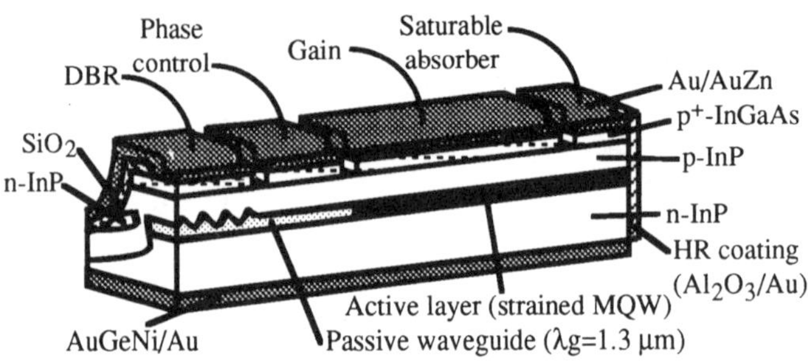

Figure 2.   Schematic of a passively mode-locked DBR-LD.

Figure 3 shows the SHG correlation trace and the time-averaged optical spectrum of passively mode-locked pulses generated from the LD in Figure 2 [10]. The absorber section was reverse biased at - 0.2 V and the gain section was biased just above the threshold (85 mA). The pulse repetition rate of 40.8 GHz agreed well with the cavity roundtrip frequency estimated from the cavity length (approximately 1.1 mm). No changes were observed in either peak intensity or shape between the autocorrelation and the cross-correlation (around 25 ps of the delay time) traces, indicating that the output pulses were continuous and coherent. The full-width at half maximum (FWHM) of the correlation trace was 5 ps. The envelope coincided well with a Gaussian waveform (open circles in the correlation trace), indicating that the actual pulse was 3.5 ps.

The spectrum was symmetric and no broadening in each longitudinal mode was observed, suggesting the dynamic frequency chirping of output pulses is very low. The FWHM of the spectrum was 1 nm and the envelope was also in good agreement with a Gaussian waveform. The time-bandwidth product was 0.43, very close to the transform-limited value of a Gaussian waveform, indicating that the output pulses were transform-limited, verifying the mode-locking method. The pulse energy was 0.21 pJ, comparable to that of similar monolithic devices reported thus far [13].

Increasing the current in the gain section from the threshold broadened the pulse width and made the spectrum asymmetric, which caused the output pulses to have some background and deviate from the transform-limited condition, indicating that there is an optimum gain condition for passive mode-locking, as suggested by the self-consistent mode-locking model [2].

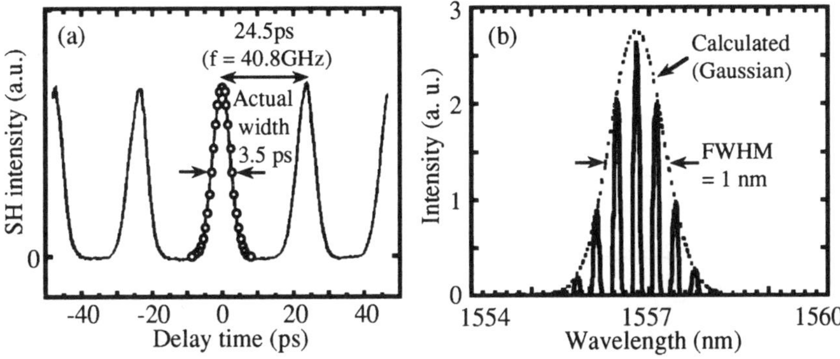

Figure 3.   (a) SHG correlation trace and (b) time-averaged optical spectrum of passively mode-locked pulses.

2. Wavelength Tuning

The formation of soliton is very sensitive to the group-velocity-dispersion (GVD) of transmission fibers [14]. To form (light) soliton, the GVD at the signal wavelength must be ranged within an anomalous region. A small GVD is necessary for forming soliton with a low pulse peak power.

The GVD also changes the soliton period, which must be larger than the period of the lumped amplifiers in soliton transmission systems considering the transmission loss of the fibers (dynamic soliton transmission) [15], and hence, the small GVD expands the maximum period of the lumped amplifiers. GVD depends on the signal wavelength of the input optical pulses, and therefore, wavelength tunability of optical pulse generator is very attractive and important in constructing soliton transmission systems.

In monolithic semiconductor DBR-LDs, wavelength-tuning can be easily achieved by current injection to the DBR section. Injection current to the heterojunction which constitutes the DBR waveguide changes the refractive index of the waveguide due to the plasma effect, resulting in a change of the Bragg wavelength.

Figure 4 shows the wavelength-tuning characteristics of the passively mode-locked DBR-LD (Figure 2) as a function of the injection current to the DBR section [16]. Increasing the injection current decreases the refractive index, and then shifts the lasing wavelength to a shorter wavelength. In our experiments, 4.2 nm of the tuning range was obtained. Fine-tuning can be achieved by current injection to the phase control section. The deviation in the pulse width through the tuning was about 2 ps. The time-bandwidth product was 0.44~0.55, close to the transform-limited value. These results indicate that wide wavelength-tuning can be achieved using DBR-type mode-locked LDs, maintaining the transform-limited condition.

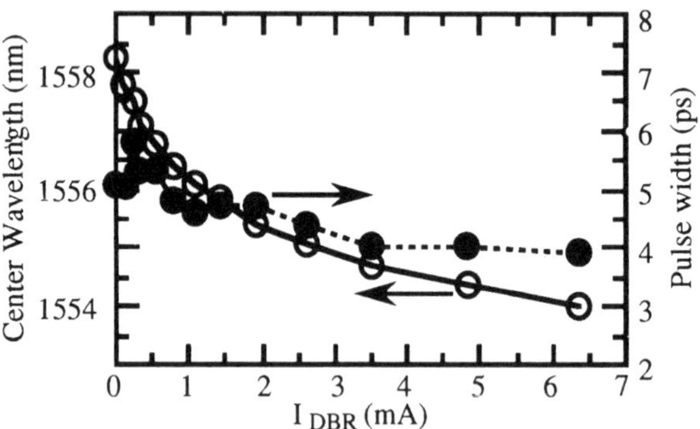

Figure 4.   Wavelength-tuning characteristics of a passively mode-locked DBR-LD.

3. Harmonic Mode-locking

Increasing the repetition rates of the optical pulses is of interest in order to increase bit-rate of optical data signals. In mode-locking methods, the repetition rate of the generating optical pulses is almost equal to the cavity roundtrip frequency, f_{rt}, and is given by

$$f_{rt} = \frac{c}{2nL},\tag{1}$$

where c is the light velocity in vacuum, n is the refractive index in the cavity and L is the cavity length.

Cavity roundtrip frequency is inversely proportional to the cavity length. Cavity lengths shorter than 50 μm are expected to produce pulses at repetition rates on the order of THz, but such a short cavity device will inevitably impose a limitation on the output power, occasionally on the lasing itself. Colliding-pulse mode-locking [9], [17,18] and harmonic mode-locking [19]-[23] techniques are   promising for generating ultrafast pulses without using a short cavity laser.

One of the problems which must be solved in harmonic mode-locked lasers is the suppression of the residual cavity modes. The residual cavity modes cause an undesirable fluctuation in pulse repetition rates [24,25]. Incorporating a frequency filter with pulse repetition rate selectivity in the laser cavity is effective for suppressing such extra cavity modes and obtaining continuous harmonic mode-locked pulses [24,25].

An intracavity Bragg reflector, DBR, with a finite facet reflectivity can operate as such a frequency filter [26,27]. Figure 5 shows examples of calculated reflectivity profiles for an intracavity Bragg reflector. In calculation, the reflectivities of the front and the rear facets of the reflector were assumed to be 0% and 27.4% (as-cleaved).

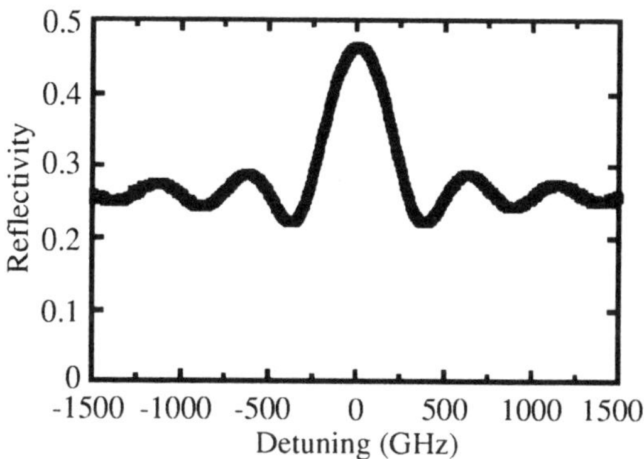

Figure 5.   Examples of the calculated reflectivity profiles for an intracavity
            Bragg reflector. The parameters used in the calculation are;
            the coupling coefficient=30 cm^{-1}, the amplitude loss coefficient
            = 5 cm^{-1}, Bragg wavelength=1,560 nm, the facet reflectivity = 0
            (front facet), 0.274 (rear facet), the refractive index = 3.25.

When the κL factor, the product of the coupling coefficient and the DBR length, is large, the penetration depth of the incident light becomes much smaller than the DBR length, resulting in little contribution of Fresnel reflection at the end facet to total reflectivity. In this case, the DBR operates as a narrow-bandwidth filter. On the other hand, when the κL factor is very small, the contribution of the grating is negligible and the DBR merely operates as an intracavity Fabry-Perot etalon.

When the κL is adequate, the total reflectivity is influenced by both the diffraction from the grating and the Fresnel reflection at the end facet. Then, the reflectivity profile has many subpeaks with intensities comparable to those of the main peak in the vicinity of the Bragg wavelength (detuning~ 0). In this case, the Bragg reflector operates as a frequency filter which enhances only the frequency components of optical pulses that coincide with the peak period in the reflectivity profile (approximately 500 GHz in the calculation), but suppresses other frequency components. The profile drastically changes with the initial phase of the grating and the DBR length. Fine-tuning of the reflectivity profile can be achieved by current injection to the DBR, because the injected current changes the refractive index of the DBR waveguide.

Figure 6 shows the SHG correlation trace and the optical spectrum of the harmonic mode-locked pulses observed in the LD shown in Figure 2 [26]. The absorber was grounded and the current in the gain section increased to 151 mA.

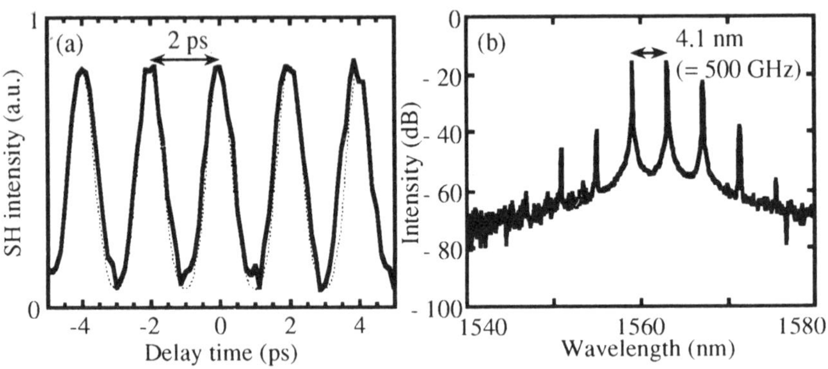

Figure 6.   (a) SHG correlation trace and (b) time-averaged optical spectrum of harmonic mode-locked pulses.

When the gain recovery time is constant, the saturation level in the gain between the pulses falls with a decrease in the pulse period. Therefore, the unsaturated gain required for the successful passive mode-locking rises with an increase in the pulse repetition rates, and hence, a higher unsaturated gain than that of the fundamental passive mode-locking (Figure 3) is required in harmonic mode-locking.

The pulse repetition rate was estimated to be 500 GHz from the peak period of the correlation trace, and it coincided well with the mode-spacing of the optical spectrum. This pulse repetition rate agreed well with the peak separation of the DBR reflectivity (Figure 5), confirming that the frequency-selectivity function of the DBR section determines the pulse repetition rates of the harmonic mode-locking.

The dotted line in the correlation trace is the theoretical curve calculated from the Fourier transformation of the optical spectrum. From the excellent agreement between the observed and the calculated curves, we concluded that the harmonic mode-locked pulses are also transform-limited. From the calculation, the actual pulse width was estimated to be 580 fs. The averaged output power was 15 mW, corresponding to 51.7 mW of the peak power. The pulse energy was 0.03 pJ.

It should be noted that, to obtain a spectrum with such good purity, fine-tuning of the injection currents to the phase control and/or DBR section is necessary. When such fine-tuning is not achieved, the suppression of the extra cavity modes is insufficient and the output pulses are modulated at the fundamental cavity roundtrip frequency. This indicates that an effective resonance in the Bragg reflectivity matching the harmonics of the extended cavity is required to ensure the stable and continuous generation of harmonic mode-locked pulses.

The correlation trace has some background due to the poor duty ratio between the pulse width and the pulse period (approximately 30%). To shorten the pulse width and obtain low-duty pulses, the absorber should be biased by a larger reverse voltage. Increasing the reverse voltage causes a faster absorption recovery time and a larger pulse energy, resulting in shortened pulses, because the pulse energy is inversely proportional to the square of the pulse width in the self-consistent mode-locking model. Rapid absorption recovery, however, causes a high saturated absorption and, hence, the unsaturated gain required for generating harmonic mode-locked pulses rises. When a large (0.2 V~) reverse voltage was applied to the absorber, the unsaturated gain obtained for 151 mA of the bias current became insufficient for stable harmonic mode-locked pulses, and mode-locking occurred at its fundamental roundtrip frequency. To obtain harmonic mode-locked pulses under a large reverse voltage, i.e. faster absorption recovery, design of a laser structure with higher unsaturated gain is necessary.

For some specific devices, higher order harmonic mode-locking was also observed [27]. Figure 7 shows the SHG correlation trace and the optical spectrum of higher order harmonic mode-locked pulses observed in a laser with a 115 μm DBR section. The pulse repetition rate was estimated to be 1.54 THz, the highest value reported to date. This is approximately the fourth harmonic of the peak period in the reflectivity profile of a 115 μm DBR, indicating that the selectivity of harmonic numbers in the Bragg reflector also determines the repetition rates of higher order harmonic mode-locking.

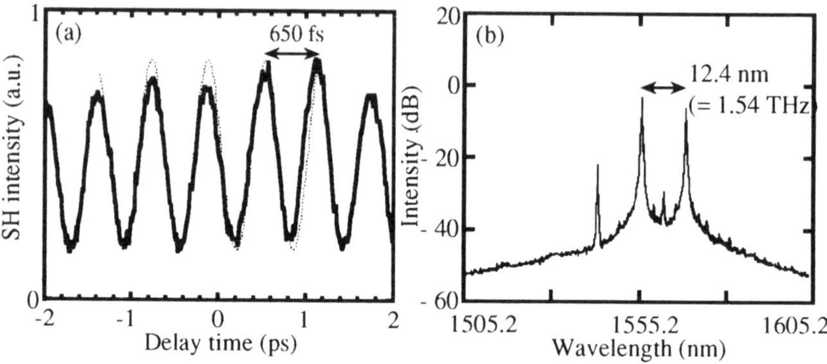

Figure 7.   (a) SHG correlation trace and (b) time-averaged optical spectrum of higher order harmonic mode-locked pulses.

4. Hybrid Mode-locking

One of the urgent problems to be solved for passive mode-locking is to control the timing of pulse generation and to reduce timing jitter. Pulse repetition rates and pulse widths fluctuate randomly in passive mode-locking because of carrier density fluctuation in the active waveguides, and the timing jitter exceeds several picoseconds [28].

When the jitter is large, the time slot in the receiver must be broadened and this gives the upper limitation of bit-rates in the transmission systems. Therefore, reducing the timing jitter of the optical signal pulses is very important in high speed optical communication systems. Hybrid and synchronous mode-locking techniques, where controllable electrical modulation or optical-pulse pumping is superimposed on passively mode-locked lasers, are solutions for reducing the jitter.

Hybrid mode-locking in our laser was achieved by superimposing a microwave signal on the absorber section. In hybrid mode-locking experiments, 50 Ω load resistance was connected parallel to the absorber in order to obtain an impedance matching to the microwave synthesizer. A laser with a 300 μm DBR was used in the hybrid mode-locking experiment. The repetition rate of the passively mode-locked pulses generated from this laser was approximately 32.765 GHz. The output pulses were converted to electrical signals by a high-speed (45 GHz) p-i-n photodiode, then the microwave spectra of the converted electrical signals were measured to study the jitter characteristics.

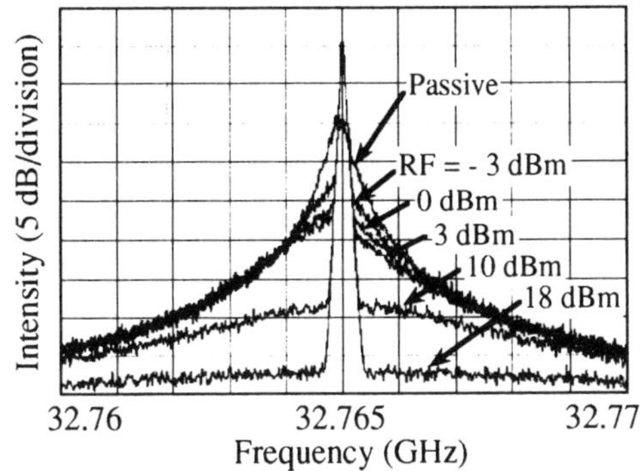

Figure 8.   Change of the microwave spectra of the hybrid mode-locked pulses at different microwave signal powers applied to the absorber. The resolution bandwidth was 100 kHz.

Figure 8 shows the change of the microwave spectra of the mode-locked pulses at different microwave signal powers. When no microwave signal was applied to the absorber, the microwave spectrum was broadened, indicating a large timing jitter. The FWHM of the spectrum was 600 kHz and the envelope corresponded well to the Lorentzian shape. When the microwave signal was applied to the absorber, the repetition frequency of the mode-locked pulses was "locked" to that of the microwave signal, and a single discrete peak appeared in the spectra. Increasing the microwave signal power suppressed the phase-noise levels in the spectra, indicating that the timing jitter was reduced.

Root-mean-squared (rms) timing jitter was obtained by integrating the single sideband phase noise level relative to the carrier per 1 Hz bandwidth [28]-[30]. The rms jitter with 18 dBm of microwave signal power was estimated to be 0.45 ps (5°). Further reduction in jitter is expected by reducing parasitic capacitance of the absorber section (10 pF in this experiment) and improving frequency response of the laser and the electrical assembly.

5. All-optical Clock Extraction from High Bit-rate Optical Data Signals Using Mode-locked Laser Diodes

All-optical signal processing is the key technology to avoid bottle-necking in electrical signal processing schemes. All-optical clock extraction from the incoming optical data signals is an urgent problem because it is necessary element for demultiplexers and regererators for time-division-multiplexing (TDM) transmission, which is a promising scheme for increasing bit rate in large-capacity optical communication systems.

The method of timing recovery uses the incoming data stream to drive a mode-locked laser [31,32] or a self-pulsating laser [33], which in turn generates an optical pulse train which has the same repetition rate as the bit rate of driving data signals; an optical cock. This pulse train can be then directly employed in further optical processing of the transmitted data. A monolithic mode-locked LD is very attractive as an optical clock-extraction device at an ultrahigh bit-rate, because it is very small in size and the pulse repetition rate is well over several ten GHz.

Figure 9 shows the experimental setup for clock extraction. A slave laser, which worked as a clock-extraction device, was a multi-electrode passively mode-locked DBR-LD with a long cavity length of approximately 5 mm [34]. When an absorber was grounded and two-divided gain sections were nonuniformly pumped, passive mode-locking occurred at 8.65 GHz of the repetition rate. The pulse width was 8.7 ps and the center wavelength was 1.555 μm.

The master laser was a gain-switched distributed feedback (DFB) laser. The repetition rate was 8.65 GHz and the center wavelength was 1.554 μm. Timing jitter was estimated to be 0.38 ps (@10 Hz~50 MHz). Gain-switched pulses were first compressed using 4-km long dispersion-shifted-fiber (D = -19 ps/km/nm). Pulse width after compression was estimated to be 8.6 ps from an SHG autocorrelation trace. Then, the pulses were coded by a $LiNbO_3$ intensity modulator driven by a pulse pattern generator. After amplification by an Er^{3+}-doped fiber amplifier, they were coupled to the mode-locked LD at the absorber facet through a 3 dB coupler. The coupling loss between the coupler output and the laser facet was approximately 6 dB. The mode-locked LD output was filtered once by a bandpass filter (λc = 1.555 μm, bandwidth = 1 nm) to remove the reflected data pulses from the absorber facet. The output from a p-i-n photodiode was measured by a sampling oscilloscope and a microwave spectrum analyzer.

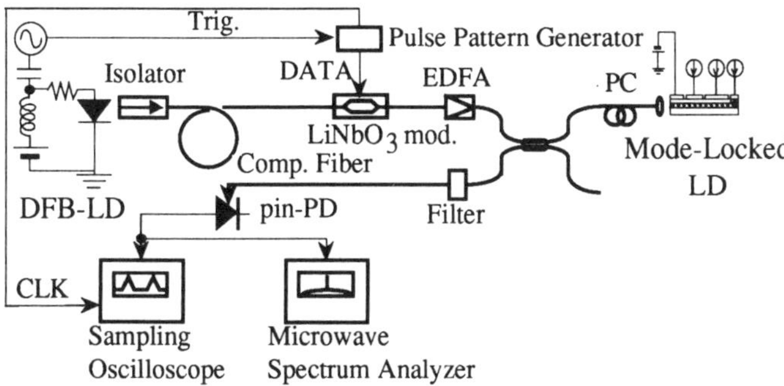

Figure 9.   Experimental setup for all-optical clock extraction using mode-locked LD.

First, we measured the change of output pulses for 1111... input data signals at different input pulse powers. Increasing the input power locked the mode-locked laser's frequency to the repetition frequency of the input data signals, and also decreased the phase noise levels in the output mode-locked pulses, indicating that the jitter was reduced by synchronous mode-locking. When the input power was 0.8 mW, the jitter was estimated to be 0.44 ps.

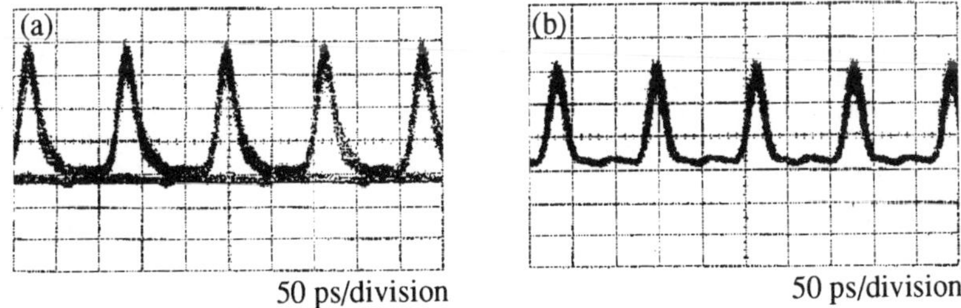

Figure 10.   Sampling oscilloscope traces of (a) the dye diagram of the input 8.65Gbit/s data signals and (b) the extracted optical clock pulses.

Figure 10 shows the sampling oscilloscope traces of the eye diagram of the input 2^{15}-1 pseudorandam-bit-stream data signals and the extracted optical clock pulses. The averaged input power was 1.2 mW. A continuous train of picosecond pulses at 8.65 GHz was clearly observed. The jitter was estimated to be 0.51 ps. The excess timing jitter was 0.34 ps (1°), as low as the jitter in the electrical clock extraction.

6. Conclusions

Ultrafast pulse generation using passively mode-locked LDs was reviewed in this chapter. Passively mode-locked LDs can provide transform-limited pulses at repetition rates higher than terahertz, verifying their usefulness as transmitters for large capacity soliton transmission systems. Wide wavelength-tunability, and jitter reduction in hybrid mode-locking operation are attractive and important features for constructing soliton transmission systems. All-optical clock extraction was also demonstrated using mode-locked LDs, expanding the application area of mode-locked LDs in large capacity optical communication systems; clock extraction device in all-optical signal processing.

References

1. Haus, H. A.: Modelocking of semiconductor laser diodes, *Jpn. J. Appl. Phys.* **20** (1981),1007-1020.
2. Haus, H. A.: Theory of mode locking with a slow saturable absorber, *IEEE J. Quantum Electron.* **QE-11** (1975), 736-746.
3. Lau, K. Y.: Narrow-band modulation of semiconductor lasers at millimeter wave frequencies (>100 GHz) by mode locking, *IEEE J. Quantum Electron.* 26 (1990), 250-261.
4. Van Der Ziel, J. P., Tsang, W. T., Logan, R. A., Mikulyak, R. M. and Augustyniak, W. M.: Subpicosecond pulses from passively mode-locked GaAs buried optical guide semiconductor lasers, *Appl. Phys. Lett.* **39** (1981) 525-527.
5. Deryagin, A. G., Kuksenkov, D. V., Kuchinskii, V. I., Portnoi, E. L. and Khrushchev, I. Yu.: Generation of 110 GHz train of subpicosecond pulses in 1.535 μm spectral region by passively modelocked InGaAsP/InP laser diodes, *Electron. Lett.* **3 0** (1994), 309-311.
6. Harder, C., Smith, J. S., Lau, K. L. and Yariv, A.: Passive mode-locking of buried heterostructure lasers with non-uniform current injection, *Appl. Phys. Lett.* **42** (1983), 772-774.
7. Vasilev, P. P. and Sergeev, A. B.: Generation of bandwidth-limited 2 ps pulses with 100 GHz repetition rate from multi-segmented injection laser, *Electron. Lett.* **25** (1989), 1049-1050.
8. Sanders, S., Eng, L., Paslaski, J. and Yariv, A.: 108 GHz passive mode locking of a multiple quantum well semiconductor laser with an intracavity absorber, *Appl. Phys. Lett.* **5 6** (1990), 310-312.
9. Chen, Y. K., Wu, M. C., Tanbun-Ek, T., Logan, R. A. and Chin, M. A.: Subpico second monolithic colliding-pulse mode-locked multiple quantum well lasers, *Appl. Phys. Let.* **58** (1991), 1253-1255.

10. Arahira, S., Matsui, Y., Kunii, T., Oshiba, S. and Ogawa, Y.: Transform-limited optical shortpulse generation at high repetition rate over 40 GHz from a monolithic passively mode-locked DBR laser diode, *IEEE Photon. Technol. Lett.* **5** (1993), 1362-1365.
11. Karin, J. R., Helkey, R. J., Derickson, D. J., Nagarajan, R., Allin, D. S., Bowers, J. E. and Thornton, R. L.: Ultrafast dynamics in field-enhanced saturable absorbers, *Appl. Phys. Lett.* **64** (1994), 676-678.
12. Chen, Y. K. and Wu, M. C.: Monolithic colliding-pulse mode-locked quantum-well lasers, *IEEE J. Quantum Electron.* **28** (1992), 2176-2185.
13. Derickson, D. J., Helkey, R. J., Mar, A., Karin, J. R., Wasserbauer, J. G. and Bowers, J. E.: Short pulse generation using multisegment mode-locked semiconductor lasers, *IEEE J. Quantum Electron.* **2 8** (1992), 2186-2202.
14. Agrawal, G. P.: *Nonlinear fiber optics*, Academic, San Diego, CA., 1989.
15. Kubota, H. and Nakazawa, M.: Long-distance optical soliton transmission with lumped amplifiers, *IEEE J. Quantum Electron.* **26** (1990), 692-700.
16. Arahira, S., Oshiba, S., Matsui, Y., Kunii, T. and Ogawa, Y.: Wavelength-tunable passively mode-locked DBR-LD, *Optics* **23** (1994), 268-274 (in Japanese).
17. Martins-Filho, J. F. and Ironside, C. N.: Multiple colliding pulse mode-locked operation of a semiconductor laser, *Appl. Phys. Lett.* **65** (1994), 1894-1896.
18. Shimizu, T., Wang, X. and Yokoyama, H.: Subterahertz-rate optical pulse generation with asymmetric colliding-pulse mode-locked semiconductor lasers, Tech. Dig. *Pacific rim conference on lasers and electro-optics* (1995), paper FV3.
19. Hirano, J. and Kimura, T.: Multiple mode locking of lasers, *IEEE J. Quantum Electron.* **QE-5** (1969), 219-225.
20. Takada, A. and Miyazawa, H.: 30 GHz picosecond pulse generation from actively mode-locked erbium-doped fibre laser, *Electron. Lett.* **2 6** (1990), 216-217.
21. Kuznetsov, M., Tsang, D. Z., Walpole, J. N., Liau, Z. L. and Ippen, E. P.: Multistable mode-locking of InGaAsP semiconductor lasers, *Appl. Phys. Lett.* **5 1** (1987), 895-897.
22. Sanders, S., Yariv, A., Paslaski, J., Ungar, J. E. and Zarem, H. A.: Passive mode locking of a two-section multiple quantum well laser at harmonics of the cavity round trip frequency, *Appl. Phys. Lett.* **58** (1991), 681-683.
23. Tucker, R. S., Koren, U., Raybon, G., Burrus, C. A., Miller, B. I., Koch, T. L. and Eisenstein, G.: 40 GHz active mode-locking in a 1.5 μm monolithic extended-cavity laser, *Electron. Lett.* **2 5** (1989), 621-622.
24. Yoshida, E., Kimura, Y. and Nakazawa, M.: Laser diode-pumped femtosecond erbium-doped fiber laser with a sub-ring cavity for repetition rate control, *Appl. Phys. Lett.* **6 0** (1992), 932-934.
25. Harvey, G. T. and Mollenauer, L. F.: Harmonically mode-locked fiber ring laser with an internal Fabry-Perot stabilizer for soliton transmission, *Opt. Lett.* **18** (1993), 107-109.
26. Arahira, S., Oshiba, S., Matsui, Y., Kunii, T., and Ogawa, Y.: 500 GHz optical short pulse generation from a monolithic passively mode-locked distributed Bragg reflector laser diode, *Appl. Phys. Lett.* **6 4** (1994), 1917-1919.
27. Arahira, S., Oshiba, S., Matsui, Y., Kunii, T. and Ogawa, Y.: Terahertz-rate optical pulse generation from a passively mode-locked semiconductor laser diode, *Opt. Lett.* **19** (1994), 834-836.
28. Derickson, D. J., Morton, P. A., Bowers, J. E. and Thornton, R. L.: Comparison of timing jitter in external and monolithic cavity mode-locked semiconductor lasers, *Appl. Phys. Lett.* **59** (1991), 3372-3374.
29. Von der Linde, D.: Characterization of the noise in continuously operating mode locked lasers, *Appl. Phys.* **B39** (1986), 201-217.
30. Rodwell, M. J. W., Bloom, D. M. and Weingarten, K. J.: Subpicosecond laser timing stabilization, *IEEE J. Quantum Electron.* **2 5** (1989), 817-827.
31. Ellis, A. D., Smith, K. and Patrick, D. M.: All optical clock recovery at bit rates up to 40 Gbit/s, *Electron. Lett.* **29** (1993), 1323-1324.

32. Ono, T., Shimizu, T., Yano, Y. and Yokoyama, H.: Optical clock extraction from 10-Gbit/s data pulses by using monolithic mode-locked laser diodes, Tech. Dig. *Optical fiber communication* (1995), paper ThL4.
33. Jinno, M. and Matsumoto, T.: All-optical timing extraction using a 1.5 μm self pulsating multielectrode DFB LD, *Electron. Lett.* **24** (1988), 1426-1427.
34. Arahira, S. and Ogawa, Y.: Passive and hybrid modelockings in a multi-electrode DBR laser with two gain sections, *Electron. Lett.* **31** (1995), 808-809.

GENERATION OF SOLITONIC PULSES IN CW-PUMPED BRILLOUIN OR RAMAN FIBER-RING LASERS

CARLOS MONTES

Laboratoire de Physique de la Matière Condensée
Centre National de la Recherche Scientifique
Université de Nice - Sophia Antipolis, Parc Valrose
F-06108 Nice Cedex 2, France

Abstract. Generation of solitonic pulses from stimulated Brillouin backscattering (SBBS) or stimulated Raman backscattering (SRBS) in a cw-pumped fiber-ring laser of large gain is only controlled by the feedback Stokes intensity R. The nonlinear dynamics of SBBS or SRBS in monomode optical fibers, including the optical Kerr effect, is governed by the 1-D coherent three-wave model, which resonantly couples two optical waves (pump and backscattered Stokes) with a material wave. Stability analysis around the steady solution presents a regular Hopf bifurcation with R as control parameter. For $R < R_{crit}$ the unstable amplitude modes merge together into a localized pulse, giving rise to soliton morphogenesis. Numerical simulation of the dynamical model starting from any initial conditions (material noise or steady state), and experiments in an optical fiber-ring cavity confirm this bifurcation and the generation of an asymptotically stable train of solitonic pulses of nanosecond width for SBBS and picosecond width for SRBS.

1. Introduction

High amplitude optical Stokes pulses have been obtained through stimulated Brillouin backscattering (SBBS) experiments in liquids [1], gases [2], plasmas [3], and in fiber-ring resonators [4]-[9], and through stimulated Raman backscattering (SRBS) in liquids [10, 11].

In monomode optical fibers the dynamics is well modelized by the 1-D 3-wave nonlinear interaction between the pump wave E_p, the material wave E_m, and the backscattered Stokes wave E_S. The resonant 3-wave interaction

A. Hasegawa (ed.), Physics and Applications of Optical Solitons in Fibres '95, 145–161.
© *1996 Kluwer Academic Publishers. Printed in the Netherlands.*

problem described by the nonlinear partial differential equations (PDE) model has been the object of many theoretical studies and numerical simulations. Within the slowly varying envelope approximation (SVEA), and for interacting waves whose profiles vanish at infinity (bounded envelopes for all three waves), the problem has been integrated by the inverse scattering transform (IST) in the non dissipative case [12]-[14], and numerically in the non integrable dissipative case [15], yielding no soliton solution for the backscattering problem : the collision of the pump wave with the material wave generates only a radiative E_S wave [14, 15]. However, we are interested in the non conservative problem of stimulated backscattering (SBBS or SRBS) in the presence of a cw-pump and strong material-wave dissipation, when the long interaction of the coherent pump may sustain Stokes localized structures. Moreover, we are concerned here with the spontaneous generation of these structures in an actual fiber-ring laser as a result of the instability of the steady-state optical distribution.

The propagative problem of convective amplification, when the cw-pump interacts with an initially bounded counterpropagating E_S signal, has been already considered for a damped material wave, in the absence of Stokes damping, yielding backward compressional E_S pulses of continuously growing amplitude and narrowing width [16, 17]. More recently, by taking also into account the Stokes damping, it has been shown that this initial value problem yields steady backscattered traveling shock pulses, whose subluminous velocity is uniquely determined by the damping coefficients and the cw-pump level [18].

Our aim here is to present the periodic boundary problem where the nonlinear 3-wave dissipative structure results from the unlimited interaction between the cw-pump and the E_S wave envelope *via* the material response. It differs from the propagative problem initiated by a bounded E_S envelope, since now, this solitary structure, issued from the instability of 3-wave overlapped envelopes can *move* - in the sense of *convective deformation* - at any velocity sub- or super-luminous. In the actual physical problem this unlimited interaction is achieved in a long-length ring-cavity, where the Stokes feedback continuously maintains all three wave envelopes in interaction. Indeed, it has been shown in a Brillouin-fiber-ring cavity, long enough to contain a large number of longitudinal modes beneath the gain curve, that spontaneous structuration of super-luminous solitary Stokes pulses [5] takes place below a critical feedback [6]-[9]. They are associated to the 3-wave soliton solution in an unbounded line [21]-[23], but modified by the strong dissipation of the material wave [24, 5, 9] and the periodic amplification in the low finesse ring-cavity.

2. Dissipative Three-wave Model For Stimurated Backscattering

In stimulated optical scattering the optical field $\widetilde{E}$ is coupled to the material vibrations *via* the nonlinear polarizability $\widetilde{P^{NL}}$:

$$[\partial_t^2 + 2\gamma_e\partial_t - (c/n)^2\partial_x^2]\ \widetilde{E} = \mu_0\partial_t^2\widetilde{P^{NL}}$$

$$\widetilde{P^{NL}} = \widetilde{P_{stim}} + \widetilde{P_{Kerr}} = \varepsilon_0\chi\widetilde{E} + \varepsilon_0\chi^{(3)}|\widetilde{E}|^2\widetilde{E}.$$

where γ_e is the optical damping.

In stimulated Brillouin scattering (SBS), $\widetilde{P_{stim}}$ is associated to the variation of the susceptibility (dielectric constant) with the hyper-sound density fluctuations $\widetilde{\rho}$:

$$\widetilde{P_{SBS}} = \varepsilon_0[\chi_0 + (\partial\chi/\partial\rho)_T\ \widetilde{\rho}]\ \widetilde{E} \simeq \varepsilon_0[\chi_0 + (\partial\varepsilon/\partial\rho)_T\ \widetilde{\rho}]\ \widetilde{E},$$

$\widetilde{\rho}$ obeying the fluid equation in the presence of the electrostrictive pressure [25]

$$[\partial_t^2 + 2\gamma_B\partial_t - c_a^2\partial_x^2]\ \widetilde{\rho} = -(\rho_0\varepsilon_0/2)(\partial\varepsilon/\partial\rho)_T\ \partial_t^2\widetilde{E^2},$$

where c_a is the acoustic velocity, and $\gamma_B = \pi\Delta\nu_B$ the acoustic damping associated to the spontaneous gainwidth $\Delta\nu_B$ [4].

In stimulated Raman scattering (SRS), $\widetilde{P_{stim}}$ is associated to the polarizability variation due to molecular vibrations

$$\widetilde{P_{SRS}} = \varepsilon_0 N[\alpha_0 + (\partial\alpha/\partial q) < q >]\widetilde{E},$$

the quantum expectation (mean) value of the vibrational coordinate $< q >$, for a constant population difference Δ, obeying the oscillatory equation [25, 26]

$$[\partial_t^2 + 2\gamma_R\partial_t - \omega_R^2] < q >= (\Delta/2m)\widetilde{E^2},$$

where ω_R is the resonance frequency and γ_R is the damping.

We shall only consider stimulated backscattering in the Stokes recoupled ring-cavity, which is the resonant direction for SBS and must be selected for SRS in order to prevent forward SRS after a simple passage. Splitting the optical field $\widetilde{E} = \widetilde{E_p} + \widetilde{E_S}$, the resonant 3-wave system couples a forward-propagating pump wave $\widetilde{E_p} = E_p(x,t)\exp[i(k_px - \omega_pt)]$ at frequency $\omega_p = k_pc/n$ with the material fluctuations of the medium $\widetilde{E_m} = E_m(x,t)\exp[i(k_mx - \omega_mt)]$ at frequency $\omega_m = k_mc_m \simeq 2c_m\omega_pn/c$, by stimulating a counterpropagating Stokes wave $\widetilde{E_S}(x,t) = E_S\exp[-i(k_Sx+\omega_St)]$ at frequency $\omega_S = \omega_p$ - ω_m; the resonance condition imposing $\mathbf{k}_m = \mathbf{k}_p$ - $\mathbf{k}_S$ yields $k_m = k_p + k_S \simeq 2k_p$. In the absence of dispersion, both SBBS and SRBS resonant processes may be described by the same 1-D 3-wave PDE system for the three complex amplitudes, relevant for single-mode optical fibers, by neglecting the wave material propagation [17, 23], but by taking into account the optical Kerr effect [4, 8], and the full dynamical equation for the material wave E_m without the SVEA [27] :

$$(\partial_t + \frac{c}{n}\partial_x + \gamma_e)E_p = -K_m E_S E_m + iK_r[|E_p|^2 + 2|E_S|^2]E_p,$$

$$(\partial_t - \frac{c}{n}\partial_x + \gamma_e)E_S = K_m E_p E_m^* + iK_r[2|E_p|^2 + |E_S|^2]E_S, \qquad (1)$$

$$[(1 + i\frac{\gamma_m}{\omega_m})\partial_t + i\frac{1}{2\omega_m}\partial_t^2 + \gamma_m]E_m = K_m E_p E_S^*,$$

where γ_m $(m = B, R)$ is the respective material damping coefficient, K_m $(m = B, R)$ is the SBS or SRS coupling constant, related to the respective steady gain coefficient [28, 29, 4, 30],

$$g_B = \frac{4K_B^2}{\gamma_B \varepsilon_0 c^2} = \frac{2\pi n^7 p_{12}^2}{c\lambda^2 \rho_0 c_s \Delta\nu_B} = 4.6 \times 10^{-11} \text{ m W}^{-1}, \qquad (2)$$

$$g_R = \frac{4K_R^2}{\gamma_R \varepsilon_0 c^2} = \frac{N(\partial\alpha/\partial q)_0}{\gamma_R n\varepsilon_0 c}\sqrt{\frac{\pi\omega_p\omega_S}{2m\omega_R cn}} = \frac{\lambda(1\mu m)}{\lambda} \times 10^{-13} \text{ m W}^{-1}, \quad (3)$$

and $K_r = n_2\omega_{p,S}/(2n)$ is the optical Kerr coefficient.

The gain coefficient g_m enters the instantanous material response model, which is obtained by neglecting the material dynamics $(\partial_t E_m \ll \mu_m E_m)$; E_m becomes a slave variable, and yields the intensity equations $[I_{p,S} = (n\varepsilon_0 c/2)|E_{p,S}|^2]$:

$$[\partial_x + (n/c)(\partial_t + 2\gamma_e)]I_p = -g_m I_p I_S,$$
$$[\partial_x - (n/c)(\partial_t + 2\gamma_e)]I_B = -g_m I_p I_S. \qquad (4)$$

We consider a cw-pumped fiber-ring cavity of length L (Figure 1), in which the recoupling of the pump wave is avoided by an intracavity isolator. The ring is then independent of the recoupling phases [7] and the problem is reduced to one control parameter, namely the Stokes intensity feedback efficiency $R < 1$. The boundary conditions are

$$I_p(0, t) = I_{cw} , \quad I_S(L, t) = RI_S(0, t). \qquad (5)$$

In a Brillouin fiber-ring laser the solitonic Stokes structure generated by the SBBS process [7] is large enough (some ns) to allow the SVEA for the three waves throughout the whole nonlinear interaction. This is also valid for SRBS during the initial stage of Stokes structuration. Moreover, since the optical Kerr effect is a slightly perturbative effect which does not

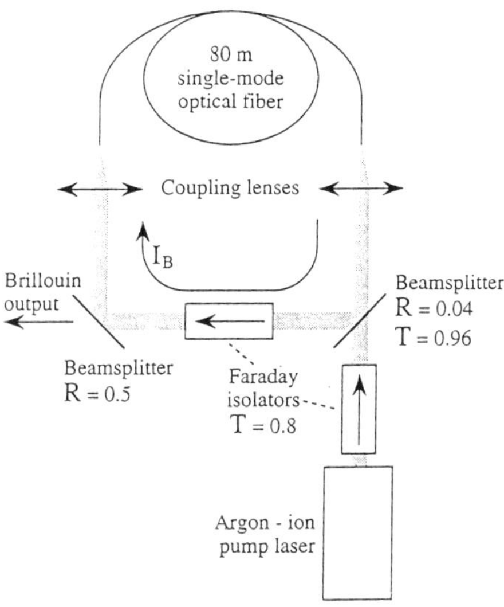

Figure 1.   Experimental setup

modify the bifurcation process [8, 9], the dynamical system (1) may be first approximated by the classical 3-wave SVEA model:

$$
\begin{aligned}
(\partial_t + \partial_x + \mu_e)E_p &= -E_S E_m, \\
(\partial_t - \partial_x + \mu_e)E_S &= E_p E_m^*, \\
(\partial_t + \mu_m)E_m &= E_p E_S^*,
\end{aligned}
\tag{6}
$$

where the envelope amplitudes, the time, and the space variables are normalized to the constant pump input $E_p(x_0, t) = E_{cw}$ and to the SBBS or SRBS coupling constant K_m (respectively $E_i/E_{cw} \to E_i$; $tK_m E_{cw} \to t$; $xcK_m E_{cw}/n \to x$), and where $\mu_{e,m} = \gamma_{e,m}(K_m E_{cw})^{-1}$ are dimensionless damping coefficients. The resulting dynamics will then be tested by computing the whole dynamical model (1).

In a Brillouin fiber-ring laser we could successfully observe the super-luminous soliton-like pulsed regime [5], predicted by numerical simulations of the model equations (6). However, we became able to actually control a stable train of solitonic pulses - experimentally and by the numerical treatment of (6) - only after having performed the stability analysis of the steady solution with the feedback R as control parameter [7] - [9]. It explains the steady mirror regime (for $R > R_{crit}$), the oscillatory regime (for $R < R_{crit}$) *via* a Hopf bifurcation, and morphogenesis of solitonic structures in the cw-pumped cavity (for $R < R_{pulse} < R_{crit}$). This general scenario of the dynamical model (6) is confirmed by numerical simulations starting from any initial material noise conditions, and remains essentially unchanged in the presence of optical Kerr effect and by accounting for complete material wave response as included in model (1).

2.1. STABILITY ANALYSIS

Stability analysis is performed for $\mu_e = 0$ by solving the linear perturbative equations around the nonlinear steady "Stokes mirror" solution where monotonic backscattered amplification of the Stokes wave in the optical medium is saturated by the monotonic depletion of the forward propagating cw-pump; by scaling $2x/\mu_m \to x$ the steady solution, common to model Equations (6) and (4), yields

$$I_p^{st}(x) = \frac{2DI_p^0}{(2D - I_p^0)\exp(-2Dx) + I_p^0} \ , \ I_S^{st}(x) = I_p^{st}(x) - 2D \qquad (7)$$

where $I_{p,S}$ are normalized to I_{cw} (i.e. $I_p^0 = 1$), where $L = g_m L I_{cw}$ stands for the dimensionless gain-length, and $D = const$ is related to R and L through (5):

$$R \exp(2DL) = R + 2(1 - R)D. \qquad (8)$$

By taking into account the time dependent optical and material responses of (6) through $I_i(x,t) = I_i^{st}(x) + \delta I_i(x) \exp(-i\omega t)$, $[i = p, S, m]$, where $I_m^{st} = I_p^{st}(x)I_S^{st}(x)/\mu_m^2$, the stability analysis is reduced to study the eigenvalue problem for complex ω

$$\frac{d^2Y}{dx^2} + [(\omega - ifD)^2 + \frac{f(1 - f)D^2}{\sinh^2[D(x_0 + x)]}] Y = 0 \ , \ f(\omega) = \frac{1 - i(\omega/\mu_m^2)}{1 - 2i(\omega/\mu_m^2)}, \qquad (9)$$

where

$$Y = [(1 - D)\sinh Dx + D\cosh Dx]^{f(\omega)} (\delta I_p - \delta I_S)/2$$

$$\frac{d(\delta I_p - \delta I_S)}{dx} = i\omega(\delta I_p + \delta I_S),$$

and the characteristic equation for ω is determined by the boundary conditions. The discrete set of frequencies Re ω which are solutions of the eigenvalue problem define the longitudinal modes of the ring cavity which are unstable for Im $\omega > 0$. By assuming in the potential term of Equation (9) $f(\omega)$ real and $D \to 0$, so that $\sinh^2[D(x_0 + x)] \sim [D(x_0 + x)]^2$, the solution is given by a combination of Airy functions Ai and Bi (C_1 and C_2 being constants):

$$Y(x) = z^{1/4}[C_1 Ai(-z) + C_2 Bi(-z)], z = [3\Omega(x + 1)/2]^{2/3}, \Omega = \omega - ifD.$$

The intensity model (4) corresponds to $f = 1$, and from Equation (9) it yields a harmonic oscillator of frequency $\Omega = \omega - iD$. But it breaks down at some critical value $R = R_{crit}^{int}$: the infinite frequency mode becoming unstable, closely followed by all other cavity modes, and the SVEA does not longer hold [7]. However this intensity model yields simple approximate expressions for the critical parameters at the bifurcation.

$$D_{crit} \simeq \frac{1}{6} \ , \ R_{crit}^{int} \simeq \frac{1}{3 \exp(L/3) - 2}.$$

The coherent 3-wave model of Equations (6), which takes into account the finite time material response, overcomes the high-frequency divergence of the instantaneous intensity model. Now, by decreasing the feedback R from the steady mirror solution, at $R = R_{crit} > R_{crit}^{int}$ appears a time dependent oscillatory regime through a regular Hopf bifurcation. Figure 2 shows the marginal stability (Im $\omega = 0$): the fundamental mode $\mathbf{M_1}$ of period close to the optical round-trip time $t_r = nL/c$ is the first unstable mode below a certain gain $L \simeq 6$, and above that it is the second perturbative mode $\mathbf{M_2}$ of period $t_r/2$, associated to the "collision" frequency of counterpropagating pump and backscattered perturbations, which is destabilized at the bifurcation. Thus, just below R_{crit}, either mode $\mathbf{M_1}$ or mode $\mathbf{M_2}$ which satisfy the SVEA, become unstable, the stability of higher order perturbative modes being successively broken for $R < R_{crit}^{int}$. For lower values of R the unstable amplitude modes laying beneath the gain curve of the stimulated process merge together into a pulse of full width at half maximum (FWHM) δt simply given by the inverse of the spontaneous gain width $\Delta\nu_m = \gamma_m/\pi$, itself narrowed by the exponential gain fact or $\sqrt{(ln2/L)}$ as was determined by Boyd et al. [31] for $L \gg 1$,

$$\delta t = \left\{ \frac{\gamma_m}{\pi} \sqrt{\frac{ln2}{L}} \right\}^{-1}. \tag{10}$$

Numerical simulation of the space-time 3-wave model (6), starting from any initial material noise conditions, show the three asymptotic regimes only depending on the feedback R (Figure 3 left h.s.). Figure 4 shows the mean asymptotic reflectivities obtained after long numerical simulation of Equations (6) and verified by the treatment of the complete dynamical model (1). It is interesting to note that, although the bifurcation causes a dramatic change in the mean amplitude of the backscattered Stokes wave, the mean energetic efficiency variation of the stimulated process is smooth. The morphogenesis of the 3-wave soliton starting from an unstable "Stokes mirror" ($R < R_{pulse} < R_{crit}$) is shown in Figure 5.

Pulse width : The dynamical treatment yields SBBS solitonic pulses in

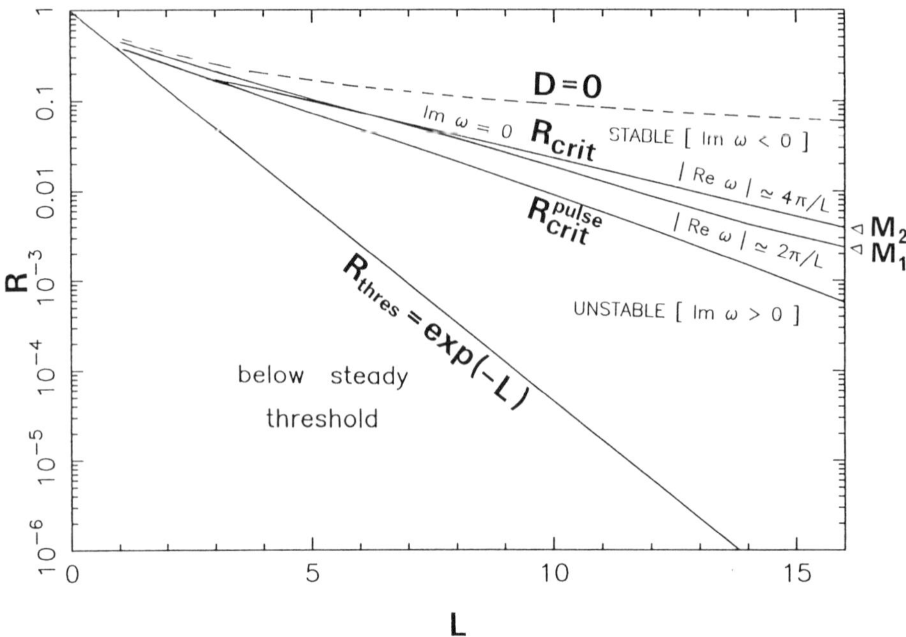

Figure 2. Marginal stability (Im $\omega = 0$) of modes $\mathbf{M_1}$ at $|R_e\omega| \simeq 2\pi/L$ and $\mathbf{M_2}$ at $|R_e\omega| \simeq 4\pi/L$. $R > R_{crit}$: steady Stokes mirror; $R_{crit} > R > R_{crit}^{pulse}$: oscillatory regime; $R_{thres} > R > R_{crit}^{pulse}$: solitonic regime.

the ~ 10 ns range [4] - [9] as shows Figure 6(a), in excellent agreement with expression (10). It would yield sub-ps width solitonic pulses for SRBS if the Kerr effect did not significantly spread them. The robustness of the solitonic pulses has been tested by numerical simulation of the coherent model (1), which includes the optical Kerr effect and the whole material dynamics. Computations of (1) in the intermediate SBBS-SRBS parameter range show that the Kerr effect broadens the pulses more than twice, as can be seen by comparing Figures 6(b) and (c). In an actual Raman-fiber-ring-laser the Kerr effect will probably limit the width to some tens of ps, therefore preventing dispersion, which is absent in model (1), from playing an important role. Typically, we obtain tens of ns width pulses for SBBS and we shall obtain tens of ps pulses for SRBS.

SBBS experiments support the bifurcation scenario, the cw-pump power of the laser being the control parameter (Figure 3 right h.s.). The setup is the basic SBS fiber ring laser (Figure 1), with a 80 m long single-mode polarization-maintaining silica fiber, cw-pumped by an argon-ion laser ($\lambda = 514.5$ nm , $\Delta\nu_B = 34$ GHz) and closed by two external beam splitters ($R = 10^{-2}$). High pump power ($P_{cw} > 400$ mW at the entrance of the

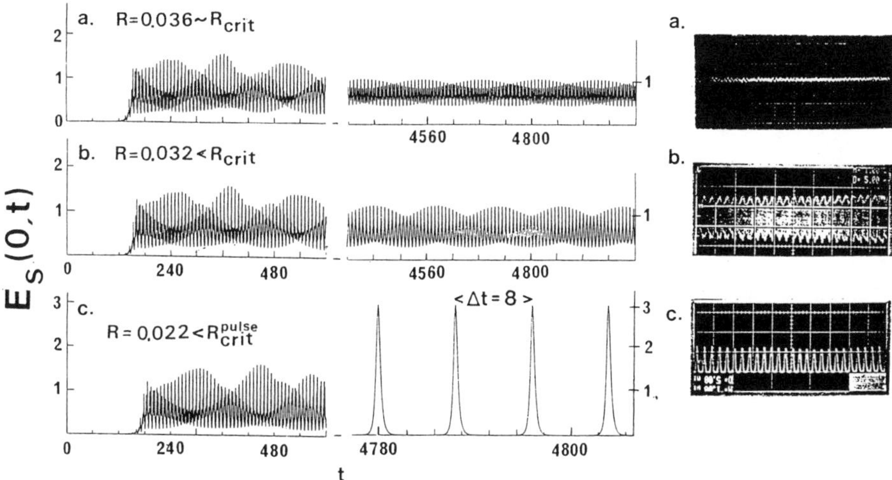

Figure 3. Temporal behaviour of the output Stokes amplitude $E_s(0,t)$, **left h.s.** solution of Equation (6) for $L = 8$, $\mu_m = 7$ and $\mu_e/\mu_m = 10^{-3}$; **right h.s.** experimental: (a) $R \sim R_{crit}$ evolution towards the mirror regime : mode $\mathbf{M_2}$ is still excited; (b) $R < R_{crit}$ oscillatory regime (mode $\mathbf{M_1}$); (c) $R < R_{crit}^{pulse}$ solitonic regime (from Reference [7]).

fiber) yields a rather stable cw Stokes output. Large pulses and low amplitude oscillations around the mirror coexist in the intermediate power range ($P_{cw} \sim 350$ mW), with a persistance of $\sim 10^3$ roundtrips. A stable pulsed regime at the roundtrip frequency is obtained for $P_{cw} < 300$ mW.

3. Asymptotic Traveling Wave Pulses

3.1. 3-WAVE DISSIPATIVE SOLITON

These solitonic pulses may be associated to the asymptotic 3-wave soliton solution in an unbounded line. Let us look for soliton solutions of Equations (6) which involve the coherent dynamics of a self-similar backward-propagating 3-wave (pump-Stokes-material) structure. In the nondissipative case (i.e. $\mu_e = \mu_m = 0$), specific traveling wave solutions have been studied [21] - [23], which persist in the dissipative case [24, 5]. These solutions belong to a broader class of solitons previously described through inverse scattering transform [12] - [14]. In the case of optical fibers, it is not possible to neglect the dampings ($\mu_i \neq 0$). Analytic traveling-wave solutions

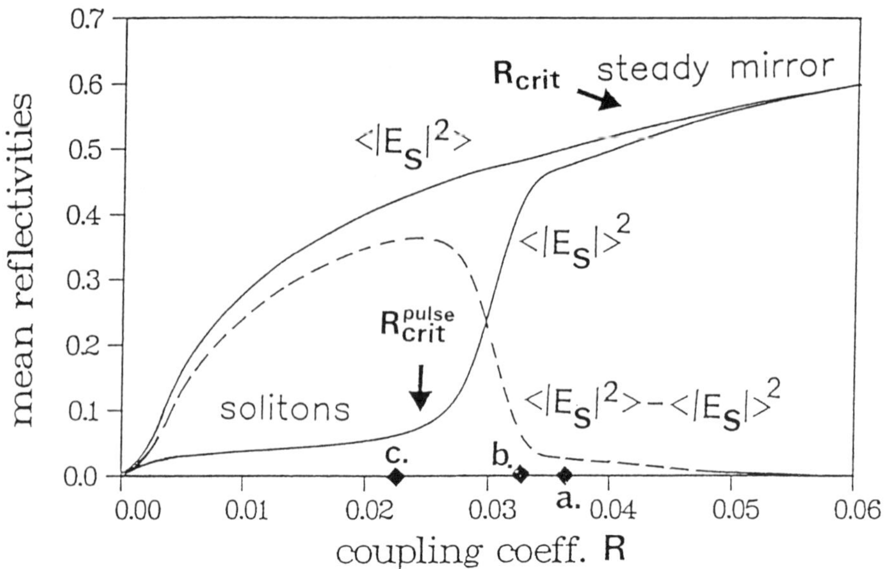

Figure 4.   Mean reflected Stokes intensities, amplitude and variance around the bifurcation vs. feedback R. Cases a., b. and c. of Figure 3, being marked on the axis (from Reference [7]).

are still available if the pump attenuation is neglected, which is locally legitimate as long as $\mu_e/\mu_m \ll 1$. Starting from Equations (6) we perform the change of frame moving in the backscattered direction ($\xi = x + vt$; $\tau = t$),

$$
\begin{aligned}
[\partial_\tau + (1+v)\partial_\xi]E_p &= -E_S E_m - \mu_e E_p, \\
[\partial_\tau + (v-1)\partial_\xi]E_S &= E_p E_m^* - \mu_e E_S, \\
[\partial_\tau + v\partial_\xi]E_m &= E_p E_S^* - \mu_m E_m.
\end{aligned}
\tag{11}
$$

Defining the A_i's fields as

$$
A_1 = |1+v|^{1/2}E_p \ , \ \ A_2 = |v-1|^{1/2}E_S \ , \ \ A_3 = |v|^{1/2}E_m \ ,
\tag{12}
$$

and looking for stationary solutions in the new frame, we have

$$
\begin{aligned}
\partial_X A_1 &= -s_1 A_2 A_3 - s_1 \rho_1 A_1, \\
\partial_X A_2 &= s_2 A_1 A_3^* - s_2 \rho_2 A_2, \\
\partial_X A_3 &= s_3 A_1 A_2^* - s_3 \rho_3 A_3,
\end{aligned}
\tag{13}
$$

where $X = \xi/|(v-1)(1+v)v|^{1/2}$, $\rho_1 = \mu_e|v-1|^{1/2}|v|^{1/2}/|1+v|^{1/2}$, $\rho_2 = \mu_e|1+v|^{1/2}|v|^{1/2}/|v-1|^{1/2}$, $\rho_3 = \mu_m|1+v|^{1/2}|v-1|^{1/2}/|v|^{1/2}$ and

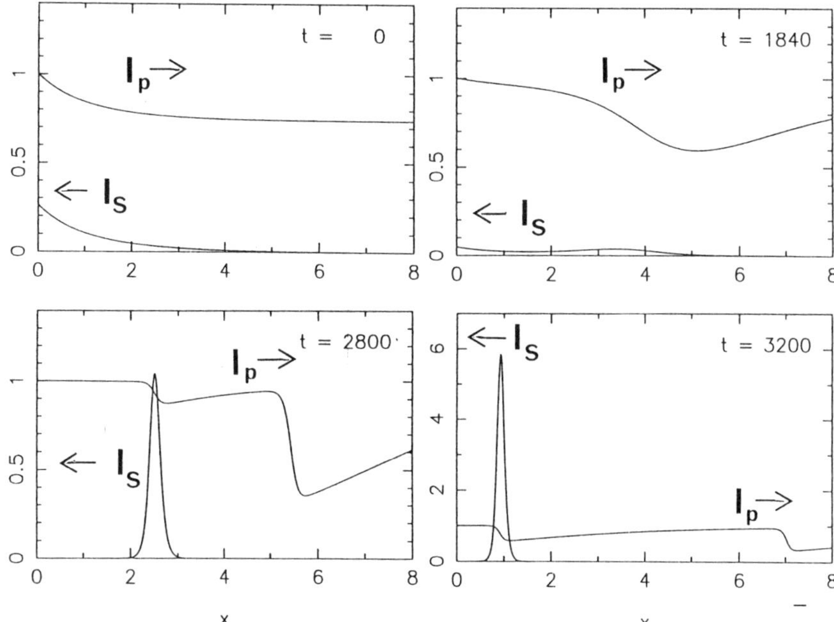

Figure 5.   Morphogenesis of the 3-wave soliton : for $R < R_{crit}^{pulse}$ the steady "Stokes mirror" is unstable and the system bifurcates towards the localized pulse regime for the backscattered Stokes wave : spatial distributions of I_p and I_s inside the optical medium at different times (from Reference [7]).

$s_1 = sgn(1+v)$, $s_2 = sgn(v-1)$, $s_3 = sgn(v)$. Taking into account stability arguments [19], we shall be only concerned with the solutions for which $v > 1$, i.e. a superluminous structure. Since $\mu_e \ll \mu_m$ and $v > 1$ we also have $\rho_1 \ll \rho_3$ but ρ_2 and ρ_3 are of the same order, i.e. $\mu_e \sim |v - 1|\mu_m$. Thus, we are able to neglect the pump damping; setting $\rho_1 = 0$ and $s_1 = s_2 = s_3 = 1$, and

$$A_0 = \rho_2 = \rho_3, \tag{14}$$

we obtain

$$A_1 = -A \tanh AX + A_0 \ , \ A_2 = A_3 = A \ \text{sech} \ AX. \tag{15}$$

Therefore:

$$\frac{\mu_e}{\mu_m} = \left| \frac{v - 1}{v} \right| \quad \Rightarrow \quad v = \frac{1}{1 - \mu_e/\mu_m}, \tag{16}$$

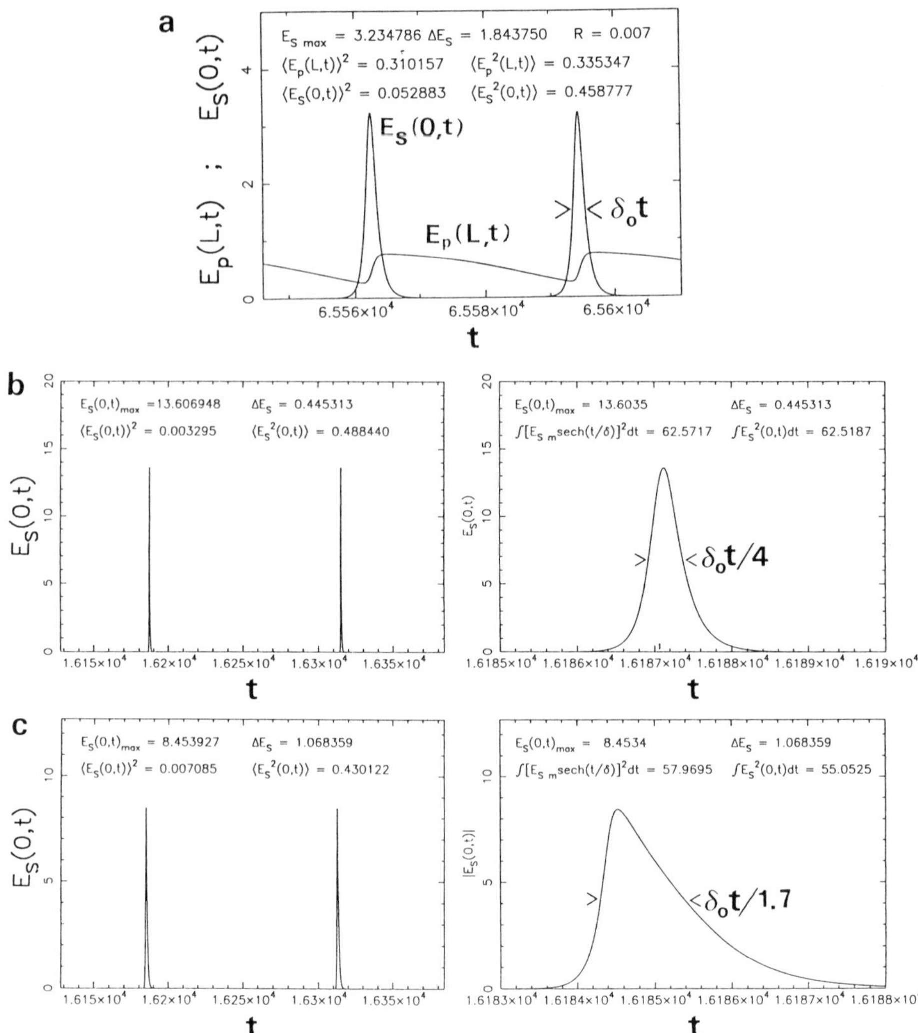

Figure 6.   (a) : Stable train of SBBS solitons generated in a low finesse ($R = 0.007$) fiber-ring cavity of gain-length $L = 10$; $\mu_m = 6.4$; and $\mu_m/\mu_e = 10^{-3}$, e.g. the experiment (cf. Figures 1 and 3) corresponding to a 80 m length silica fiber and 20 mW pump power ($\times 4$ due to the antiguiding sound factor, cf. Table 1 of Reference [7]), yields an amplitude width of $\delta_0 t \simeq 20$ ns.

Towards the SRBS soliton laser ? (b) : Numerical simulations of Equations (6) for the same gain $L = 10$, but for a $\times 4$ greater material wave damping than (a) ($\mu_m = 25.6$), and a smaller feedback ($R = 0.003$), yield a stable train of solitonic pulses of 1/4 width. (c) : For a strong pump power the Kerr effect becomes important. Computing Equations (1) with a Kerr/SRBS rate $\kappa_r = E_p K_r/K_m = 10^{-3}$ (i.e. $P_{cw} \simeq 1\ W$) yields a pulse only compressed by a factor 1/1.7 and presenting strong asymmetry. For higher pump power, higher values of μ_m are required to prevent the pulse from breaking down and to obtain considerable compression, which would correspond to actual SRBS lasers.

which yields for the wave amplitudes :

$$\begin{aligned}
E_p &= P_0 - P\,\tanh[(x+vt)/\Delta], \\
E_S &= S\,\mathrm{sech}[(x+vt)/\Delta], \\
E_m &= SP[2/(S^2+P^2)]^{1/2}\mathrm{sech}[(x+vt)/\Delta],
\end{aligned} \tag{17}$$

with the extra relationships :

$$P_0 = (\mu_e\mu_m)^{1/2}, \quad S/P = (2\mu_m/\mu_e - 1)^{1/2}, \quad \Delta = \frac{(\mu_e/\mu_m)^{1/2}}{P(1-\mu_e/\mu_m)}. \tag{18}$$

The only one free parameter left will be taken as the undepleted pump amplitude at the far front end of the localized structure, namely $P_0 + P = E_p(x+vt \to -\infty) = 1$.

This superluminous self-similar Stokes pulse does not contradict by any means the special theory of relativity. Its motion can be viewed as the result of a convective amplification of the leading edges of the Stokes and material pulses, whereas their rears are depleted, the pump wave being partially restored after the interaction. No transportation of information can be obtained *via* this process which can only occur if a sufficiently extended background of Stokes light is available so that the superluminous self-similar amplification takes place. Without imposing the symmetrical constraint (14), we can obtain a large class of sub- and super-luminous backscattered soliton-like solutions, the velocity depending on the front slope of the localized envelope [32].

Ring-cavity : In a long-length but nevertheless confined ring-cavity, the small value of R required to be in the unstable domain of soliton generation, renders the periodical evolution in the ring-laser deeply different from the infinite unbounded interaction in the line. At each round-trip, the outcoming Stokes pulse is re-injected at the other cavity end with the small recoupling factor R. Therefore this quasi-soliton pulse is periodically amplified and accelerated in the cavity and never reaches its invariant asymptotic form.

3.2. 2-WAVE ADIABATIC RAMAN PULSE

A new solitary wave solution may be obtained for the instantaneous material response model governed by the two coupled Equations (4) for the optical intensities. This model is commonly applied to stimulated Raman scattering (SRS) since it implies $\partial_t E_R \ll \gamma_R E_R$; it points out the possibility of picosecond range pulse generation for optical telecommunications. Starting from Equations (4), and after normalization, we again perform the change of frame moving in the backscattered direction ($\xi = x + vt$; $\tau = t$),

$$[\partial_\tau + (1 + v)\partial_\xi]I_p = -I_p I_S - \alpha I_p,$$
$$[\partial_\tau + (v - 1)\partial_\xi]I_S = I_p I_S - \alpha I_S, \tag{19}$$

where $\alpha = \mu_e \mu_m$. Then, by defining the J_i's intensities as

$$J_1 = |1 + v|I_p, \quad J_2 = |v - 1|I_S, \tag{20}$$

and looking for stationary solutions in the new frame, we have :

$$\partial_X J_1 = -J_1 J_2 - \beta_1 J_1,$$
$$\partial_X J_2 = J_1 J_2 - \beta_2 J_2, \tag{21}$$

where $X = \xi/|(v - 1)(1 + v)|$, $\beta_1 = \alpha|v - 1|$, $\beta_2 = \alpha|1 + v|$. For the same reasons as before, we shall be only concerned with the solutions for which $v > 1$. Since $v \simeq 1 + \Delta v$ ($|\Delta v| \ll 1$) and $\beta_1 \ll \beta_2$ we shall also neglect the pump damping. Setting $\beta_1 = 0$ and $\beta_2 = \beta$, we obtain the following set of coupled equations

$$J_1' = -J_1 J_2 \, , \ J_2' = J_1 J_2 - \beta J_2, \tag{22}$$

yielding for J_1 the equation

$$J_1 J_1'' - (J_1')^2 - J_1^2 J_1' + \beta J_1 J_1' = 0. \tag{23}$$

Introducing the change $J_1 = \exp U$ which satisfies $dU/dX = J_1'/J_1 = -J_2$, Equation (23) yields

$$dU/dX = \exp U - \beta U + C, \tag{24}$$

where C is a constant that can be removed with the change [$W = U - C/\beta$, $X = \eta \exp(C/\beta)$, $a = \alpha \exp(1/\alpha)$]

$$\frac{dW}{d\eta} = \exp W - aW, \tag{25}$$

the constants C, β and a being directly related to v and the dissipation α by taking $I_p(\xi \to -\infty) = 1$. For $a > e$, (i.e. for the localization threshold condition $\alpha < 1$), we obtain two finite values of W, say W_1 and W_2, where all the derivatives of W vanish when $\eta \to \pm\infty$. Therefore Equation (25) yields a traveling localized structure of the kink form for I_p and of the pulse form for I_S, e.g. for $I_p(\xi \to -\infty) = 1$ and $\alpha = \mu_e \mu_m = 0.625$ we have $a = \alpha \exp(1/\alpha) = 3.09564$, $W_2 = 1/\alpha = 1.6$, $W_1 = 0.573$, $I_p(\xi \to +\infty) =$

$\exp(W_1 - W_2) = 0.358$. The amplitude I_S and the width Δ_S of the Stokes structure are related to the velocity $v = 1 + \delta v$ through

$$I_S = [1 - \alpha(1 - log\alpha)](v + 1)/\delta v, \tag{26}$$

$$\Delta_S = \delta v/(1 + \alpha), \tag{27}$$

giving extremely high amplitude and narrow Raman pulses in the ps range, which should be useful for optical telecommunications. However, the intensity model (4) or (19) is singular [7] and is not appropriate for describing the nonlinear dynamics in a resonator, since no stable saturated regime can be reached. The instantaneous material response is responsible for unlimited compression and amplification during the feedback evolution. For a small finesse resonator, even for high μ_m values, which could justify the adiabatic approximation, only the coherent models (1) or (6) lead to a stable periodic pulsed regime.

4. Coherence Of The Stokes Solitonic Pulse

The amplitude of the material wave being always much lower than the optical ones, it can be shown that its phase may also rotate much faster than the optical ones [17]. An additive interesting feature is thus obtained when strong phase fluctuations are numerically imposed to the pump: the 3-wave structure then evolves to cancel this perturbation through complementary material phase fluctuations and the Stokes pulse remains nearly unaffected [5, 9]. A SBS Stokes wave with a coherence much higher than that of the pump has been experimentally observed [33], as well as pulses with very stable shapes [5, 9]. This mechanism can be regarded as the (1-D space $\times$ 1-D time) equivalent of the stationary 2-D phase conjugation [34].

5. Acknowledgement

The author thanks Professor A. Hasegawa for his kind invitation to the Kyoto "International Symposium on Physics & Applications of Optical Solitons in Fibers", and the "Groupe de Dynamique Non Linéaire en Optique" of the LPMC laboratory, J. Botineau, O. Legrand, C. Leycuras, A. Mamhoud, A. Picozzi and E. Picholle for their contribution to the work presented here.

References

1. Pohl, D. and Kaiser, W.: Time-resolved Investigations of Stimulated Brillouin Scattering in transparent and absorbing media : Determination of phonon lifetimes, *Phys. Rev. B* **1** (1970), 31-43.

2. Gorbunov, V. A., Papernyi, S. B. and Startsev, V. R.: Time compression of pulses in the course of stimulated Brillouin scattering in gases, *Sov. J. Quantum Electron* **13** (1983), 900-905.

3. Gellert, B. and Kronast, B.: Investigation of Stimulated Brillouin Scattering Under Well-Defined Interaction Conditions, *Appl. Phys. B* **32** (1983), 175-186; *Appl. Phys. B* **33** (1984), 29-41.

4. Botineau, J., Leycuras, C., Montes, C. and Picholle, E.: Stabilization of a stimulated Brillouin fiber ring laser by strong pump modulation, *J. Opt. Soc. Am. B* **6** (1989), 300-312.

5. Picholle, E., Montes, C., Leycuras, C., Legrand, O. and Botineau, J.: Observation of Dissipative Superluminous Solitons in a Brillouin Fiber Ring Laser, *Phys. Rev. Lett.* **66** (1991), 1454-1457.

6. Montes, C., Picholle, E., Botineau, J., Legrand, O. and Leycuras, C.: in *Nonlinear Coherent Structures in Physics and Biology*, Lectures Notes in Physics **393**, ed. M. Remoissenet and M. Peyrard, Springer-Verlag, Berlin, (1991), 44-51.

7. Montes, C., Mamhoud, A. and Picholle, E.: Bifurcation in a cw-pumped Brillouin fiber-ring-laser: Coherent soliton morphogenesis, *Phys. Rev. A* **49** (1994), 1344-1349.

8. Montes, C., Mamhoud, A. and Picholle, E.: Hopf bifurcation in cw-pumped fiber resonators: generation of stimulated Brillouin solitons, in *Nonlinear CoherentStructures in Physics and Biology*, ed. by K.H Spatschek and F.G. Mertens, Plenum Press, New-York, (1994), 357-363.

9. Montes, C., Mamhoud, A., Botineau, J. and Picholle, E.: Three-wave soliton morphogenesis in a cw-pumped Brillouin fiber-ring-laser, in *Fluctuation Phenomena: Disorder and Nonlinearity*, ed. by A.R. Bishop, S. Jim énez and L. V ázquez, World Scientific, Singapore, (1995), 397-405.

10. Maier, M., Kaiser, W. and Giordmaine, J. A.: Intense light bursts in the stimulated Raman effect, *Phys. Rev. Lett.* **17** (1966), 1275-1277.

11. Arrivo, S. A., Spears, K. G. and Sipior, J.: Picosecond transient backward stimulated Raman scattering and pumping of femtosecond dye lasers, *Opt. Comm.* **116** (1995), 377-381.

12. Zakharov, V. E. and Manakov, S. V.: Resonant interaction of wave packets in nonlinear media, *Sov. Phys. JETP Lett.* **18** (1973), 243-247 ; *Sov. Phys. JETP Lett.* **42** (1976), 842-845.

13. Kaup, D. J.: The three-wave interaction, a nondispersive phenomenon, *Stud. Appl. Math.* **55** (1976), 9-44.

14. Kaup, D. J., Reiman, A. and Bers, A.: Space-time evolution of nonlinear three-wave interactions. Interaction in a homogeneous medium, *Rev. Mod. Phys.* **51** (1979), 275-309; *Erratum* 915-917.

15. Chow, C. C.: Spatiotemporal chaos in nonintegrable three-wave interactions, *Physica D* **81** (1995), 237-270.

16. V. A. Gorbunov: Formation and amplification of ultrashort optical pulses as a result of stimulated scattering in opposite directions, *Sov. J. Quant. Electron.* **14** (1984) 1066-1073.

17. Coste, J. and Montes, C.: Asymptotic evolution of stimulated Brillouin scattering: Implications for optical fibers, *Phys. Rev. A* **34** (1986), 3940-3949.

18. Mikhailov, A., Montes, C., Picozzi, A. and Ginovart, F.: Subluminous solitary attractor in dissipative 3-Wave stimulated backscattering, submitted to Phys. Rev. A

19. Chiu, S. C.: On the self-induced transparency effect of the three-wave resonance process, *J. Math. Phys.* **19** (1978), 168-176.

20. Kaup, D. J.: The First-order Perturbed SBS Equations, *Nonlinear Sci.* **3** (1993), 427-443.

21. Armstrong, J. A., Jha, S. S. and Shiren, N. S.: Some Effects of Group-Velocity Dispersion on Parametric Interactions, *IEEE J. Quant. Elect.* **QE-6** (1970), 123-129.

22. Nozaki, K. and Taniuti, T.: Propagation of Solitary Pulses in Interactions of Plasma Waves, *J. Phys. Soc. Jpn.* **34** (1973), 796-800; Oshawa, Y. and Nozaki, K.: Propagation of Solitary Pulses in Interaction of Plasma Waves. II, *J. Phys. Soc. Jpn.* **36** (1974), 591-595; Nozaki, K.: Stability of Triple Solitary Waves and Effects of Inhomogeneities, *J. Phys. Soc. Jpn.* **37** (1974), 1124-1128.

23. Montes, C. and Legrand, O.: Nonstationary stimulated Brillouin backscattering, in *Electromagnetic and acoustic scattering: detection and inverse problem*, ed. World Scientific, Singapore, (1989), 209-221; Legrand, O. and Montes, C.: Apparent superluminous quasi-solitons in stimulated Brillouin backscattering, *J. Phys. Colloq. France* **50** (1989), C3-147-155.

24. Morosov, S. F., Piskunova, L. V., Sushik, M. M. and Freidman, G.I.: Formation and amplification of quasi-soliton pulses in head-on stimulated scattering, *Sov. J. Quant. Electron.* **8** (1978), 576-580.

25. Kaiser, W. and Maier, M.: Stimulated Rayleigh, Brillouin and Raman Spectroscopy, in *Laser Handbook* **2**, edit. by T.F. Arecchi and F.O. Schulz-Dubois, North-Holland, Amsterdam, (1972), 1078-1150.

26. Newell, A. C. and Moloney, J. V.: Stimulated Raman Scattering, in *Nonlinear Optics*, 253-259, Addison-Wesley, Redwood City, (1992), 253-259

27. Montes, C. and Pellat, R.: Inertial response to nonstationary stimulated Brillouin backscattering: Damage of optical and plasma fibers, *Phys. Rev. A* **36** (1987), 2976-2979.

28. Ippen, E. P. and Stolen, R. H.: Stimulated Brillouin scattering in optical fibers, *Appl. Phys. Lett.* **21** (1972), 539-541.

29. Stolen, R. B. and Ippen, E. P.: Raman gain in glass optical waveguides, *Appl. Phys. Lett.* **22** (1973), 276-278.

30. Agrawal, G. P.: in *Nonlinear fiber optics*, AT&T Bell Laboratories and Academic Press, San Diego (1989), 218-288.

31. Boyd, R. W., Rzazewsky, K. and Narum, P.: Noise initiation of stimulated Brillouin scattering, *Phys. Rev. A* **42** (1990), 5514-5521.

32. Montes, C. and Picozzi, A.: Solitary 3-wave dissipative structures in cw-pumped stimulated backscattering, *in preparation*

33. Smith, S. P., Zarinetci, F. and Ezekiel, S. E.: Narrow-linewidth stimulated Brillouin fiber laser and applications, *Opt. Lett.* **16** (1991), 393-395.

34. Fisher, R. A., in *Optical Phase Conjugation*, Academic Press, Orlando (1983), chapt. 6 & 7.

HIGH-SPEED FIBER TRANSMISSION USING OPTICAL PHASE CONJUGATION

SHIGEKI WATANABE AND TERUMI CHIKAMA
Fujitsu Laboratories Ltd.
1015 Kamikodanaka, Nakahara-ku, Kawasaki 211, Japan

Abstract. We have shown the effectiveness and possible application of optical phase conjugation (OPC) in high-speed optical fiber transmission. By compensating for both chromatic dispersion and the optical Kerr effect in a transmission fiber by OPC, we can increase the transmission distance and capacity in future high-speed fiber transmission systems. Waveform pre-compensation using a fiber compensator and a phase conjugator at the transmitting terminal can potentially upgrade the conventional wavelength-division multiplexed transmission systems.

1. Introduction

In long-haul and high-speed optical fiber communication systems using Er^{3+}-doped fiber amplifiers (EDFAs), group velocity dispersion (GVD) and self-phase modulation (SPM) in the fiber limits the achievable bit-rate-distance (BL) product [1]. This also limits the signal bit-rate in wavelength-division multiplexing (WDM) transmission system. To further increase the capacity in future, the system will be required to compensate for the distortion due to GVD and SPM.

Many dispersion compensation techniques have been proposed and demonstrated based on either optical or electrical processes. Optical approaches include the use of in-line optical filters [2], pre-chirping of the signal wave [3], the use of two-mode fiber near the cutoff [4], and the use of an equalizing fiber [5, 6]. However, the distortion caused by SPM cannot be compensated for with these techniques. In optical soliton transmission, SPM cancels the pulse broadening caused by GVD [7, 8].

Optical phase conjugation (OPC) [9] is a candidate to compensate for both GVD and SPM [10, 11]. Using OPC, waveforms distorted by GVD in

A. Hasegawa (ed.), Physics and Applications of Optical Solitons in Fibres '95, 163–176.

one fiber are reshaped by the following OPC and subsequent propagation through the second fiber with the same GVD. The effectiveness of this method in high-speed fiber transmissions has been demonstrated with a phase conjugator using parametric four-wave mixing (FWM) [12]-[14]. OPC compensates for both positive and negative GVD and the effect of OPC does not depend on the type of optical modulation format [15]. Demonstration of ultra-short pulse recovery using OPC [16] shows that the method is applicable in ultra-high speed systems [17]. Recently, a soliton transmission system using OPC was shown, in which the soliton interaction due to the fiber nonlinearity was suppressed [18]-[20]. Theoretical investigations indicate that OPC can be effectively applied to long-distance transmissions through both dispersion-shifted fiber (DSF) with normal dispersion [10], [21] and standard single-mode fiber (SMF) [15]. SPM compensation by OPC, however, is limited by the asymmetry of strength of the optical Kerr effect along the fiber with respect to the OPC position due to fiber loss and amplifier gain.

A method for the exact compensation of GVD and SPM using OPC has been proposed [22]. Here, GVD and nonlinearity along the fibers are controlled so as to substantially cancel the effects of fiber loss and amplifier gain. The effect of the method was demonstrated by a transmission using waveform pre-compensation by a fiber compensator and OPC [23].

When applied to wavelength-division multiplexed (WDM) systems, OPC simultaneously compensates for waveform distortion in each channel [24] by converting WDM signal waves to WDM phase-conjugate waves. For long-distance WDM transmission, pre-compensation has the potential to exactly compensate for the channels' distortion [22], even when each channel is distorted by a different dispersion due to the second-order dispersion. The possibility of OPC compensating the nonlinear interaction among adjacent channels [25, 26] allows for further expansion in capacity.

We describe the application of OPC to conventional high-speed fiber transmission systems including WDM systems and the resultant increase in their performance.

2. Waveform Distortion Compensation by OPC

Figure 1 illustrates compensation for waveform distortion due to GVD and SPM using OPC. The transmission fiber employs optical amplifiers to compensate for fiber loss. The signal wave E_S with optical power P_1 is transmitted through Fiber-1 of length L_1, which has a loss, α_1, a dispersion, D_1, and an optical nonlinearity coefficient, γ_1. A phase conjugator then converts the wave to a phase-conjugate wave E_C ($\propto E_S^*$). E_C with optical power P_2 is then transmitted through Fiber-2 with parameters L_2, α_2, D_2, and γ_2. By taking the complex-conjugate of the wave, OPC inverts the phase shift (modulation) accumulated during transmission over Fiber-1. Waveform distortion is compensated for by giving E_C the same distortion in Fiber-2.

The condition for compensation is approximately achieved when the total GVD in both fibers is identical [9]-[12], that is

$$D_1 \, L_1 = D_2 \, L_2, \tag{1}$$

and when the path averaged SPM-induced phase shift along both fibers is almost the same [10, 11], such that

$$\gamma_1 \, \overline{P_1} \, L_1 \simeq \gamma_2 \, \overline{P_2} \, L_2. \tag{2}$$

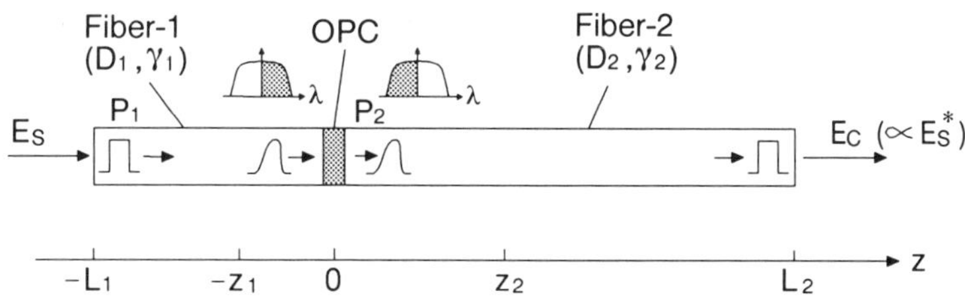

Figure 1. Compensation for GVD and SPM by OPC.

$\overline{P_j} = (1/L_j) \int P_j(z)dz$ ($j = 1, 2$) indicates the pass-averaged power in Fiber-j. Equations (1) and (2) show that the fiber length should be proportional to the inverse of the GVD or the nonlinearity. The above equations determine the lengths of Fiber-1 and Fiber-2 and the optical signal powers. Note that once the transmission fiber is used in the OPC system for a certain bit-rate, it can thereafter be used in a higher bit-rate system as the compensation condition does not depend on the bit-rate.

SPM compensation by OPC, however, is limited by the asymmetric distribution of the Kerr effect along the fiber with respect to the OPC position. This is caused by the asymmetric light intensity change due to fiber loss and amplifier gain. Therefore, the above approximation applies only when the change of SPM along the fiber is small.

A method for the exact compensation of GVD and SPM using OPC has been proposed [22], in which the ratio of dispersion to the strength of the optical Kerr effect is set to be the same at two positions, $-z_1$ and z_2, with respect to OPC as follows:

$$\frac{D_1(-z_1)}{\gamma_1(-z_1)P_1(-z_1)} = \frac{D_2(z_2)}{\gamma_2(z_2)P_2(z_2)}. \tag{3}$$

Positions $-z_1$ and z_2 are the points to which the dispersion (or the optical Kerr effect) from the OPC position ($z = 0$) are the same as

$$-\int_0^{-z_1} D_1(z)dz = \int_0^{z_2} D_2(z)dz. \tag{4}$$

Under the conditions in Equations (3) and (4), the output wave from the second fiber is the exact complex conjugate of the input wave into the first fiber. By providing the equal ratio of the dispersion and the nonlinearity at the corresponding positions throughout the fibers, we can thus exactly compensate for the waveform distortion due to GVD and SPM.

From Equations (3) and (4), the relation between dz_1 and dz_2 is given by

$$\frac{dz_1}{dz_2} = \frac{D_2(z_2)}{D_1(-z_1)} = \frac{\gamma_2(z_2)P_2(z_2)}{\gamma_1(-z_1)P_1(-z_1)}. \tag{5}$$

As shown by Equation (5), the larger the dispersion or the nonlinearity, the smaller the change of corresponding z.

When the fiber parameters are constant as α_j, D_j, γ_j in Fiber-j, and if the change of SPM in the fibers is small, we obtain the approximate formula (1) and (2) by integrating Equation (5) along the fibers.

By providing an equal ratio of nonlinearity and dispersion at corresponding positions in the two fibers, the effects of fiber loss and amplifier gain are substantially canceled. We can then increase the BL-product until compensation efficiency is limited by the second-order GVD, the polarization mode dispersion (PMD) and a reduction in the signal-to-noise (S/N) ratio due to amplifier noise [27].

To exactly satisfy Equation (3), it is most practical that the dispersion in fibers changes with changes in intensity [22].

3. System Configuration

Figure 2 shows typical configurations of the transmission system using OPC. Usually, two fibers of almost the same length and dispersion are used, and the phase conjugator is placed midway (Figure 2(a)). Two windows of dispersion are opened for long-distance transmission in the 1.55-µm wavelength range [28]. One is an anomalous dispersion of +16~18 ps/nm/km as in a standard SMF, and the other is a small normal dispersion in DSF. In both windows, the influence of the modulation instability is efficiently suppressed [28]. By controlling the ratio of dispersion and nonlinearity, we can effectively increase the amplifier spacing in the system [22]. An alternative would be to use a fiber with a small dispersion for the transmission line and a fiber with a large dispersion for concentrated compensation in the transmitting or receiving terminal [15]. In the system shown in Figure 2(b), the waveform is pre-compensated for in the transmitting terminal, whereas Figure 2(c) shows a system where the distorted waveform is post-compensated for at the receiving terminal. In these configurations, systematic parameters should be set to satisfy Equation (3) for an ideal compensation. Equations (1) and (2) can also be applied approximately when the amplifier spacing is not too long and the change in the Kerr shift due to intensity change is small [10, 11].

The pre-compensation configuration shown in Figure 2(b) has various advantages. First, we can use conventional DSFs and EDFAs, even when wavelength shift occurs by OPC, since only Fiber-2 is used as the transmission line. Note that we can use the self-filtering effect of EDFAs' chain by tuning the wavelength of E_C (λ_C) to the gain peak wavelength of EDFAs (λ_g). Second, by locating the phase conjugator in the transmitting terminal, we can achieve a stable OPC performance. It is much easier to realize a polarization insensitive OPC. Finally, the system can be used in WDM schemes by preparing a fiber pre-compensator at every WDM channel, the dispersion of which is adjusted for each channel. By using this configuration, we can compensate for GVD and SPM in every channel even when the GVD is different from channel to channel due to the second-order dispersion of the transmission fiber.

To implement the pre-compensation and satisfy the compensation condition, a dispersion-compensating fiber (DCF) is a promising candidate. It is possible to obtain DCFs with a large normal dispersion [5, 6], and having a smaller mode-field diameter (MFD) and a larger nonlinearity compared with DSFs [29]. This suggests that by making a dispersion-decreasing DCF (DD-DCF), in which the dispersion of DCF changes to follow an intensity change, we can implement fiber pre-compensation.

A short DD-DCF and OPC can thus act as a pre-compensator that compensates for GVD and SPM in a longer DSF transmission line with smaller normal dispersion. This method was recently demonstrated in a 20 Gb/s signal transmission [23].

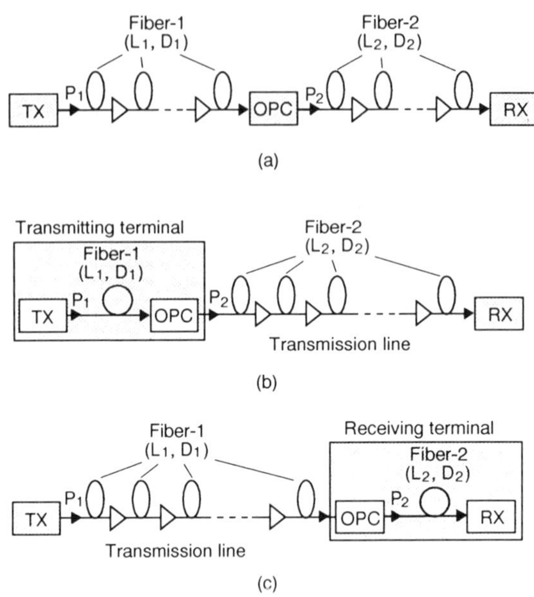

Figure 2. Optical transmission system using OPC.

4. Experiment

4.1. EXPERIMENTAL SETUP

Figure 3 shows the experimental setup [23]. As light sources, we used three-electrode $\lambda/4$-shifted DFB-LDs [30]. A time-division multiplexed 20 Gb/s signal wave, E_S, at wavelength $\lambda_S = 1{,}551$ nm was generated by time-division multiplexing two 10 Gb/s RZ signals with ~40 ps pulse width (FWHM). To generate 10 Gb/s RZ pulses, we intensity-modulated E_S using the first LiNbO$_3$ modulator (LN-1) with a 10-GHz sinusoidal signal followed by the second LiNbO$_3$ modulator (LN-2) intensity-modulating with a 10 Gb/s NRZ data stream (PN: $2^{23} - 1$). The modulated E_S with a power of P_1 was input to two-stage DD-DCFs and the waveform was pre-compensated.

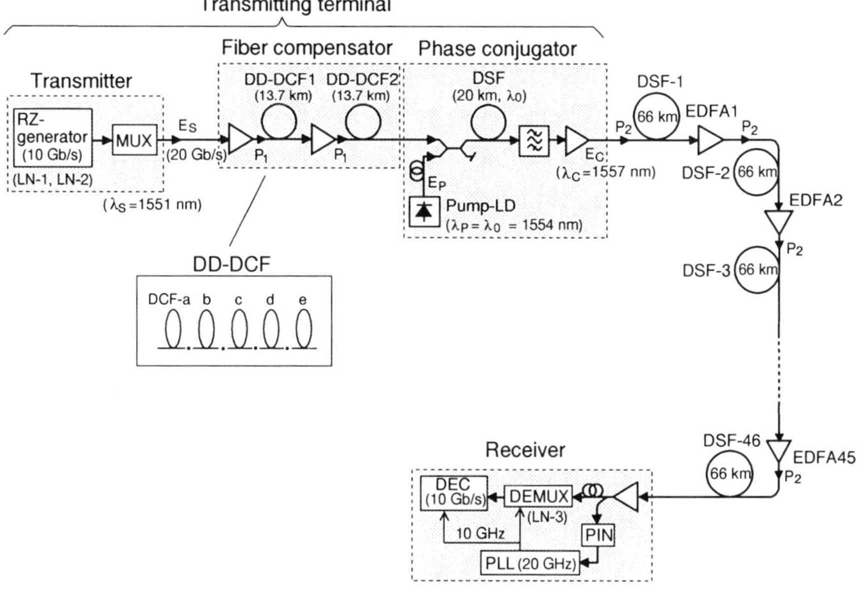

Figure 3. Experimental setup for 20 Gb/s-3,000 km transmission. DD-DCF consisting of five-stages DCF spliced with 0.1 dB splice loss (inset).

TABLE 1. The length and dispersion of each DCF in DD-DCF.

DCF	Length (km)	D_1 (ps/nm/km)
a	2.8	-80.6
b	2.7	-57.9
c	2.8	-43.7
d	2.7	-32.1
e	2.7	-27.0

Each DD-DCF consisted of five DCFs (Figure 3 inset) being spliced together. The loss of each DD-DCF was 0.46 dB/km and the MFD of each DCF was set to ~4 μm. To approximately satisfy the condition in Equation (3), D_1 should decrease along with the average optical power in each DCF. We therefore set the length and D_1 of five DCFs, as shown in Table 1, and spliced these five DCFs together. The length of each DD-DCF was 13.7 km and the total dispersion was –662.8 ps/nm.

An optical phase conjugator then converted the pre-compensated E_S to a copropagating phase-conjugate wave, E_C, at $\lambda_C = 1{,}557$ nm by non-degenerate forward FWM in a 20 km DSF [12, 15] using a pump wave, E_P, at $\lambda_P = 1{,}554$ nm. The conversion efficiency from E_S to E_C was -12 dB. The generated E_C was input to the 3,036 km-long transmission line consisting of 46 DSF sections (0.21 dB/km loss) and 45 EDFAs, each with a noise figure of ~6 dB. The averaged dispersion of the line at λ_C was -0.44 ps/nm/km. Therefore, the difference in the total dispersion between 2-stage DD-DCFs and the transmission lines was ~10 ps/nm. The length of each fiber span was 66 km and an input optical power, P_2, of E_C to each span was set to +6 dBm. The optimum value of P_1 under the above conditions was +16 dBm. The nonlinear constant of the DD-DCF was estimated to be $\gamma_1 \sim 18.0$ W^{-1}km^{-1}.

To suppress the stimulated Brillouin scattering (SBS), E_S and E_P were frequency modulated [31] by 500 kHz and 150 kHz sinusoidal signals. At the receiver, E_C was time-division demultiplexed to 10 Gb/s using the third LiNbO$_3$ modulator (LN-3) and a phase-lock loop (PLL), and the bit error rate (BER) was measured.

For comparison, we tested for 1,518 km transmission using one DD-DCF and 23 fiber spans.

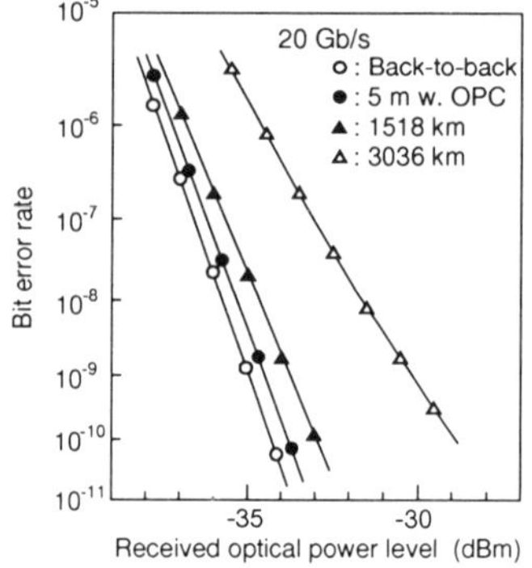

Figure 4. BER characteristics.

4.2. RESULTS AND DISCUSSIONS

Figure 4 shows the BER characteristics. Even after transmission over 3,036 km, we could detect the signal at a BER of less than 10^{-9}. The power penalty of 4.8 dB at a BER of 10^{-9} was mainly due to a degradation of the S/N ratio from the accumulated EDFA noise and intersymbol interference caused by residual waveform distortion. In the experiment, λ_C was detuned by about 1.5 nm from $\lambda_g \sim 1{,}558.5$ nm. If we can tune λ_C to λ_g, we can achieve a better performance with a higher S/N ratio. In 1,518 km transmission, the waveform was almost completely recovered with a 1.2 dB penalty.

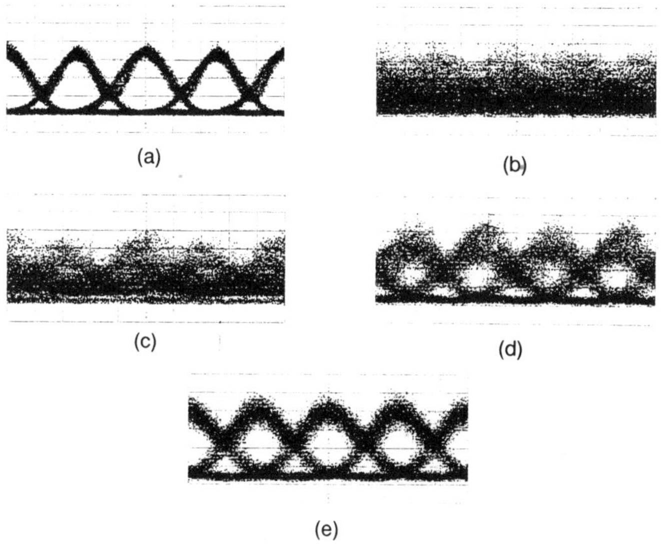

Figure 5. Detected waveforms at (a) transmitter out, (b) OPC out, (c) at 1,518 km, (d) at 2,706 km and (e) after 3,036 km. (V: 12 mV/div., H: 20 ps/div.)

Figure 5 shows the detected waveforms for 3,036 km transmission. Figures 5(a) to 5(e) show the waveforms at the transmitter and OPC outputs, at 1,518 km, 2,706 km, and after 3,036 km, respectively.   The pre-distorted waveforms were gradually compensated for as E_C was transmitted, and were well recovered at 3,036 km. The residual distortion in Figure 5(e) was caused by an imperfect compensation condition. This was mainly because the fiber span was not sufficiently shorter than the nonlinear length [28]. That is, $(\gamma_2 \overline{P_2})^{-1}$ was ~120 km for a P_2 of +6 dBm (especially $(\gamma_2 P_2)^{-1}$ ~50 km at near the input point). If we shorten the amplifier spacing without changing the path-averaged power, we can obtain better compensation. Dividing the DD-DCF into more than five will further improve the compensation.

The results show the effectiveness of OPC compensating for both GVD and SPM.   Note that only five-stages of stepped changes in DCF dispersion were shown to be effective in the DD-DCF, whereas conventional DSFs and EDFAs were used in the transmission line without control over the dispersion. The only requirement was that the total GVD before and after OPC be almost identical. This method can thus be used to upgrade any conventional system by simply inserting the pre-compensator and the phase conjugator between the transmitter and the transmission line.

5. Application to WDM Transmission

To increase the bit-rate of WDM systems, we have to consider the difference in dispersion from channel to channel. The higher the bit-rate, the stricter the required compensation of the dispersion in each channel. The second-order dispersion makes it difficult to compensate for every channel's dispersion. As shown in Figure 6, the pre-compensation system can be effectively applied to high-speed WDM systems only by fitting a pre-compensator at every WDM channel [22]. By adjusting dispersion of the compensator for each channel, we can exactly compensate for GVD and SPM at every channel. When necessary, it is possible to use plural phase conjugators before WDM.

To test the feasibility, we simulated 40 Gb/s IM signal transmission. We assumed the transmission fiber (Fiber-2) to be a DSF of length $L_2 = 2,000$ km with a loss of $\alpha_2 = 0.2$ dB/km, and with a constant normal dispersion $D_2 < 0$ ps/nm/km and a second-order dispersion of $D_2' = 0.08$ ps/nm^2/km. We set the nonlinearity constant of Fiber-2 to $\gamma_2 = 2.6$ W^{-1}km^{-1}. The average input signal power P_2 was +4 dBm and the fiber loss was compensated for by EDFAs at 50 km spacing. The fiber compensator (Fiber-1) was DD-DCF with a length of $L_1 = 20$ km. DD-DCF consisted of five DCFs, each of length 4 km, with a constant loss of $\alpha_1 = 0.35$ dB/km [6]. Its nonlinearity was assumed to be $\gamma_1 = 15.0$ W^{-1}km^{-1}, which was six times larger than γ_2. This value is reasonable using the nonlinear refractive index recently reported [29]. Larger value may be expected considering the value of 18.0 W^{-1}km^{-1} which we obtained for the DD-DCF used in the experiment. We set the dispersion of the k-th (k = 1, 2, --, 5) DCF, $D_1(k)$, in proportion to its average power with the average value of $D_1 \equiv \Sigma_k \{D_1(k)\}/5 = 100{\times}D_2$. Large values of D_1 are possible by designing the refractive-index profile in the core of the DCF [5, 6]. The optimum input optical power P_1 for each DD-DCF at the above settings was +15.2 dBm. We can decrease the necessary D_1 value and P_1 by increasing L_1 using cascaded Fiber-1s with optical amplifiers for loss compensation. The second-order dispersion of Fiber-1 was assumed to be $D_1' = 0.16$ ps/nm^2/km. The bandwidth of the lowpass filter in the receiver was set to 24 GHz.

Figure 7 shows the eye-opening penalty dependence on D_2. For $D_2 > -0.1$ ps/nm/km, a power penalty due to the nonlinear effect appeared. So long as we used the normal dispersion of $D_2 < -0.1$ ps/nm/km, a significant penalty did not appear. The result shows the dispersion independence or the wideband characteristics of the system.

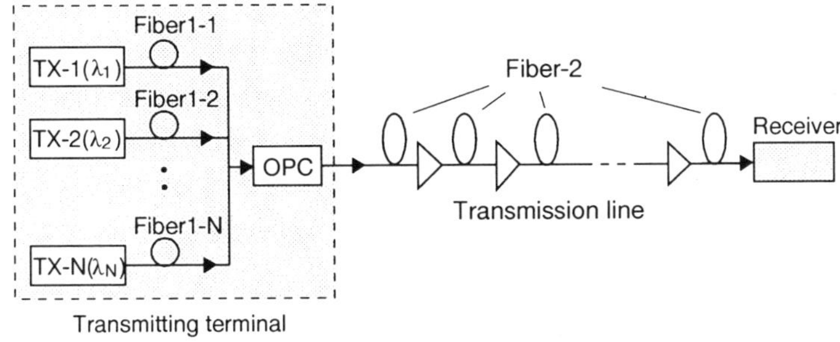

Figure 6. Application of pre-compensation by OPC to WDM transmission.

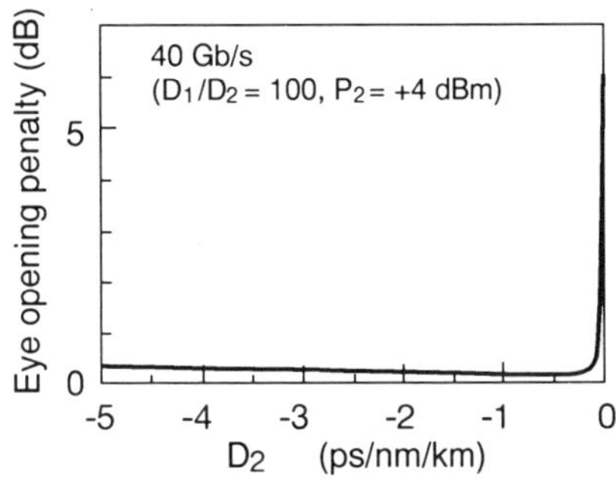

Figure 7. Calculated power penalty dependence on D_2.

The permissible range for D_2 shown in Figure 7 indicates the potential applicability of our method to ultra-wideband WDM optical transmission systems. Considering that $D'_2 = 0.08$ ps/nm^2/km and assuming that we use the dispersion range $-4.0 < D_2 < -0.5$ ps/nm/km, we can utilize more than 40 nm of bandwidth. Such a wide bandwidth will be sufficient to meet the requirements for future WDM systems, and a large value of D_2 is desirable in suppressing the nonlinear interactions between channels [32].

By providing a suitable DD-DCF for each channel and taking the phase conjugate of the waves, we can upgrade conventional WDM systems to higher bit-rate systems. A transparency of OPC that its effect does not depend on the bit-rate and the modulation format further extends the flexibility and capacity of the system.

6. Conclusion

OPC is effective when applied in a high-speed fiber transmission system by compensating for both chromatic dispersion and the optical Kerr effect. By designing the ratio of the dispersion to the optical Kerr effect equal at each positions, OPC exactly compensates for waveform distortion due to GVD and SPM. OPC allows increased transmission distances and capacity for future high-speed fiber transmission systems. Waveform pre-compensation by OPC at the transmitting terminal has the potential to increase the capacity in WDM systems by providing a pre-compensator at every WDM channel, the dispersion of which is adjusted for each channel. By transmission through a fiber with a large normal dispersion which suppresses nonlinear crosstalk, OPC has the potential capacity of several hundred Gb/s WDM transmission.

7. Acknowledgment

The authors acknowledge S. Kaneko and M. Shirasaki for useful discussions, G. Ishikawa, A. Sugata and H. Ooi for their support during experiments, and H. Kuwahara for encouragement.

References

1. Agrawal, G. P.: *Nonlinear Fiber Optics.*, Academic Press, New York, 1989.
2. Quellette, F.: Dispersion cancellation using linearly chirped Bragg grating filters in optical waveguides, *Opt. Lett.,* **12** (1987), 847-849.
3. Koch, T. L. and Alferness, R. C.: Dispersion compensation by active predistorted signal synthesis, *J. Lightwave Technol.,* **LT-3** (1985), 800-805.
4. Poole, C. D., Wiesenfeld, J. M. and McCormick, A. R.: Broadband dispersion compensation by using the higher-order spatial mode in a two-mode fiber, *Opt. Lett.,* **1 7** (1992), 985-987.
5. Vengsarkar, A. M. and Reed, W. A.: Dispersion-compensating single-mode fibers: efficient designs for first- and second-order compensation, *Opt. Lett.,* **18** (1993), 924-926.
6. Onishi, M., Koyano, Y., Shigematsu, M., Kanamori, H. and Nishimura, M.: Dispersion compensating fiber with a high figure of merit of 250 ps/nm/dB, *Electron. Lett.,* **3 0** (1994), 161-163.
7. Hasegawa, A. and Tappert, F.: Transmission of stationary nonlinear optical pulses in dispersive dielectric fibers. I. Anomalous dispersion, *Appl. Phys. Lett.,* **23** (1973), 142-144.

8. Mollenauer, L. F., Stolen, R. H. and Gordon, J. P.: Experimental observation of picosecond pulse narrowing and solitons in optical fibers, *Phys. Rev. Lett.*, **45** (1980), 1095-1098.
9. Yariv, A. , Fekete, D. and Pepper, D. M.: Compensation for channel dispersion by nonlinear optical phase conjugation, *Opt. Lett.*, **4** (1979), 52-54.
10. Kikuchi, K. and Lorattanasane, C.: Compensation for pulse waveform distortion in ultra-long distance optical communication systems by using nonlinear optical phase conjugator: *OAA'93*, SuC1 (1993), 22-25.
11. Watanabe, S., Chikama, T., Ishikawa, G., Terahara, T. and Kuwahara, H.: Compensation of pulse shape distortion due to chromatic dispersion and Kerr effect by optical phase conjugation: *IEEE Photon. Technol.Lett.*, **5** (1993), 1241-1243.
12. Watanabe, S., Naito, T. and Chikama, T.: Compensation of chromatic dispersion in a single-mode fiber by optical phase conjugation, *IEEE Photon. Technol. Lett.*, **5** (1993), 92-95.
13. Jopson, R. M., Gnauck, A. H. and Derosier, R. M.: 10-Gb/s 360-km transmission over normal-dispersion fiber using mid-system spectral inversion, *OFC '93*, **PD3** (1993), 17-20.
14. Tatham, M. C., Sherlock, G. and Westbrook, L. D.: Compensation of fiber chromatic dispersion by mid-way spectral inversion in a semiconductor laser amplifier, *ECOC'93*, **ThP12.3** (1993), 61-64.
15. Watanabe, S., Ishikawa, G., Naito, T. and Chikama, T.: Generation of optical phase-conjugate waves and compensation for pulse shape distortion in a single-mode fiber, *J. Lightwave Technol.*, **12** (1994), 2139-2146.
16. Laming, R. I., Richardson, D. J., Taverner, D. and Payne, D. N.: Transmission of 6 ps linear pulses over 50 km of standard fiber using midpoint spectral inversion to eliminate dispersion, *IEEE J. Quantum Electron.*, **3** (1994), 2114-2119.
17. Ellis, A. D., Tatham, M. C., Davies, D. A. O., Nessett, D., Moodie, D. G. and Sherlock, G.: 40 Gb/s transmission over 202 km of standard fiber using midspan spectral inversion, *Electron. Lett.*, **31** (1995), 299-301.
18. Forysiak, W. and Doran, N. J.: Conjugate solitons in amplified optical fiber transmission systems, *Electron. Lett.*, **30** (1994), 154-155.
19. Wen, S. and Chi, S.: Undoing of soliton interaction by optical phase conjugation, *Electron. Lett.*, **30** (1994), 663-664.
20. Yoshioka, H. and Fujii, Y.: Soliton transmission using optical phase-conjugators-Compensation for timing jitter due to soliton-soliton interaction and SSFS-, *IOOC'95*, **FD3-3** (1995), 111-112.
21. Kurtzke, C. and Gnauck, A.: How to increase capacity beyond 200 Tbit/s km without solitons, *ECOC'93*, ThC 12.**12** (1993), 45-48.
22. Watanabe, S. and Shirasaki, M.: Exact compensation for both chromatic dispersion and Kerr effect in a transmission fiber using optical phase conjugation, *J. Lightwave Technol.*, **14** (1996), 243-248.
23. Watanabe, S., Kaneko, S., Ishikawa, G., Sugata, A., Ooi, H. and Chikama, T.: 20 Gb/s fiber transmission experiment over 3000 km by waveform pre-compensation using fiber compensator and optical phase conjugator, *IOOC'95*, **PD2-6** (1995), 31-32.
24. Jopson, R. M., Gnauck, A. H., Iannone, P. P. and Derosier, R. M.: Polarization-independent mid-system spectral inversion in a 2-channel, 10-Gb/s, 560-km transmission system, *OFC'94*, **PD22** (1994), 104-107.
25. Watanabe, S.: Cancellation of four-wave mixing in a single-mode fiber by midway optical phasec onjugation, *Opt. Lett.*, **19** (1994), 1308-1310.
26. Gnauck, A. H., Jopson, R. M. and Derosier, R. M.: Demonstration of nonlinearity compensation by spectral inversion in a three-channel WDM system, *ECOC'94*, PD (1994), 91-94.
27. Desurvire, E.: *Erbium-doped fiber amplifiers*, John Wiley & Sons, New York (1994).
28. Lorattanasane, C. and Kikuchi, K.: Design of long-distance optical transmission systems using midway optical phase conjugation, *IEEE Photon. Technol. Lett.*, **7** (1995), 1375-1377.
29. Kato, T., Suetsugu, Y., Takagi, M., Sasaoka, E. and Nishimura, M.: Measurement of the nonlinear refractive index in optical fiber by the cross-phase-modulation method with depolarized pump light, *Opt. Lett.*, **20** (1995), 988-990.

30. Ogita, S., Kotaki, Y., Matsuda, M., Kuwahara, Y., Onaka, H., Miyata, H. and Ishikawa, H.: FM response of narrow-linewidth, multielectrode $\lambda/4$ shift DFB laser, *IEEE Photon. Technol. Lett.*, **2** (1990), 165-166.
31. Cotter, D.: Stimulated Brillouin scattering in monomode optical fiber, *J. Opt. Commun.*, **4** (1983), 10-19.
32. Bergano, N. S., Davidson, C. R., Vengsarker, A. M., Nyman, B. M., Evangelides, S. G., Darcie, J. M., Ma, M., Evankow, J. D., Corbett, P. C., Mills, M. A., Ferguson, G. A., Pedrazzani, J. R., Nagel, J. A., Zyskind, J. L., Sulhoff, J. W. and Lucero, A. J.: 100 Gb/s WDM transmission of twenty 5 Gb/s NRZ data channels over transoceanic distances using a gain flattened amplifier chain, *ECOC'95*, **Th.A.3.1**, (1995), 967-970.

DESIGN AND CHARACTERIZATION OF OPTICAL FIBERS FOR HIGH CAPACITY TRANSMISSION

MASAYUKI NISHIMURA
Sumitomo Electric Industries, Ltd.
1, Taya-cho, Sakae-ku, Yokohama 244, Japan

Abstract. Various new designs of optical fibers are being developed in pursuing the ultimate performance of optical fiber communications. The current major concerns with optical fiber design are control of chromatic dispersion and nonlinearities. This paper reviews the current status of new optical fiber development, such as improved designs of dispersion-shifted fibers, dispersion-flattened fibers, and so on. Topics related to measurement techniques for nonlinear refractive index and chromatic dispersion are also discussed.

1. Introduction

The history of the development of optical fiber transmission systems has been a never-ending expansion of the transmission distance and capacity. In the interim, optical fibers have also been evolved in terms of optical attenuation and transmission bandwidth. Recent dramatic success of erbium-doped fiber amplifiers (EDFAs) has further accelerated the development toward ultra-long distance and/or ultra-high capacity transmission, such as soliton transmission, wavelength-division-multiplexed (WDM) transmission, etc. Since attenuation of optical fibers can be easily compensated for by use of EDFAs, one of the current major concerns with optical fibers is control of dispersion characteristics. Relatively high power levels of optical signals traveling in very long optical fibers have also increased the importance of nonlinearity of optical fibers.

In order to pursue the ultimate performance of optical fiber communications, various new designs of optical fibers are being developed. Main design issues are the control of chromatic dispersion and nonlinearity. This paper

177

A. Hasegawa (ed.), Physics and Applications of Optical Solitons in Fibres '95, 177–189.
© 1996 *Kluwer Academic Publishers. Printed in the Netherlands.*

reviews several topics related to those new design optical fibers. Measurement techniques for nonlinear refractive index and longitudinal distribution of chromatic dispersion will also be discussed.

2. New Designs of Optical Fibers

2.1. DISPERSION-SHIFTED FIBER WITH REDUCED MFD

Standard single mode fibers (SMFs) have their minimum chromatic dispersion in the 1.3 μm wavelength window because of the dispersion characteristics of silica glass. Dispersion-shifted fibers (DSFs) whose zero-dispersion wavelengths are shifted to the 1.55 μm wavelength window have been developed to realize both the lowest attenuation and the minimum dispersion around 1.55 μm. They have already been introduced into commercial telecommunication systems on a considerable scale.

When DSFs are to be used in soliton transmission, not only does dispersion need to be controlled within a certain range (positive, or anomalous, with small absolute values, typically), but also nonlinearity of the optical fiber is another important parameter determining the system design or performance. Since the optical peak power of soliton pulses required for stable transmission is directly proportional to the inverse of the nonlinear coefficient, it may be beneficial under certain conditions to enhance the nonlinear coefficient of the optical fiber. One of the direct and effective approaches for increasing the nonlinear coefficient is to reduce the mode field diameter (MFD), because the nonlinear coefficient is defined as n_2/A_{eff}, where n_2 is the nonlinear refractive index and A_{eff} is the effective core area, and A_{eff} is generally proportional to (MFD)^2. Typical MFDs of the standard DSFs range from 7 to 8 μm. DSFs with reduced MFDs have been explored for such soliton transmission or Raman amplification [1]. Figure 1 illustrates an example of the refractive index profiles of the fibers investigated. In general, in order to reduce the MFD, it is necessary to increase the refractive index difference (Δ) between the core and the cladding typically by increasing the concentration of germanium doped in the core. This, however, would cause higher attenuation of the fiber because of larger Rayleigh scattering induced by germanium. To overcome this difficulty, in this fiber, fluorine is added in the cladding for lowering the refractive index, which makes it possible to achieve large Δ without significantly increasing the germanium concentration in the core.

Two types of DSFs have been designed and manufactured by the VAD method. Structural parameters and typical results of measurement are summarized in Table 1. MFDs smaller than 6 μm have been realized. Measured attenuation at 1.55 μm is 0.227 dB/km for Type A, and 0.265 dB/km for Type B. Although these values are relatively large compared with those

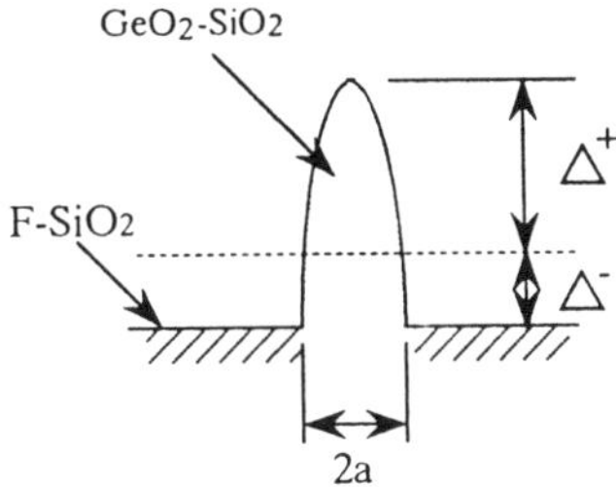

Figure 1.   Refractive index profile of dispersion-shifted fiber with reduced MFD.

TABLE 1. Structural parameters and characteristics of
fabricated DSFs with reduced MFDs

Parameter	Type A	Type B
Δ^+ [%]	1.0	1.2
Δ^- [%]	0.35	0.65
Core diameter [μm]	5.3	5.2
MFD [μm]	5.56	4.88
Cutoff wavelength [μm]	1.22	1.40
Zero dispersion wavelength [μm]	1.50	1.55

of the standard DSFs (typically 0.20 to 0.21 dB/km), they are considerably small for such high NA (small MFD) fibers. Further improvement of attenuation would be possible by fine tuning of manufacturing conditions.

2.2. DISPERSION-SHIFTED FIBER WITH ENLARGED MFD

In conventional transmission schemes using the non-return-to-zero (NRZ) format, nonlinearities of optical fibers are usually aspects which need to be avoided or suppressed.

Especially when DSFs are used in dense WDM transmission systems, the four wave mixing (FWM) becomes one of the major factors limiting the system performance. Much effort is being made to suppress the FWM effect in WDM transmission. For example, appropriate allocation of signal wavelengths with unequal spacing has been found to be highly effective [2]. As for DSFs, it has been pointed out [3],[4] that proper management of chromatic dispersion along the transmission line is essential. It is also effective to design DSFs so as to reduce the nonlinear coefficient. This can

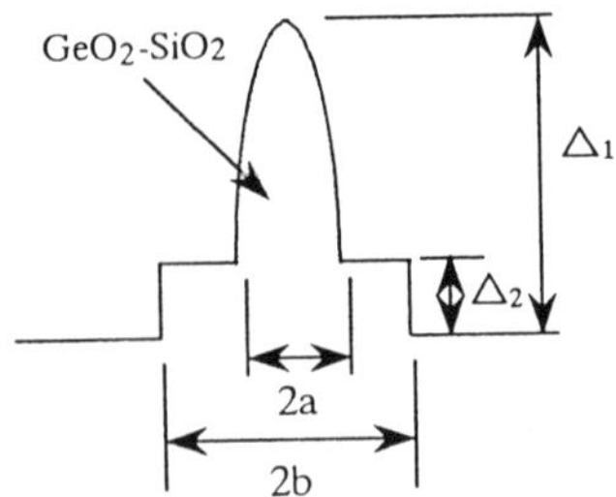

Figure 2.   Refractive index profile of the dual-shape-core dispersion-shifted fiber.

be done primarily by enlarging the MFD in this case.

It is not very difficult to design DSFs having large MFDs, if the bending loss performance is not a concern. In general, there is a trade-off between enlarging the MFD and suppressing the bending loss. The bending loss performance must be kept to a certain level in order to avoid any attenuation increase during the cabling process or installation. This makes the design optimization of DSFs somewhat complicated. In the early stage of DSF development, various designs of DSFs exhibiting good bending loss performance were developed, for example, the segmented core profile [5], the dual shape core profile [6], etc. As research of the dense WDM transmission systems progresses, it has become increasingly important to further optimize the DSF design so as to realize as large an MFD as possible without deteriorating the bending loss [7]-[9].

Figure 2 illustrates the dual shape core profile. This profile has four major design parameters, the refractive index difference of the inner core (Δ_1), that of the outer core (Δ_2), the inner core diameter (2a) and the outer core diameter (2b). It has been found through intensive design analysis [9] that, by properly choosing those parameters, it is possible to expand MFDs without deteriorating the bending loss performance. Figure 3 shows the relationship between the MFD and the bending loss for a bend diameter of 30 mm, which has been observed with the conventional DSFs and the newly designed DSFs. It is clear that MFDs can be enlarged beyond 8.5 μm while keeping the bending loss performance equivalent to that of the conventional DSFs. It should also be mentioned, however, that there is another trade-off between enlarging the MFD and suppressing the dispersion slope [9]. This needs to be taken into account in the design optimization.

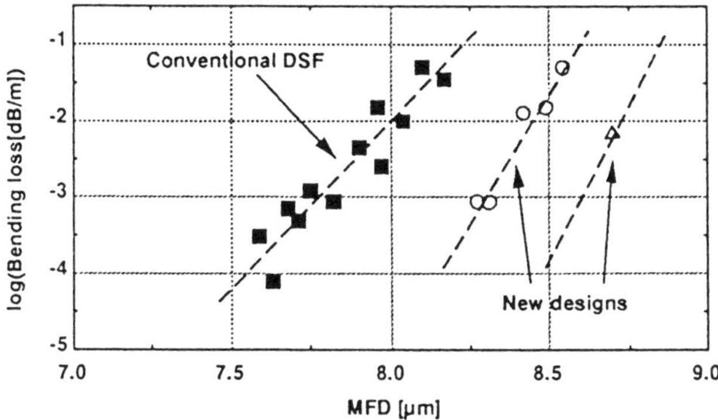

Figure 3.   Relationship between MFD and 30 mmφ bending loss for dispersion-shifted fibers.

2.3. DISPERSION-FLATTENED FIBER

As mentioned in the previous subsection, DSFs generally have some dispersion slope in the vicinity of the zero dispersion wavelength. When the bit-rate is extremely high ($>$ 100 Gbit/s, for example), this second order dispersion may affect the transmission even if the signal wavelength coincides perfectly with the zero dispersion wavelength of the whole transmission line consisting of DSFs. In WDM transmission, the effect of the dispersion slope is more apparently detrimental, since the residual chromatic dispersion differs for each signal wavelength and the difference is accumulated along the transmission line.

In order to realize low chromatic dispersion in a relatively wide wavelength range, dispersion-flattened fibers (DFFs) have been developed [10]-[11]. Various designs of DFF have been proposed, and some of them are intended for use in the broadband WDM transmission (e.g. 1.3/1.55 μm). In this section, however, DFFs specifically designed for use in the 1.55 μm window, which exhibit nearly zero dispersion slopes around 1.55 μm, are discussed.

Figure 4 illustrates the refractive index profile of the DFF with the double-cladding structure [12]. In mass-production of optical fibers, some degree of variation of structural parameters are inevitable. It is therefore important in designing DFFs to choose the target profile so that statistical cancellation of the dispersion slopes is realized when many fibers of different lots, each of which deviates slightly from the design target, are concatenated. Also important is the bending loss characteristics. Table 2 lists the structural parameters of the DFF optimized with regard to the above [12].

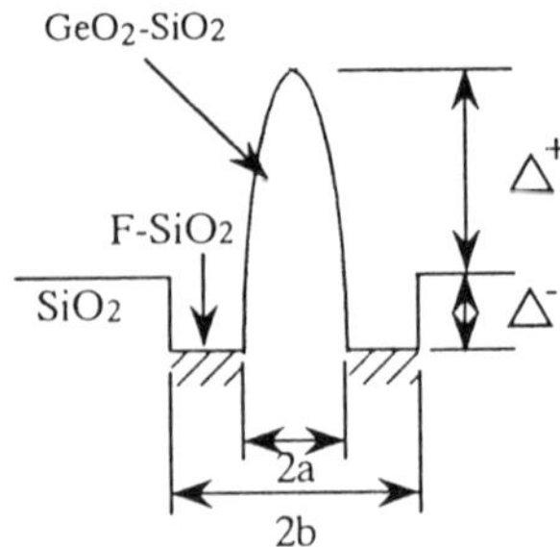

Figure 4. Refractive index profile of dispersion-flattened fiber.

DFFs with the design shown in Figure 4 and Table 2 have been fab-

TABLE 2. Structural parameters of the optimized DFF with the double-cladding structure.

Δ^+ [%]	1.0
Δ^- [%]	0.5
2a/2b	0.5

ricated by the VAD method. Several fibers with slightly different cladding diameters were drawn from the same preform in order to examine the effect of fluctuation of the core diameter. The maximum variation of the core diameter was $+/-3\%$. Figure 5 shows chromatic dispersion characteristics of the fabricated fibers. It should be noted that only a few percent deviation of the core diameter caused a significant scatter of chromatic dispersion (from -4.65 to $+3.99$ ps/km/nm at 1.55 μm). Dispersion slopes at 1.55 μm ranged from -0.037 to $+0.016$ ps/km/nm^2. Attenuation was 0.30 dB/km at 1.55 μm. In order to see the cancellation effect of the dispersion slope, these fibers were spliced and overall chromatic dispersion was measured. The result is also shown in Figure 5. Almost zero dispersion slope was demonstrated in a wide wavelength range around 1.55 μm.

Although DFFs have been successfully developed, they have not yet been introduced into commercial telecommunication systems, mainly because chromatic dispersion characteristics are so sensitive to structural variation in manufacturing which makes stable and high-yield production very difficult, and also because attenuation tends to be relatively higher than

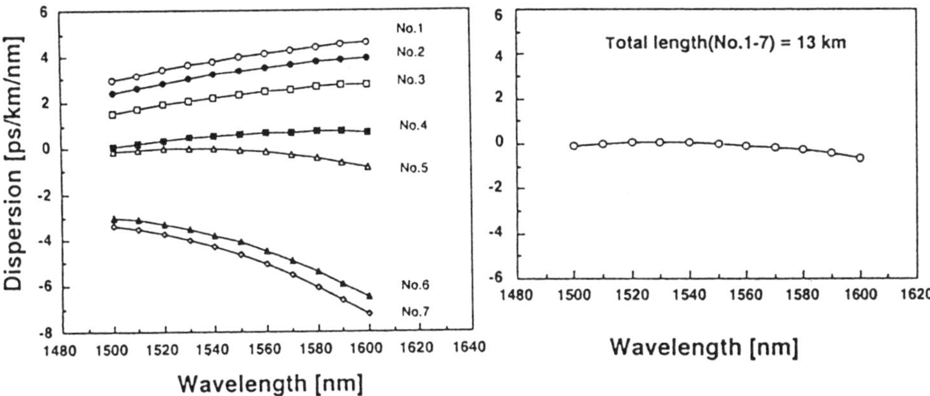

Figure 5. Dispersion characteristics of dispersion-flattened fibers fabricated with slightly different diameters (left) and a spliced link (right).

standard DSFs due to their complicated profiles and high concentrations of dopants.

2.4. DISPERSION COMPENSATING FIBER

As EDFAs have been demonstrating dramatic success, a major concern has emerged on how to upgrade the current transmission systems operating in the 1.3 μm window using standard SMFs. A relatively large chromatic dispersion, typically 16 to 18 ps/km/nm at 1.55 μm, must be dealt with in some way to utilize the embedded SMFs for high bitrate and long distance transmission using EDFAs. Various techniques have been proposed and developed to compensate for the chromatic dispersion of SMFs. One of the most promising techniques is believed to be dispersion compensating fibers (DCF's) which exhibit large negative dispersion in the 1.55 μm window [13]-[15].

As a dispersion compensating device, the primary characteristics of DCFs are, of course, chromatic dispersion and attenuation. A figure of merit (FOM) has been defined as the ratio of the absolute value of dispersion and attenuation. One of the refractive index profiles of the DCFs which have realized very high FOMs close to 300 ps/km/dB [15],[16], is very similar to that of the reduced MFD DSF shown in Figure 1. The fiber structure and manufacturing conditions have been optimized to maximize the FOM. It has been revealed experimentally that doping fluorine in the cladding is very effective for reducing attenuation with this profile [15].

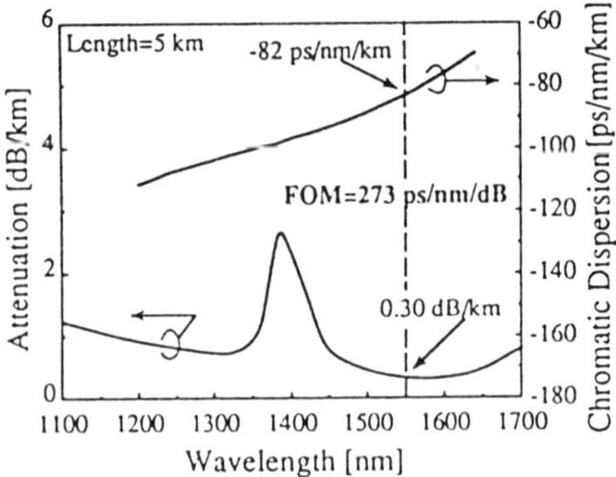

Figure 6. Spectral attenuation and dispersion of a high FOM dispersion compensating fiber.

Typical dispersion and attenuation characteristics of such high FOM DCF are shown in Figure 6.

Especially when DCFs are employed in the 1.55 μm band dense WDM transmission systems, dispersion slope becomes essential to expand the usable bandwidth in which residual dispersion can be minimized by DCFs. It is possible by utilizing elaborate refractive index profiles, such as the double-cladding structure [17],[18], the triple-cladding structure [19], etc., to tailor the dispersion characteristics of DCFs so that both dispersion and dispersion slope are negative and the ratio of the two is comparable to that of SMFs.

3. New Characterization Techniques for Optical Fibers

3.1. NONLINEAR REFRACTIVE INDEX

As the importance of the nonlinearity of optical fibers has increased, substantial efforts have been made recently to precisely determine the nonlinear refractive index (n_2) of optical fibers. The self-phase-modulation (SPM) method [20]-[22] and the cross-phase-modulation (XPM) method [23],[24] are the two major techniques which are widely used. At this moment, some inconsistency is indicated among the various measurement results of those methods. Further study appears to be necessary to determine n_2 with high absolute accuracy. As an example of the n_2 measurement technique, the improved XPM method with depolarized pump light [24] is described in this section. Figure 7 illustrates a setup of the measurement system. High power pump light and relatively low power probe light are combined and launched into the optical fiber under testing. When the pump light is modulated

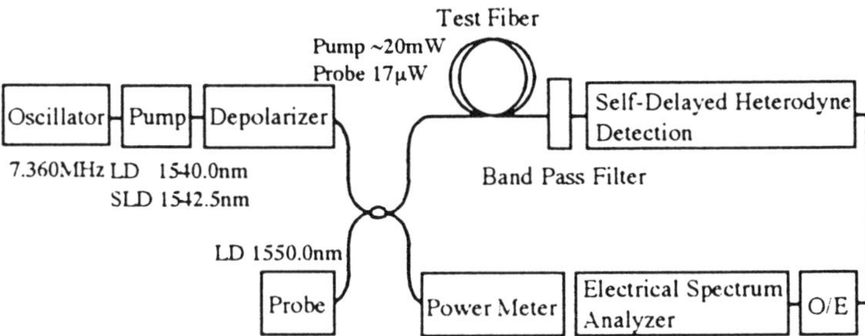

Figure 7.   Measurement setup for nonlinear refractive index by the improved XPM method.

in its intensity at a certain frequency, the probe light is modulated in its phase through the XPM. The phase shift induced by the phase modulation can be measured by detecting the probe signal at the output end of the test fiber with the self-delayed heterodyne technique and analyzing its frequency components. One can determine n_2 from the phase shift measured as a function of the optical pump power launched into the test fiber. Since the XPM effect depends largely on the polarization states of the pump and the probe, it is essential for accurate measurement to control the polarization states inside the test fiber. Depolarizing the pump light has been found to be very effective for improving the measurement accuracy. Repeatability (standard deviation) within $+/-1\%$ has been achieved.

It is well-known that some additives (such as germanium) doped in silica glass modify n_2 of the glass. It is useful for optical fiber designing to know precisely how much each dopant material affects n_2 of the silica glass. In order to do this, various optical fibers with different designs have been measured by using the improved XPM method described above, and the effects of germanium and fluorine on n_2 have been quantitatively determined [25]. Figure 8 illustrates refractive index profiles of the fiber used in the experiment.

The measured n_2 represents the cross-sectional average of local n_2 weighted by the optical power distribution. Assuming that n_2 varies proportionally to the concentration of each dopant, it is possible to determine the contribution of each dopant to n_2 by calculating the optical power distribution of each fiber and relating the assumed n_2 profiles (corresponding to the dopant profiles) to the measured results. The empirical equations giving n_2 of germanium-doped silica and fluorine-doped silica have been derived as

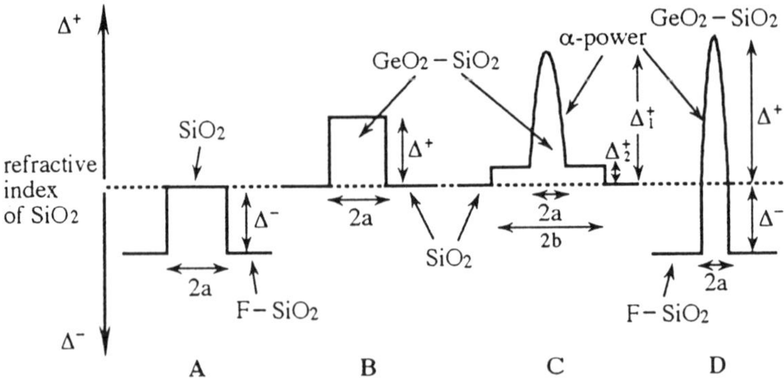

Figure 8. Refractive index profiles of optical fibers tested for nonlinear refractive index.

follows:

$$n_2(\text{GeO}_2 - \text{SiO}_2)[10^{-20}\text{m}^2/\text{W}] = 2.76 + 0.0974\text{X} \tag{1}$$

and

$$n_2(\text{F} - \text{SiO}_2)[10^{-20}\text{m}^2/\text{W}] = 2.76 + 1.03\text{Y}, \tag{2}$$

where X is the concentration of GeO_2 [mol%] and Y is the concentration of F [wt%].

3.2. DISTRIBUTION OF CHROMATIC DISPERSION

Chromatic dispersion measured by most of the current standard techniques represents the value averaged along the whole optical fiber under testing. Longitudinal distribution or variation of chromatic dispersion along long optical fibers or transmission lines becomes essential in ultra high capacity transmission, such as soliton or WDM. Several new measurement techniques have been developed. A multi-wavelength optical time domain reflectometry (OTDR) technique [26] has been proposed to measure longitudinal distribution of chromatic dispersion. A nonlinear OTDR technique base on the modulational instability [27] has also been developed for the same purpose. A four-wave-mixing (FWM) technique [28] is a simple alternative for estimating longitudinal variation of chromatic dispersion (not as a function of the location) in optical fibers or spliced links.

Figure 9 shows the measurement setup of the FWM technique. High power probe light and pump light are combined and launched into the test fiber to induce the FWM and the resultant optical spectrum is observed at the output of the test fiber. When the pump wavelength, which is swept in the measurement wavelength range, coincides with the zero-dispersion

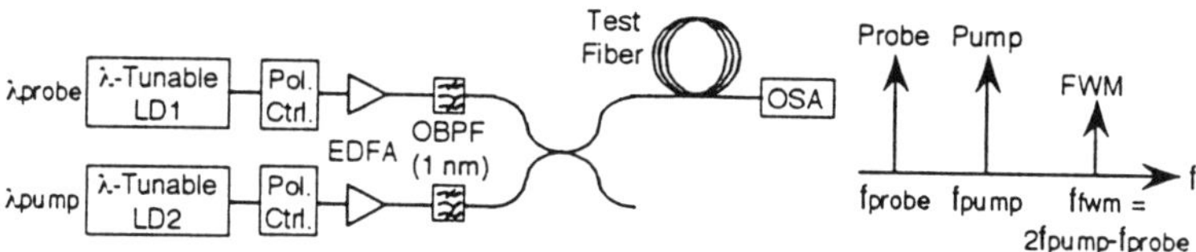

Figure 9.   Measurement setup for dispersion variation by the FWM technique.

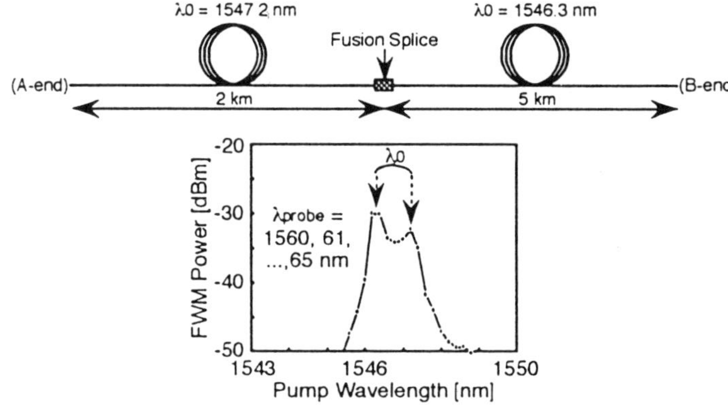

Figure 10.   An example of the measurement by the improved FWM method.

wavelength of the fiber, the efficiency of the FWM is maximized. By carefully observing the FWM efficiency as a function of the pump wavelength, variation of the zero-dispersion wavelength can be estimated [28]. Since the FWM efficiency exhibits some oscillatory dependence on wavelength, however, it is difficult to clearly determine the zero-dispersion wavelengths in many cases. By repeating the measurement of the FWM efficiency with different probe wavelengths and averaging those results, we can significantly improve the technique [29]. Figure 10 demonstrates an example of the measurement by the improved FWM method. A spliced fiber link consisting of two DSFs with the zero-dispersion wavelengths of 1547.2 nm and 1546.3 nm was tested. The FWM power measured as a function of the pump wavelength can be seen to clearly indicate the zero-dispersion wavelengths of the link.

4. Summary

In order to meet the requirements posed by the pursuit of higher capacity and longer transmission distance in optical communication, various optical fibers of new designs are being developed. These new optical fibers are often accompanied with highly complicated refractive index profiles and tight tolerances in production. It should be pointed out that not only optical performance but also production cost must be taken into consideration in development of those fibers. It appears that measurement techniques also need to be further improved or newly developed to satisfy the increasing demands for higher accuracy or new specifications. All of these are challenging tasks for optical fiber manufacturers.

5. Acknowledgement

The author would like to thank Y. Suetsugu, Y. Terasawa, M. Onishi, T. Kato and all the other colleagues who have contributed to the work presented in this paper.

References

1. Kanamori, H., *et al.*: Fabrication of high-NA single-mode fibers for nonlinear transmission systems by the VAD method, *Technical Digest of 15th European Conference on Optical Communication (ECOC '89)* **TuP-4**, (1989), 458-461.
2. Forghieri, F., *et al.*: Reduction of four-wave mixing crosstalk in WDM systems using unequally spaced channels, *IEEE Photon. Technol. Lett.*, **6** (1994), 754-756.
3. Chraplyvy, A. R., *et al.*: 160-Gb/s (8*20 Gb/s WDM) 300-km transmission with 50-km amplifier spacing and span-by-span dispersion reversal, *Technical Digest of Optical Fiber Communication Conference (OFC '94)*, **PD-19**, (1994),
4. Bergano, N. S., *et al.*: Four-channel WDM transmission experiment over transoceanic distances, *Technical Digest of Optical Amplifiers and Their Applications Conference (OAA '94)*, **PD-7**, (1994).
5. Bhagavatula, V. A. and Blaszyk, P. E.: Single-mode fiber with segmented core, *Technical Digest of Topical Meeting on Optical Fiber Communication (OFC "83)*, **MF5**, (1983), 10-12.
6. Kuwaki, N., *et al.*: Transmission characteristics of dispersion-shifted dual shape core single-mode fibers, *IEEE J. Lightwave Technol.*, **LT-5** (1987), 792-797.
7. Bhagavatula, V. A., *et al.*: Dispersion-shifted single-mode fiber for high-bit-rate and multiwavelength systems, *Technical Digest of Optical Fiber Communication Conference (OFC '95)*, **ThH1**, (1995), 259-260.
8. Nouchi, P., *et al.*: Low-loss single-mode fiber with high nonlinear effective area, *Technical Digest of Optical Fiber Communication Conference (OFC '95)*, **ThH2**, (1995), 260-261.
9. Terasawa, Y., *et al.*: Design optimization of dispersion shifted fiber with enlarged mode field diameter for WDM transmission, *Technical Digest of Tenth International Conference on Integrated Optics and Optical Fibre Communication (IOOC '95)*, **FA2-2**, (1995).
10. Miya, T., *et al.*: Fabrication of low dispersion single-mode fibers over a wide spectral range, *IEEE J. Quantum Electron*, **QE-17** (1981), 858-861.

11. Cohen, L. G., *et al.*: Ultrabroadband single-mode fibers, *Technical Digest of Topical Meeting on Optical Fiber Communication (OFC '83)*, **MF4**, (1983), 10-11.

12. Kubo, Y., *et al.*: Dispersion flattened single-mode fiber for 10,000km transmission system, *Technical Digest of 16th European Conference on Optical Communication (ECOC '90)*, **WeB3.1**, (1990), 505-508.

13. Dugan, J. M., *et al.*: All-optical, fiber-based 1550nm dispersion compensation in a 10 Gbit/s, 150km transmission experiment over 1310nm optimized fiber, *Technical Digest of Optical Fiber Communication Conference (OFC '92)*, **PD-14**, (1992).

14. Izadpanah, H., *et al.*: Dispersion compensation for upgrading interoffice networks built with 1310nm optimized SMFs using an equalizer fiber, EDFAs and 1310/1550nm WDM, *Technical Digest of Optical Fiber Communication Conference (OFC '92)*, **PD-15**, (1992).

15. Onishi, M., *et al.*: Dispersion compensating fibre with a high figure of merit of 250ps/nm/dB, *Electron. Lett.*, **30**, (1994), 161-162.

16. Onishi, M., *et al.*: Dispersion compensating fiber with a figure of merit of 273ps/nm/dB and its compact packaging, *Technical Digest of Fifth Optoelectronics Conference (OEC '94)*, **14B1-3**, (1994), 126-127.

17. Onishi, M., *et al.*: High NA double-cladding dispersion compensating fiber for WDM systems, *Technical Digest of 20th European Conference on Optical Communication (ECOC '94)*, (1994), 681-684.

18. Akasaka, Y., *et al.*: Dispersion-compensating fiber with W-shaped index profile, *Technical Digest of Optical Fiber Communication Conference (OFC '95)*, bf ThH3, (1995), 261-262.

19. Vengsarkar, A. M., *et al.*: Fundamental-mode dispersion-compensating fibers: design considerations and experiments, *Technical Digest of Optical Fiber Communication Conference (OFC '94)*, **ThK2**, (1994), 225-227.

20. Stolen, R. H. and Lin, C.: Self-phase-modulation in silica optical fibers, *Physical Review*, **17**, (1978), 1448-1453.

21. Kim, K. S., *et al.*: Measurement of the nonlinear index of silica-core and dispersion-shifted fibers, *Opt. Lett.*, **19**, (1994), 257-259.

22. Namihira, Y., *et al.*: Nonlinear coefficient measurements for dispersion shifted fibres using self-phase modulation method at 1.55mm, *Electron. Lett.*, **30**, (1994), 1171-1172.

23. Wada, A., *et al.*: Measurement of nonlinear-index coefficients of optical fibers through the cross-phase modulation using delayed-self-heterodyne technique, *Technical Digest of 18th European Conference on Optical Communication (ECOC '92)*, **MoB1**, (1992), 45-48.

24. Kato, T., *et al.*: Measurement of the nonlinear refractive index in optical fiber by the cross-phase-modulation method with depolarized pump light, *Opt. Lett.*, **20**, (1995), 988-990.

25. Kato, T., *et al.*: Estimation of nonlinear refractive index in various silica based glasses for optical fibers, submitted to *Opt. Lett.*, for publication.

26. Ohashi, M. and Tateda, M.: Novel technique for measuring longitudinal chromatic dispersion distribution in singlemode fibres, *Electron. Lett.*, **29**, (1993), 426-428.

27. Nishi, S. and Saruwatari, M.: A technique to measure distributed zero dispersion wavelength of optical fibers using pulse amplification due to modulation instability, *Technical Digest of Optical Amplifiers and their Applications (OAA '94)*, **ThC4-1**, (1994).

28. Onaka, H., *et al.*: Measuring the longitudinal distribution of four-wave-mixing efficiency in dispersion-shifted fibers, *IEEE Photon. Technol. Lett.*, **6**, (1994), 1454-1456.

29. Suetsugu, Y., *et al.*: Measurement of zero-dispersion wavelength variation in concatenated dispersion-shifted fibers by improved four-wave-mixing technique, to be published in *IEEE Photon. Technol. Lett.*.

APPLICATION OF OPTICAL SOLITON FOR ULTRA-FAST LONG-HAUL TRANSMISSION AND MEASUREMENT TECHNOLOGY

K. IWATSUKI

NTT Optical Network Systems Laboratories
1-2356 Take, Yokosuka, Kanagawa 238, Japan

Abstract. This paper describes some issues and their countermeasures on applying solitons to ultra-high speed and/or ultra-long distance transmission. The application of adiabatic soliton compression to an ultra-fast measurement based on electro-optic sampling is also described.

1. Introduction

The use of optical solitons is one of the attractive way to compensate for the pulse distortion due to fiber dispersion. Their potential for an ultra-high speed and/or ultra-long distance transmission were intensively investigated. As for the ultra-long distance soliton transmission, soliton control techniques both in frequency domain [1] and the time domain [2], [3] have been recently demonstrated with recirculating loop, and resulted in dramatically reducing the timing jitter and the accumulation of amplified stimulated emission (ASE) noise. The recirculating loop offers the exact uniform transmission line even in large variation of fiber dispersion in the loop as long as the loop length is short relative to the soliton period. Straight transmission lines, however, are not uniform in practice, thus leads to soliton pulse distortion accompanied with dispersive wave radiation due to the distribution in fiber dispersion along the line. In the next section, we discussed the impact of distributed fiber dispersion with and without the soliton control. As for the ultra-high speed soliton transmission, the soliton period becomes extremely short, thus shorten the allowable amplifier spacing in practice as well as affected with the higher order fiber dispersion. In Section 3, we demonstrated two approaches [4], [5] to overcome the above issues in the ultra-high speed soliton transmission. The ultra-fast measurement technique is also needed to realize ultra-high speed soliton transmission. In Section 4, the adiabatic soliton compression technique [6] was applied to the electro-optic sampler [7].

191

A. Hasegawa (ed.), Physics and Applications of Optical Solitons in Fibres '95, 191–207.

2. Ultra-long Distance Transmission

The impact of distributed fiber dispersion was discussed in this section to foresee the practical potential for ultra-long distance soliton transmission. We theoretically and experimentally discussed the SNR due to the timing jitter and the ASE noise accumulation with the signal frequency sliding under the uniform transmission line. In the practical transmission line, the fiber dispersion is distributed along the line. The impact of distributed fiber dispersion was discussed with and without the soliton control.

2.1. SIGNAL FREQUENCY SLIDING IN UNIFORM TRANSMISSION LINE

The signal frequency sliding [8], recently proposed as the soliton control in a frequency domain, gives the advantage of soliton propagation without the change in the average fiber dispersion along the line, and thus avoids the higher order dispersion effect: a significant advantage over the sliding frequency filtering [3]. The signal frequency sliding is realized by a chain of optical bandpass filters and frequency shifters. This technique periodically slides and cut the ASE noise with a frequency shifters and the filters, thus saturating the ASE noise accumulation. The soliton itself is guided with the fixed frequency filters, and thus the timing jitter is reduced.

The timing jitter $\sqrt{\langle \delta t^2 \rangle}$ with the signal frequency sliding is derived as [9],

$$\langle \delta t^2 \rangle = \langle \delta t^2 \rangle_{CL}\, f_{CL}(4\gamma_t z) + \langle \delta t^2 \rangle_{GH}\, f_{GH}(4\gamma_t z)) \tag{1}$$

$$\langle \delta t^2 \rangle_{CL} = 0.0935\, \frac{\lambda^4 D^2\, \delta f}{c^2\, t_s}\, z^2 \tag{2}$$

$$\langle \delta t^2 \rangle_{GH} = 0.1959\, \left(\frac{2\, \alpha\, n_2\, n_{sp}\, h\, D\, Q}{t_s\, A_{eff}} \right) z^3 \tag{3}$$

$$f_{CL}(x) = \left(\frac{1 - e^{-x}}{x} \right)^2 \tag{4}$$

$$f_{GH}(x) = \frac{3}{2x^3}\, (2x - 3 + 4e^{-x} - e^{-2x}) \tag{5}$$

$$\gamma_t = \frac{\rho}{6\, \left(\frac{2\pi t_s}{1.763} \right)^2\, l_{amp}} \tag{6}$$

$$Q = \left\{ \frac{\sinh(\alpha l_{amp})}{\alpha l_{amp}} \right\}^2 \tag{7}$$

where, $\langle \delta t^2 \rangle_{CL}$ and $\langle \delta t^2 \rangle_{GH}$ are the timing jitter due to the carrier linewidth of pulse source and the Gordon-Haus effect without the optical bandpass filter, respectively. The reduction factors of timing jitter with optical bandpass filter are denoted as $f_{CL}(4\gamma_t z)$ and $f_{GH}(4\gamma_t z)$, respectively. Here, λ is the central wavelength of optical soliton, D the fiber dispersion, Δf the carrier linewidth of pulse source, c the light velocity in vacuum, z the transmission distance, a the fiber loss coefficient, n_2 the fiber nonlinear coefficient, t_s is the soliton pulse width, n_{sp} the population inversion factor, h Planck's constant, and A_{eff} the effective core area, ρ the peak curvature of the optical bandpass filter, l_{amp} the amplifier spacing. Figure 2.1(a) shows $\sqrt{\langle \delta t^2 \rangle}$ as a function of z with and without

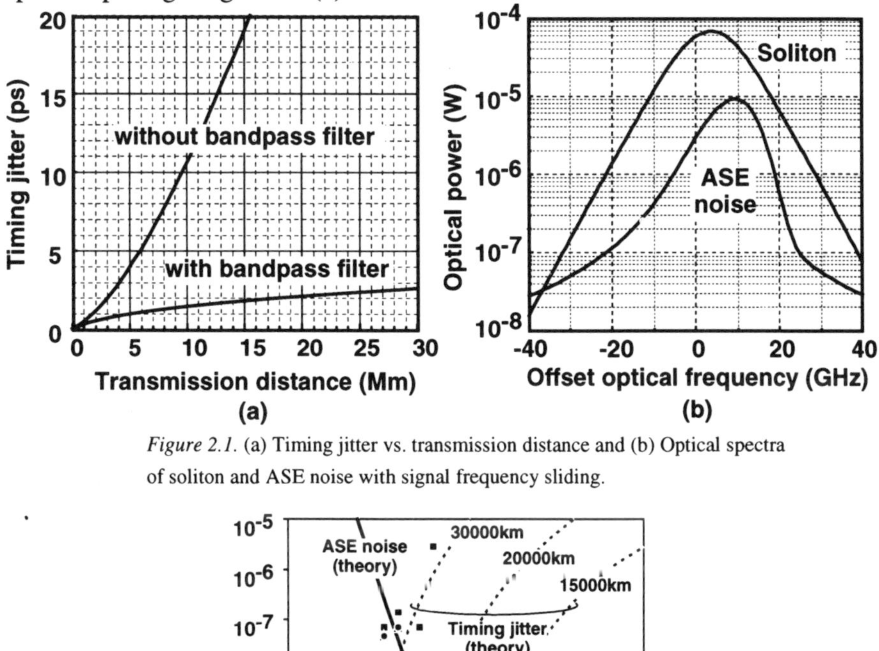

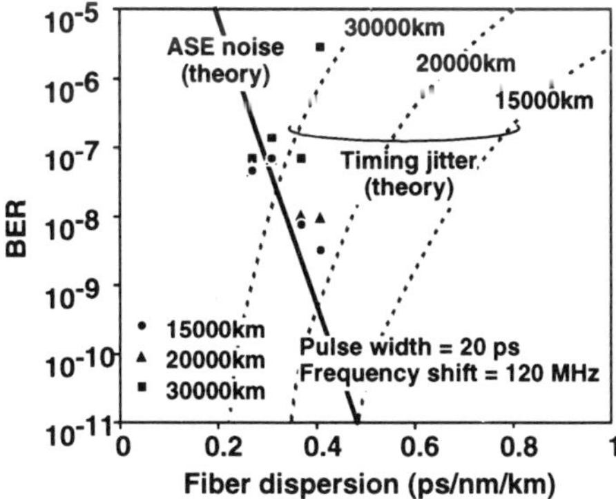

Figure 2.1. (a) Timing jitter vs. transmission distance and (b) Optical spectra of soliton and ASE noise with signal frequency sliding.

Figure 2.2. Bit error rate as a function of fiber dispersion with signal frequency sliding.

the optical bandpass filter, where λ, t_s, D, l_{amp}, α, n_{sp} and Δf are 1553 nm, 20 ps, 0.3 ps, 30 km, 0.25 dB/km, 1.5, 1 MHz, respectively. The 1 nm Lorentzian-shaped optical bandpass filter was used. The optical bandpass filter effectively reduces the timing jitter.

The ASE noise spectrum density with the signal frequency sliding is expressed as the following equations.

$$N(f,z) = 2n_{sp}h\nu \, (G \, e^{2\gamma l_{amp}} - 1) \sum_{k=1}^{N} e^{2\gamma l_{amp}(k-1)} \prod_{m=1}^{k} H(f - mF) \qquad (8)$$

$$\gamma = \gamma_t + \frac{9}{8\rho \, l_{amp}} \frac{F^2}{} \left(\frac{2\pi t_s}{1.763}\right)^4 \qquad (9)$$

where $N(f,z)$ is the ASE noise spectrum density, f the optical frequency, N the number of optical amplifiers, F the frequency shift of one span and $H(f)$ the Lorentzian-shaped filter transfer function. Equation (9) expresses the condition of stable soliton propagation [8] in the signal frequency sliding. Figure 2.1(b) shows the optical spectra of optical soliton and ASE noise calculated with Equations (8) and (9), where F and z are 120 MHz and 10,000 km, respectively. The SNR can be estimated with these spectra as shown in Figure 2.1(b).

We demonstrated the signal frequency sliding soliton transmission utilizing the recirculating loops [10], [11]. A pulse source consisted of a DBR-LD with CW operation and a MQW-EA modulator sinusoidally driven at 10 GHz. The recirculating loop was constructed with a 33 km dispersion shifted single-mode fiber (DSF) and two EDFA's. The measured population inversion factors n_{sp} of the two EDFA were ~1.2. An optical filter with 1 nm bandwidth (peak curvature $\rho = 1.48 \times 10^{-22}$ s^2) was placed in the loop to guide the soliton frequency. An acousto-optic switch (AO) and a 3 dB fiber coupler closed the loop. The AO played two roles: the recirculating switch and the frequency shifter. We employed an optical receiver having the effective window time of ~32 ps to measure the bit error rate (BER).

Figure 2.2 shows the BER vs. the fiber dispersion, when F and t_s are 120 MHz and 20 ps, respectively. Since the SNR given by the ASE noise is theoretically saturated with the frequency sliding, the theoretical BER$_{(ASE)}$ due to the ASE noise, denoted as the solid line, becomes constant with the increase of z. Therefore, the BER$_{(ASE)}$ depends on the input signal power, which is proportional to D. As the timing jitter due to the Gordon-Haus effect and the carrier linewidth of pulse source (~1MHz) theoretically increases with ~$\sqrt{Dz}$ [9], the theoretical BER$_{(jitter)}$ due to the timing jitter degrades with D and z. The dashed lines denote the theoretical BER$_{(jitter)}$ at the transmission distances of 15,000 km, 20,000 km and 30,000 km, respectively. The experimental results indicated with circles, triangles and squares, correspond to the respective theoretical curves. Therefore,

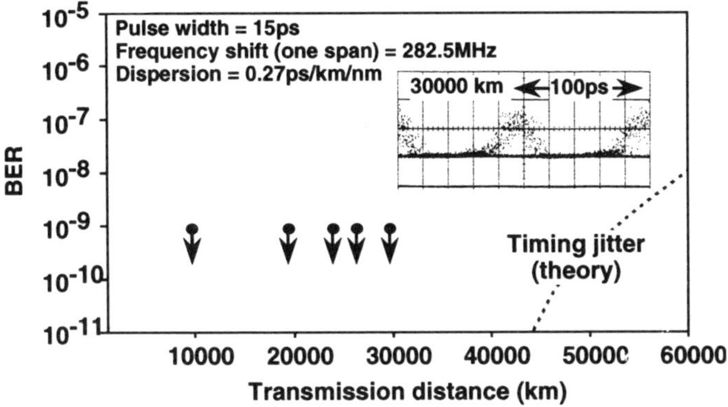

Figure 2.3. Bit error rate as a function of transmission distance with signal frequency sliding.

the fiber dispersion must be selected to minimize both $BER_{(ASE)}$ and $BER_{(jitter)}$.

To achieve the $BER_{(ASE)}$ less than 10^{-12}, we set the frequency shifts, the initial pulse width and the average fiber dispersion as 565 MHz, 15 ps and 0.27 ps/km/ nm, respectively. The resultant BER versus transmission distance is shown in Figure 2.3. Error free (BER $< 10^{-9}$) soliton transmission free of the limit of ASE noise and timing jitter can be achieved over 30,000 km with the signal frequency sliding technique. The change in the state of polarization in the loop affected stable BER measurement, thus restricting the minimum detectable BER as small as $\sim 7 \times 10^{-10}$. The measured transmission distance was limited only by our electrical delay line equipment. The theory predicts that the BER of 10^{-9} would be encountered over more than $\sim 50,000$ km. The eye pattern at 30,000 km is also depicted in the inset of Figure 2.3. Good eye opening was obtained even at 30,000 km.

2.2. IMPACT OF DISTRIBUTED FIBER DISPERSION IN SOLITON TRANS-MISSION

The potential of soliton control for an ultra-long distance transmission over more than 10,000 km was demonstrated using the recirculating loop in the previous subsection. Since the local average dispersion of recirculating loop is the same as the average dispersion, the recirculating loop offers the exact uniform transmission line for soliton even in large variation of fiber dispersion in the loop. This leads to the stable soliton transmission. Straight transmission lines, however, are not uniform in practice, thus leads to soliton pulse distortion accompanied with dispersive wave radiation due to the distributed fiber dispersion along the line.

In this subsection, the impact of distributed fiber dispersion in a soliton transmission is discussed experimentally and numerically [12]. The FSA commercial submarine optical amplifier transmission line recently embedded between Kagoshima and Okinawa was employed in the experiment. The system length and the amplifier spacing l_{amp} are 900 km and 90 km, respectively. The fiber dispersion was carefully managed within each repeater span with allocating a normal dispersion (over -0.3 ps/km/nm) at its input end and an anomalous dispersion at its output end to reduce the fiber nonlinearity induced excess noise in the NRZ signal format[13].

Figure 2.4 shows the experimental configuration. The transmission line was constructed by successively and appropriately connecting the up- and down-lines with EDFA's including 1 nm optical bandpass filters at Kagoshima and Okinawa. Two kinds of pulse sources, integrated MQW-EA/DFB-LD (EA source) [14] and gain-switched MQW-DFB-LD (GS source) [15], were employed in transmitters. The transmitters generated 20 ps pulse signals encoded using a LiNbO$_3$ modulator with a 2^{23}-1 PRBS pattern at 10 Gb/s. The central wavelengths of transmitters were adjusted at 1,555 nm to achieve the average fiber dispersion of 0.2 ps/km/nm. The soliton period Z_0 was estimated as 800 km. Receivers consisted of EDFA's including 1 nm optical bandpass filters and 10 Gb/s optical receivers [4]. The equalizer bandwidth and the window time of optical receivers were 7 GHz and 47 ps, respectively.

Figure 2.5 shows the SNR vs. average input power Pav into the fiber. The closed circles, triangle and squares indicate the experimental results of EA pulse source at 1,800 km, 2,700 km and 3,600 km, respectively. The SNR of ~23 dB (BER~10^{-12}) was experimentally achieved at 2,700 km at Pav=1.0 dBm. The solid and dashed broken lines are the numerical SNR's of the FSA and the uniform fiber-dispersion including the ASE noise. The solid lines denote theoretical SNR given by the ASE noise from the EDFA's. The numerical results well agree with the experimental ones, when the average input power of the N=1 soliton Pn is estimated as -3.0 dBm. The theoretical Pn was estimated with the experimental parameters as -4.5 dBm. We believe that the difference of Pn is attributed to the initial waveform in this experiment, which was different from the ideal ones.

In the range of Pav marked with "A", the dispersive waves associated with the N=1 soliton are generated except the value of Pn. Therefore, the excess degradation of SNR due to the dispersive waves is caused above and below Pn with the increase of the transmission distance. The waveform changes along the transmission distance in the range of Pav marked with "B" where the N=2 soliton is formed, so that the SNR drastically degraded with the increase of Pav.

The eye patterns at Pav=1.0 dBm and the transmission distance of 2,700 km was measured with a fast response p-i-n photo-detector (32 GHz bandwidth), as shown in Figure 2.6(a). The corresponding eye patterns without and with equaliz-

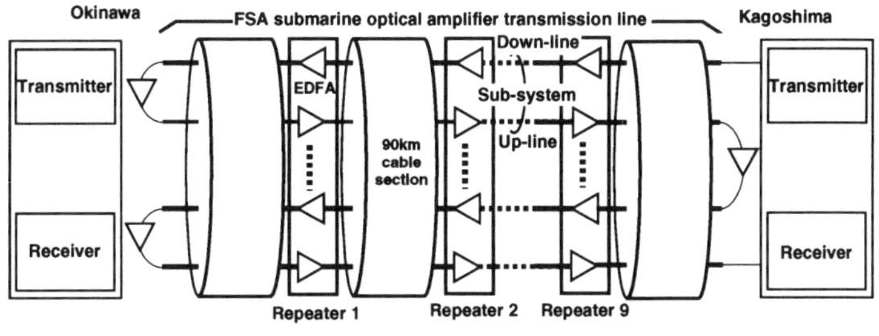

Figure 2.4. Experimental configuration.

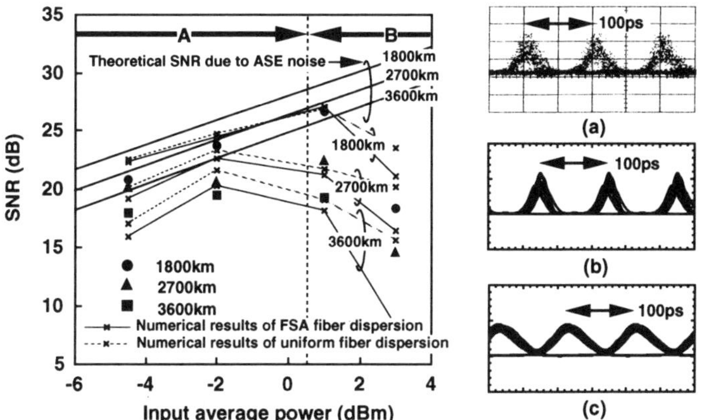

Figure 2.5. SNR vs. input average power.

Figure 2.6. Eye patterns at input power=1.0 dBm and transmission distance=2,700 km.

ing were numerically calculated including the ASE noise, as shown in Figure 2.6(b) and (c). Good eye openings were obtained both in the experimental and numerical results.

Figure 2.7 shows the SNR as a function of transmission distance. The squares and circles denote the SNR of EA and GS pulse sources at Pav=-2.0 dBm. The other symbols summarized in Figure 2.7 indicate the numerical results. The dashed and solid lines denote theoretical SNR due to the ASE and the timing jitter of the N=1 soliton. The resultant SNR's were almost the same as two pulse sources within ~$3Z_0$, and well agree with numerical ones.

As the numerical results of the FSA and uniform fiber dispersion at l_{amp} of 90 km almost coincide with the theoretical SNR's due to the ASE noise, the SNR in this experiment is dominantly restricted by the ASE noise rather than the impact of distributed fiber dispersion. To reduce the impact of the ASE noise and the timing jitter, the amplifier spacing l_{amp} was numerically shorten to 50 km, preserving the allocation of fiber dispersion. This leads to remarkable increase of the SNR degradation due to the distributed fiber-dispersion as shown in Figure

2.7. The difference between the theoretical SNR due to the ASE noise and the numerical results of uniform fiber-dispersion is attributed to the numerical back-to-back SNR limited to ~30 dB due to patterning effect at the electrical equalizer.

Figure 2.8 shows the numerical and the theoretical results of SNR vs. transmission distance with the signal frequency sliding at 10 Gb/s. The circles and squares denoted the numerical SNR of uniform and FSA fiber dispersion. The solid and dashed lines indicated the theoretical SNR due to ASE noise and timing jitter calculated by Equations (1)-(9). The parameters are summarized as follows: $l_{amp} = 50$ km, $D = 0.3$ ps/km/nm, $n_{sp} = 1.8$, $t_s = 13$ ps, $\rho = 0.64 \times 10^{-22}$ s^2, $F = 600$ MHz and $df = 1$MHz. The power of each optical amplifier was assumed to be automatically controlled to the value yielding N=0.9 soliton. The numerical results shows that the attained transmission distances at SNR of 22 dB for both

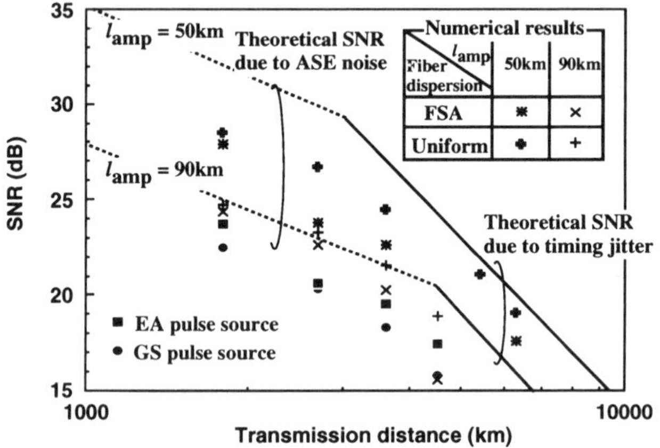

Figure 2.7. SNR vs. transmission distance.

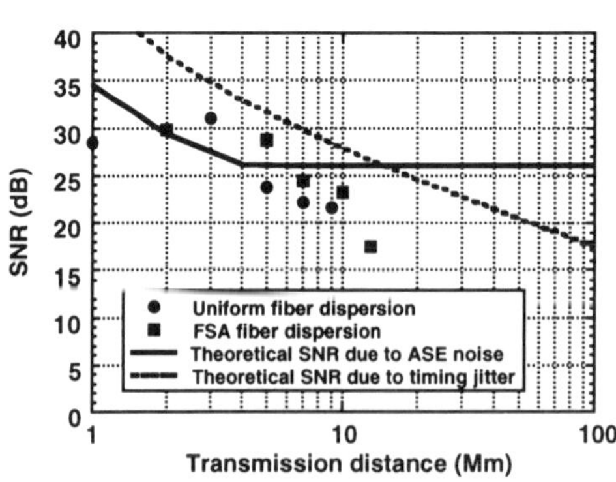

Figure 2.8. SNR vs. transmission distance with signal frequency sliding.

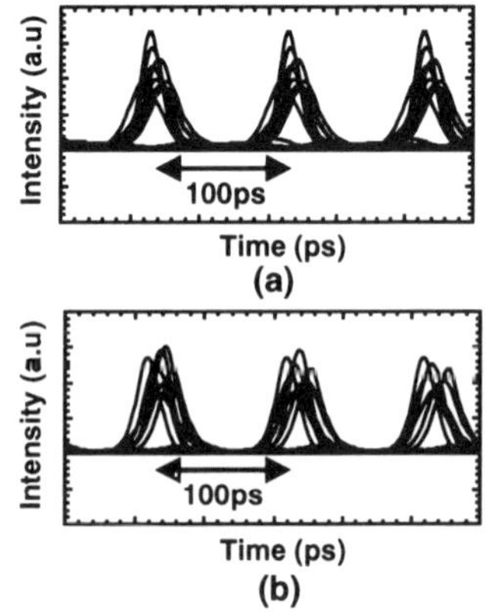

Figure 2.9. Numerical eye pattern.
(a)Uniform fiber dispersion
(b)FSA fiber dispersion

fiber dispersion are the almost 10,000 km. The transmission distance is improved with the signal frequency sliding to reduce the timing jitter and the impact of distributed fiber dispersion. The eye patterns for the uniform and the FSA fiber dispersion at 10,000 km are shown in Figure 2.9. Good eye openings are obtained for both cases.

3. Ultra-high Speed Transmission

Ultra-short transform limited (TL) pulses with a small duty factor are required for ultra-high speed optical time-division multiplexing/demultiplexing (OTDM) soliton transmission to avoid the nonlinear interaction between adjacent solitons, thus leading to the drastic reduction of the soliton period Z_0. As a result, the allowable amplifier spacing l_{amp} becomes extremely short according to the guiding-center soliton criterion of $l_{amp} \ll Z_0$ [16], [17]. The polarization division multiplexing (PDM) [4], [18] is one of the promising techniques to overcome this difficulty.

An ultra-high speed optical transmission over more than 100 Gb/s requires optical pulses of a few pico-seconds. These pulses are extremely distorted by the higher order dispersion in the transmission line. Especially in non-repeated transmission at higher bit rate, the high fiber-input power is required to achieve a sufficient SNR, thus leading to pulse degradation due to the combination effect of fiber nonlinearity and fiber dispersion. We demonstrated a new approach with optically filtering the solitary wave to overcome the pulse distortion [4].

3.1. PDM SOLITON TRANSMISSION

The PDM technique using orthogonally polarized solitons is simply achieved with passive optical components (in practice, an automatic polarization controller with relatively slow response would be required), and advantageously reduces the nonlinear interaction which is suffered by single-polarization solitons from the overlap of pulse tails. The experimental setup is shown in Figure 3.1. The pulse train was generated a gain-switched DFB-LD at a 7.5 GHz repetition rate followed by a 1.8 km polarization-maintaining linear compression fiber and a spectral filter [15]. A Ti:LiNbO$_3$ intensity modulator (LN modulator) driven with a pulse pattern generator (PPG) modulated the pulse train at 7.5 Gb/s (PRBS 2^7-1). The modulated pulses were adiabatically compressed with the use of a 2 km dispersion decreasing fiber (DDF) [6], whose dispersion gradually changed from 4.74 to 1.06 ps/km/nm. An optical multiplexer (Optical MUX) was constructed from three stage of polarization maintaining fiber-delays and a 16 m polarization maintaining fiber whose polarization axis was set at 45 degrees with respect to that of the fiber-delays. A 60 Gb/s OTDM pulse stream was generated by the fiber-delays. The following polarization maintaining fiber orthogonally split the OTDM pulses, and simultaneously one of the orthogonally polarized OTDM pulses was delayed by polarization dispersion of 25 ps, so as to attain a 120 Gb/s PDM pulse stream.

The pulse stream was injected into a 154 km transmission line consisting of five spans of ~30 km dispersion shifted fibers (DSF's). The average dispersion and loss coefficient of the DSF's were 0.026 ps/km/nm and 0.21 dB/km, respectively. The length and input power of each section were adjusted to propagate a fundamental guiding-center soliton. The average input power into each section was 4.5 dBm.

Figure 3.2(a) and (b) show the streak image of the 120 Gb/s PDM fixed pulse patterns and the SHG autocorrelation traces of 60 Gb/s OTDM pulse patterns before and after transmission, respectively. The SHG autocorrelator traces have the weaker peaks at both sides than those at the center. This is because the optical pulses interfere themselves at the center of SHG autocorrelator traces, while the weaker peaks are the cross-correlation between the randomly modulated optical pulses. The transmitted pulse width and time-bandwidth product were measured as 3.7 ps and 0.32, respectively. The pulse broadening due to the fiber dispersion was successfully suppressed by the soliton effect. The pulse width increased to 7.7 ps due to the fiber dispersion after transmission at the average input power of 0 dBm, as shown in Figure 3.2(c).

The receiver consisted of a polarization beam splitter (PBS), three cascaded LN modulators operating as an optical demultiplexer, and a 7.5 Gb/s optical receiver (OR) following an Er-doped fiber pre-amplifier with a 1 nm OBPF. The 120 Gb/s PDM signal was first demultiplexed with the PBS. The LN modulators were driven with 15 GHz sinusoidal waves of $2V_\pi$, V_π amplitude and 7.5 GHz sinusoidal waves of V_π amplitude, respectively, from a 7.5 GHz clock recovery circuit in the OR, so as to demultiplex the 60 Gb/s pulses to 7.5 Gb/s.

Figure 3.3 shows the measured the BER after optical demultiplexing as a function of average received power. The demultiplexed eye pattern after transmission is shown in the inset of Figure 3.3. Open and closed circles correspond to the BER before and after transmission, respectively. No error-rate floor was observed. The sensitivity at a BER of 10^{-9} was -27.5 dBm before transmission and degraded to -25.5 dBm after transmission.

3.2 NONLINEAR PULSE TRANSMISSION NEAR ZERO DISPERSION

The transmitted pulses near zero dispersion region generate dispersive waves at the trailing edge which is shifted into the normal dispersion region by fiber nonlinearity, while the main part of pulse spectrum moves to the anomalous dis-

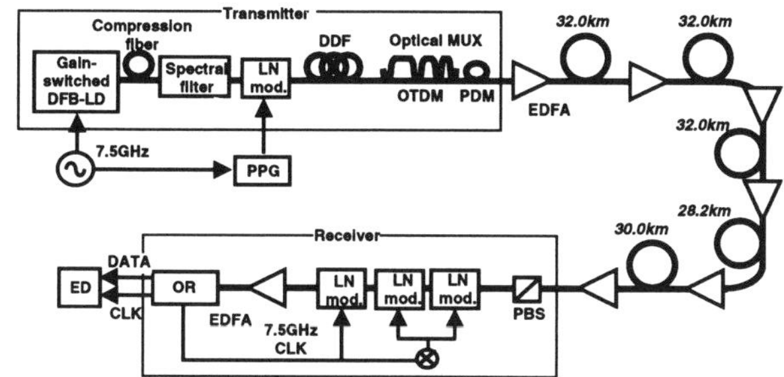

Figure 3.1. Experimental setup.

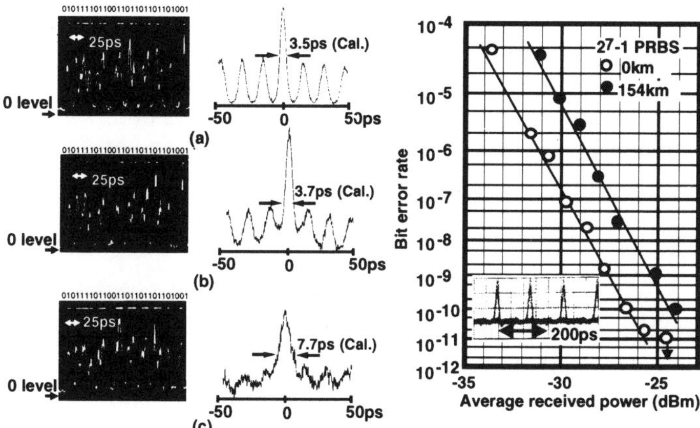

Figure 3.2. Streak images of 120 Gb/s PDM fixed pulse patterns before and after transmission. (a)before transmission, (b)after transmission (average input power = 4.5 dBm) and (c)after transmission (average input power = 0 dBm).

Figure 3.3. Bit error rate as a function of average received power.

person region and forms a soliton [19]. Therefore, we can remove the pulse distortion at the trailing edge with optical filtering [20]. The experimental setup is shown in Figure 3.4. The transmitter was almost the same configuration in the previous subsection. The adiabatically compressed pulses were optically multiplexed with a four stage Mach-Zehnder interferometer type multiplexer fabricated with a Planer Lightwave Circuit (PLC) technology [21], to achieve a 160 Gbit/s single polarized pulse stream. To attain transform-limited pulses and also to make the central pulse wavelength approach the zero dispersion wavelength of the transmission fiber, we employed a 3 nm optical bandpass filter (OBPF1). The attained pulse width and the central wavelength were 2 ps and 1,552.5 nm, respectively.

The pulse stream was injected into a 75 km DSF. The zero dispersion wavelength of the DSF was 1,553 nm. The dispersion slope was ~0.07 ps/km/ nm^2, which corresponds to the higher order dispersion length L_D' of 12.7 km.

The receiver consisted of a timing recovery circuit, an optical demultiplexer and a 10 Gbit/s optical receiver (OR) following an Er-doped fiber pre-amplifier. The 10 GHz component was electrically extracted from the 160 Gbit/s multi-plexed signal as a clock and provided to a control pulse source for the demultiplexer. The 160 Gbit/s transmitted signal was demultiplexed to 10 Gbit/s with a nonlinear optical loop mirror, which included a traveling wave semicon-ductor laser amplifier (TW-SLA NOLM) so called TOAD [22], gated with the ~1 ps control pulse [23]. The demultiplexed pulse stream was received with the OR and the BER was measured by an error detector (ED).

Figure 3.5 shows the optical spectra of transmitted pulses for various averaged input power into the 75 km DSF. The wavelength shift, which caused by the combination of fiber nonlinearity and higher order dispersion, was defined as the wavelength difference between the spectral peak wavelength in the anomalous dispersion region of transmitted pulses and the central wavelength of the initial pulses. The wavelength shift increased with the input power as shown in Figure 3.5.

Figure 3.6 (a) shows the spectrum and streak image of initial pulses. Figures 3.6 (b) and (c) show the spectra and streak images without and with a 3 nm optical bandpass filter (OBPF2) after transmission. The input power Pin to the DSF was 15 dBm, which corresponds to the wavelength shift of around 1.5 nm. The value of Pin was experimentally determined to achieve the minimum pulse width of transmitted pulses (2.5 ps) after spectral filtering. When Pin was lower than 15 dBm, the spectrum of optical pulses was located in the zero dispersion region due to the small wavelength shift, thus leading to pulse broadening with higher order dispersion. The large wavelength shift at high Pin also caused pulse broadening due to anomalous dispersion. As shown in Figure 3.6(b), the streak image was degraded by the dispersive wave around the pulse edges in the normal dispersion region, as compared with Figure 3.6(a). As shown in Figure 3.6(c), the temporal waveform was improved by eliminating the optical spectrum in the normal dispersion region, thus suppressing the dispersive wave.

We numerically calculated the time-resolved spectra of nonlinear pulse propagation. Calculated parameters are the same as those of the experiments. Figure 3.7 shows the time-resolve spectra after 75 km propagation. The spectral resolution was 2 nm. In the normal dispersion region, the pulse broadening is caused by the fiber dispersion, whereas the pulse broadening is suppressed by the soliton effect around the wavelength of 1,554 nm in the anomalous dispersion region. Therefore, the pulse distortion and the pulse broadening can be reduced by optical filtering in the anomalous dispersion region.

Figure 3.8 shows the BER as a function of averaged received power to the OR

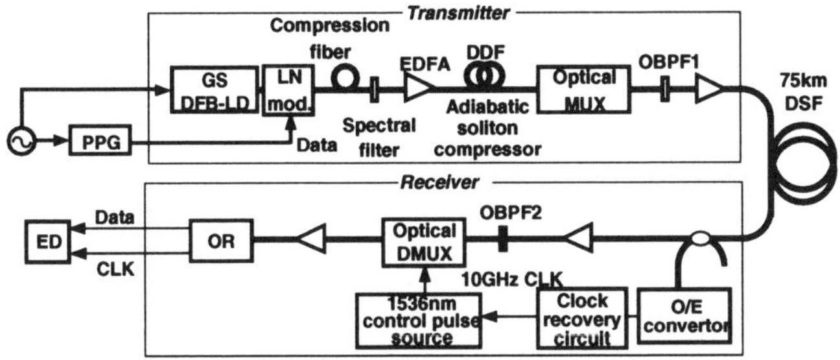

Figure 3.4. Experimental setup.

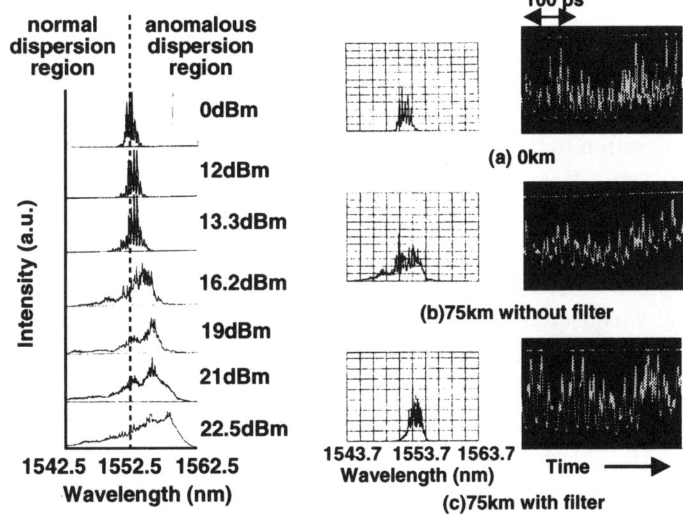

Figure 3.5. Optical spectra of transmitted pulses for various input average power.

Figure 3.6. Optical spectra and streak images.

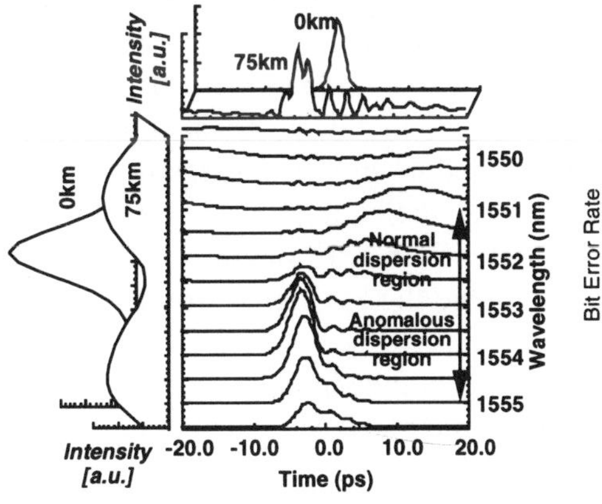

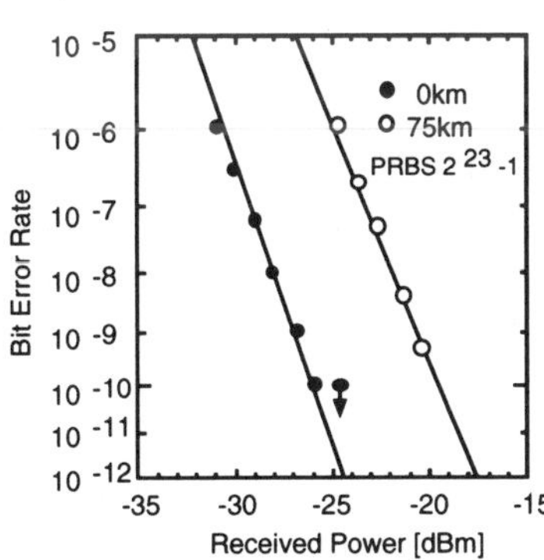

Figure 3.7. Numerical time-resolved spectra after 75 km propagation.

Figure 3.8. Bit error rate as a function of average received power.

at the optimum wavelength of 1,554 nm. The closed and open circles correspond to the BER before and after transmission. No error-rate flow was observed.

4. Application of Optical Soliton to Measurement Technology

The optical pulse-pattern signal measurement with sufficient time resolution are needed to realize ultra-high speed optical transmission. In this section, a novel waveform measurement technique based on electro-optic (EO) sampling for observing optical pulse-pattern signals more than 100 Gbit/s with high sensitivity and polarization independence is demonstrated. This technique requires a pedestal-free ultra-short optical pulses as the sampling pulse, because the pedestals and/or the wings around the sampling pulses drastically degrade the time resolution. The adiabatic soliton compression technique[6], employed as the pulse signal generation in the previous section, can be also applied to produce the sampling pulses.

Figure 4.1 is the schematic diagram of the optical signal measurement system [7]. A 1.55 μ m waveguide p-i-n photodiode with polarization independence, wide bandwidth (dc - 110 GHz , 3 dB bandwidth) and high sensitivity (> 0.63 A/ W responsivity) was used as an OE converter [24]. The EO sampling head was made of a 50 Ω coplanar waveguide on an InP substrate. The InP was an electro-optically sensitive EO transducer [25], [26]. The waveguide p-i-n photodiode was integrated on the same substrate to minimize waveform distortion. A 50 Ω resistor and two 10,000 pF capacitors were used in the bias/termination circuit to achieve a wide bandwidth (1 MHz - 100 GHz). The bias/termination circuit was integrated on an Al_2O_3 substrate and bonded to the EO sampling head.

The electric field of signal, converted with the p-i-n photodiode, propagated the 50 Ω coplanar waveguide, and penetrated into the InP substrate with changing the substrate's birefringence. The sampling beam was focused onto the center conductor of the coplanar waveguide through the backside of the substrate, and picked up the change in the birefringence induced by the electrical signal. The reflected sampling beam at the substrate changed its state of polarization due to the birefringence change, and was detected as the change in the intensity of sampling beam with the EO sampling optics. In the EO sampling optics, the differential detection scheme was used to double the EO signal strength and to reduce the laser amplitude noise.

To evaluate the bandwidth of the system, the optical impulse response of the system was measured with the pump and probe method [26]. The impulse was an adiabatically compressed soliton pulse of a 0.4 ps pulse width. The measured response is shown in Figure 4.2. The impulse response appears within a 10 ps time interval, although a small reflection (< -22 dB) occurs at the bonding section.

This result shows that the system can measure return-to-zero (RZ) optical pulse-pattern signals over more than 100 Gbit/s with an extinction ratio of more than 22 dB.

Figure 4.3 shows the measurement result of 80 Gb/s fixed pattern, which was 40 times averaged; the acquisition time was less than 0.2 s. The pulse pattern is clearly resolved. The measured pulse pattern was generated with the same technique described in the previous section. The pulse widths of the signal pulses and the sampling pulses were 2 ps and 0.4 ps, respectively, measured with a SHG autocorrelation.

5. Summary

We discussed some issues and their countermeasures on applying solitons to ultra-high speed and/or ultra-long distance transmission. The use of adiabatic soliton compression to achieve an ultra-fast measurement technique based on electro-optic sampling was also discussed.

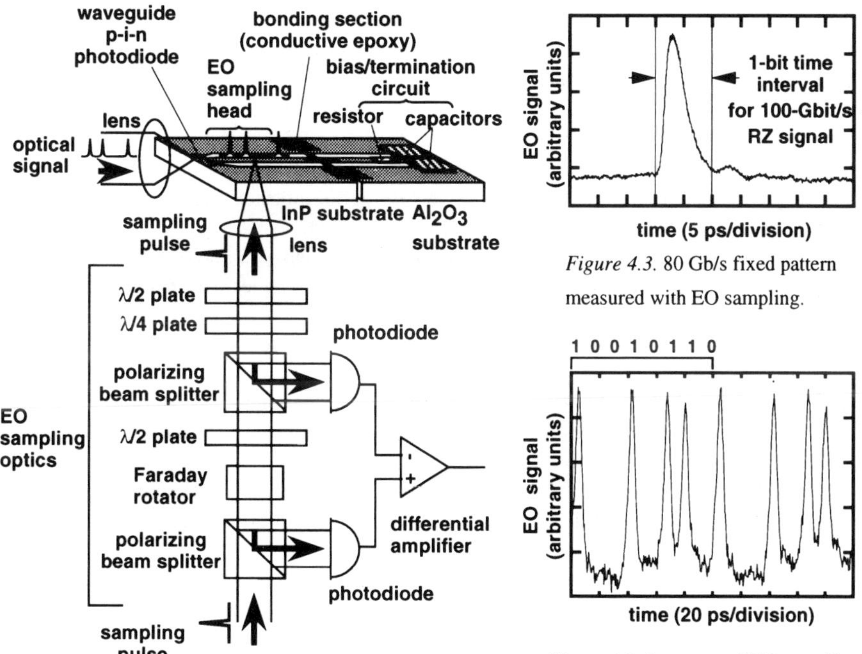

Figure 4.3. 80 Gb/s fixed pattern measured with EO sampling.

Figure 4.2. Response of EO sampling.

Figure 4.1. Schematic diagram of optical signal sampler.

6. Acknowledgment

The author would thank Drs. T. Aoyama, I. Kobayashi and M. Aiki for their encouragement, and express his thanks to Messrs. S. Nishi, K. Suzuki and S. Kawai for their fruitful and helpful discussions. He would thank Dr. S. Saito, Messrs. A. Naka and T. Matsuda for their contribution in the field demonstration. He would also thank Dr. T. Nagatsuma and Mr. M. Yaita for their collaboration in the EO sampling.

References

1. Mollenauer, L. F. Mamyshev, P. V. and Neubelt, M. J. : Measurement of timing jitter in filter-guided soliton transmission at 10 Gbits/s and achievement of 375 Gbits/s-Mm, error free, at 12.5 and 15 Gbits/s, *Opt.Lett.* **19** (1994), 704-706.
2. Nakazawa, M. Suzuki, K. Yamada, E. Kubota, H. Kimura, Y. and Takaya, M. : Experimental demonstration of soliton data transmission over unlimited distances with soliton control in time and frequency domains, *Electron.Lett.* **29** (1993), 729-730.
3. Widdowson, T. Malyon, D. J. Ellis, A. D. Smith K. and Blow, K. J. : Soliton shepherding: All-optical active soliton control over global distances, *Electron.Lett.* **30** (1994), 990-991.
4. Iwatsuki, K. Suzuki, K. Nishi, S. and Saruwatari, M.: 60Gb/s x 2ch time/polarization multiplexed soliton transmission over 154km utilizing adiabatically compressed gain-switched DFB-LD pulse source, *IEEE Photon.Technol.Lett.* (1994), 1377-1379.
5. Suzuki, K. Iwatsuki, K. and Nishi, S.: Optical filtering technique for suppressing nonlinear pulse distortion caused by long transmission distance, *OAA'95* **FB3** (1995).
6. Suzuki, K. Iwatsuki, K. Nishi, S. and Saruwatari, M.: 160Gb/s sub-picosecond transform-limited pulse data stream utilizing adiabatic soliton compressor and optical time-division multiplexer, *IEEE Photon.Technol.Lett.* **6** (1994), 352-354.
7. Yaita, M. Nagatsuma, T. Suzuki, K. Iwatsuki, K. Kato, K. and Muramoto, Y.: Ultrafast optical pulse-pattern signal measurement using optoelectronic techniques, *Electron.Lett.* **31** (1995), 1501-1502.
8. Kodama, Y. Romagnoli, M. and Wabnitz, S.: Stabilisation of optical solitons by an acousto-optic modulator and filter, *Electron.Lett.* **30** (1994), 261-262.
9. Iwatsuki, K. Kawai, S. Nishi, S. and Saruwatari, M.: Timing jitter due to carrier linewidth of laser-diode pulse sources in ultra-high speed soliton transmission, *J.Lightwave Technol.* **13** (1995), 639-649.
10. Kawai, S. Iwatsuki, K. and Nishi, S.: Demonstration of error free optical soliton transmission over 30,000km at 10Gb/s with signal frequency sliding technique, *Electron.Lett.* **31** (1995), 1463-1464.
11. Iwatsuki, K. Kawai, S. Nishi, S.: Theoretical and experimental study of timing jitter and SNR in soliton transmission with frequency sliding, *ECOC'95* **Tu.L.1.5** (1995).
12. Iwatsuki, K. Saito, S. Suzuki, K. Naka, A. Kawai, S. Matsuda, T. Nishi, S.: Field demonstration of 10Gb/s-2700km soliton transmission through commercial submarine optical amplifier system with distributed fiber dispersion and 90km amplifier spacing, *ECOC'95 Post deadline paper* **Th.A. 3.6** (1995).
13. Murakami, M. Takahashi, T. Aoyama, M. Amemiya, M. Sumida, M. Ohkawa, N. Fukada, Y.

Imai, T. and Aiki, M.: 2.5Gbit/s-9720km, 10Gbit/s-6480km transmission in the FSA commercial system with 90km spaced optical amplifier repeaters and dispersion-managed cables, *Electron.Lett.* **31** (1995), 814-815.

14. Suzuki, M. Taga, H. Tanaka, H. Edagawa, N. Utaka, K. Yamamoto, S. Sakai, K. and Wakabayashi, H.: Transform-limited optical pulse generation up to 20 GHz repetition   rate by sinusoidally driven InGaAsP electroabsorption modulator", *CLEO'92 Post deadline papers* **CPD26** (1992).

15. Iwatsuki, K. Suzuki, K. and Nishi, S.: Generation of transform limited gain-switched DFB-LD pulses < 6ps with linear compression and spectral window, *Electron.Lett.* **27** (1991), 1981-1982.

16. Hasegawa A. and Kodama, Y.: Guiding-center soliton in optical fibers, *Opt.Lett.* **15** (1990), 1443-1445.

17. Blow, K. J. and Doran, N. J.: Average soliton dynamics and operation of soliton systems with lumped amplifiers, *IEEE Photon.Technol.Lett.* **3** (1991), 369-371.

18. Evangelides Jr., S. G. Mollenauer, L. F. Gordon, J. P. and Bergano, N. S.: Polarization multiplexing with solitons, *IEEE J.Lightwave Technol.* **10** (1992), 28-35.

19. Wai, P. K. A. Menyuk, C. R. Chen, H. H. and Lee, Y. C.: Soliton at the zero-group-dispersion wavelength of a single-mode fiber, *Opt. Lett.* **vol. 12, no. 8** (1987), 628-630.

20. Suzuki, K. Iwatsuki, K. Nishi, S. and Saruwatari, M.: 160 Gb/s single polarized ultra-short pulses transmission over 62 km wiht fiber nonlinearity near zero dispersionwavelength., *OAA'94 Postdeadline papers*, **PD8** (1994).

21. Kawachi M. and Jinguji, K.: Planar Lightwave circuits for optical signal processing.", *OFC'94 Technical Digest*, **FB7** (1994).

22. Sokoloff, J. P. Prucnal, P. R. Glesk, I. and Kane, M.: A terahertz optical asymmetric demultiplexer(TOAD), *IEEE Photon. Technol. Lett.* **5** (1993), 787-790.

23. Suzuki, K. Iwatsuki, K. Nishi, S. and Saruwatari, M.: Error-free demultiplexing of 160 Gbit/s pulse signal using optical loop mirror including semiconductor laser amplifier, *Electron. Lett.* **30** (1994), 1501-1502.

24. Kato, K., Kozen, A., Muramoto, Y., Itaya, Y., Nagatsuma, T., and Yaita, M.: 110-GHz, 50%-efficiency mushroom-mesa waveguide p-i-n photodiode for a 1.55-mm wavelength, *IEEE Photon Technol. Lett.* **6** (1994), 719-721.

25. Weingarten, K. J., Rodwell, M. J. W., and Bloom, D. M.: Picosecond optical sampling of GaAs integrated circuits, *IEEE J Quantum Electron.* **24** (1988), 198-220.

26. Nagatsuma, T., Yaita, M., Shinagawa, M., Kato, K., Kozen, A., Iwatsuki, K., and Suzuki, K.: Electro-optic characterization of ultrafast photodetectors using adiabatically compressed soliton pulses, *Electron. Lett.* **30** (1994), 814-815.

ALTERNATIVES FOR SOLITON TRANSMISSION
OVER TRANSOCEANIC DISTANCES

E. DESURVIRE AND J. CHESNOY
Alcatel Corporate Research Centre,
33, Rue Emeriau, Paris, France

Abstract. Soliton transmission is a promising technology for future undersea communication systems, owing specifically to its very high capacity potential ($>$40-100 Gbit/s), the possibility of long amplifier/repeater spacings ($>$60-100 km), and increased transmission distances ($>$10,000 km). In order to achieve such performance, however, various soliton transmission control techniques are required. To date, the two main types of soliton control is either purely passive (fixed or sliding frequency-guiding filters), or active (synchronous modulation with guiding-filtering). The first approach is fully compatible with wavelength-division multiplexing, advantageously providing routing functionality. The second enables full signal regeneration, yielding enhanced system margins and virtually removing transmission distance limitations. In this paper, we first discuss the key technologies of soliton transmission, and then present recent experimental results obtained in our laboratory for both types of soliton systems.

1. Introduction

With the advent of practical diode-pumped erbium-doped fiber amplifiers (EDFA) [1], soliton communications have become a realistic alternative for transoceanic systems. In order to experience industrial developments, however, solitons systems must also meet growing market requirements in terms of bandwidth, functionality, reliability, and all this at competitive cost.

Currently, much attention is given to conventional non-return-to-zero systems (NRZ), in which various techniques of wavelength-division multiplexing, dispersion-management, signal preemphasis, gain equalization, and polarization-scrambling have led to successful laboratory demonstra-

209

A. Hasegawa (ed.), Physics and Applications of Optical Solitons in Fibres '95, 209–236.
© *1996 Kluwer Academic Publishers. Printed in the Netherlands.*

tions of 40-100 Gbit/s capacity over transoceanic distances [2]. While in such systems the routing functionality can be as high as that of twenty optical channels, optical amplifiers are closely spaced (45 km), the line bit-rates are comparatively low (5 Gbit/s), and the distances relatively limited (6,300 km). On the other hand, the rapid evolution of undersea networks is likely to provoke in the near future an unprecedented demand in capacity, which could be substantially higher than that of the past twenty years for point-to-point transoceanic links (i.e. 20% annual growth [3]). Finally, the cost of the transmission cables would have to be kept competitive, which suggests reducing the number of line components such as optical amplifiers. Such a not-so-distant future will therefore require alternative technologies for undersea communications.

The aforementioned requirements are met potentially, if not effectively yet, by soliton systems. Since their early conception in 1973 [4], and first experimental amplified loop transmission experiments in 1988-90 [5], it was found that for best performance, soliton systems require some form of in-line transmission control. To date, the two main techniques of soliton transmission control are either purely passive (i.e. sliding frequency-guiding filters [6]), or active (i.e. synchronous modulation with fixed guiding-filtering [7]). The interest of the first is to be naturally compatible with wavelength-division multiplexing, or WDM (by use of Fabry-Perot filters), which advantageously provides network routing functionality, combined with the fact that no qualification for new active components is required. With respect to conventional NRZ systems, sliding-filters soliton transmission enables higher line bit-rates (10-20G bit/s), longer amplifier spans (e.g. 63 km), and substantially longer transmission distances (10-20 Mm) [8][9].

On the other hand, soliton transmission based on synchronous modulation and fixed frequency-guiding filtering enables full signal regeneration, which yields enhanced system margins while virtually removing any transmission distance limitation [10]; amplifier spacings as long as 105-140 km have been demonstrated for 20 Gbit/s transoceanic and global distances [11]. For this reason, this technique is best adapted to point-to-point trunks applications with very high line-capacity (through optical time-division multiplexing, or OTDM) and large amplifier spacings (>100 km).

In this paper, we discuss first the key technologies of soliton transmission, and then present recent results obtained in our laboratory concerning both frequency-guiding-filter and synchronously-modulated soliton systems in WDM and OTDM transmission experiments, respectively.

2. Key Technologies

While soliton systems exhibit the aforementioned interests relative to NRZ systems, rapid progress towards reaching 100-160 Gbit/s aggregate capacity is required in order to find applications in the submarine communications market. Figure 1 summarizes the various physical and technological limitations experienced by both NRZ and soliton systems when (O)TDM and/or WDM techniques are used to increase the aggregate system capacity [12]. For soliton systems, the line capacity is primarily limited by the Gordon-Haus and interaction jitters as well as the effect of soliton self-frequency shift; on the other hand, the wavelength-channel capacity is limited by collision-induced jitter and gain flatness. Regarding amplifier spacing, it is first limited by the requirement of average (or guiding-center) soliton propagation, i.e. the soliton period (z_o) must exceed the amplification period (z_a). With passive control through sliding frequency-guiding filters, the limit is given by $z_a < z_o/2$. In WDM transmission, z_a is also limited by the detrimental effect of inelastic collisions, the condition being $z_a < z_{coll}/2$ $(z_{coll}$ = collision length) [13].

The use of large-effective-area fibers (LEA) makes possible to achieve significantly longer amplifier spacings, or alternatively, to increase the transmission distance or power margins in long amplifier-spacing systems [9], as shown in the next section. Finally, synchronous modulation makes possible to extend the amplifier spacing significantly closer to the limit $z_a \approx z_o$, which is also reviewed in the next section. Technology limits for dense OTDM concern short-pulse (ps) sources with ultra-low intrinsic jitter, as well as multiplexers/demultiplexers of high output/input aggregate capacity. Possible industrial candidates for these components are for soliton sources: gain-switched distributed-feedback (DFB) laser diodes, three-section distributed Bragg reflector (DBR) laser diodes, integrated laser-modulators or ILM (see detailled discussion in next section) and mode-locked ring fiber lasers (for ultra-short pulse generation). MUX components concern polarization-maintaining unbalanced fiber Mach-Zehnders, and their integrated-optics counterparts at ultra-high bit rates; DEMUX components can be based on either polarization-independent (PI) loop mirrors [14], fiber or semiconductor-diode four-wave-mixing, or electroabsorption modulators (PI-EAM) [11]. In these devices, clock recovery is provided by electronic narrowband circuitry including high-finesse RF cavity filters or phase-locked loops.

Concerning dense WDM, the limits are introduced by EDFA bandwidth (range of gain flatness), i.e. about 10nm with filter equalization [2], and the available EDFA output power. Automatic gain control (AGC) and passive equalization filtering are key functions in wideband WDM, due to fluc-

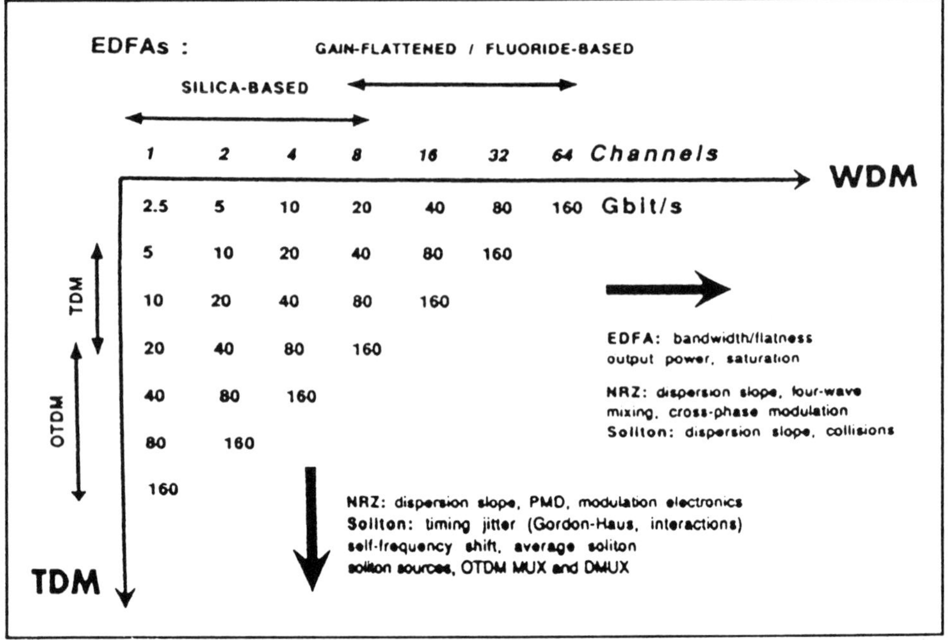

Figure 1. Time-division (TDM) and wavelength-division (WDM) multiplexing alternatives for increasing the aggregate capacity in NRZ and soliton systems, and corresponding physical/technological sources of performance limitation.

tuations in traffic and wavelength allocations. Extended bandwidth (30 nm) could also be provided by fluoride-based EDFAs [15], although these remain to be tested in long-haul transmission. Like in preceeding NRZ systems, techniques of dispersion management (collision suppression) and adaptative filtering (gain equalization) with EDFAs of improved gain flatness should permit a full implementation of soliton WDM, while operating at the higher SDH line rate of 10 Gbit/s, possibly 20 Gbit/s. Passive frequency-guided soliton systems are becoming realistic with practical photosensitive fiber Fabry-Perot (FP) filters [16], while overall performance is enhanced by use of large effective-area transmission fiber [9], as is described in the next experimental section. Concerning regenerated soliton systems, the key components to develop are 20-40 GHz polarization-independent intensity/phase modulators on one hand, and 20-40 Gbit/s to 10 Gbit/s time-division demultiplexers, on the other hand.

Ultimate line capacities at or beyond 40 Gbit/s should eventually be reached when implementing all-optical soliton regeneration, which means the development of all-optical clock recovery and modulation techniques [17]-[20]. For the clock recovery, the critical features include: polarisation insensitivity, environmental/thermal stability, control of pulse walk-off, amplitude and timing jitter, phase-locking, tolerance to long zero sequences, and operation at exact SDH frequencies. Concerning optically-controlled soliton modulation, the critical features include polarisation insensitivity, insertion loss, modulation depth (intensity modulation), modulation acceptance window and curvature, suppression of unwanted parasitic intensity or phase modulation.

3. Experimental Results

In this section, we shall describe experimental results obtained in our laboratory, concerning both WDM and OTDM transoceanic soliton systems, and based on frequency-guiding filtering and synchronous modulation control, respectively. The transmission experiments were carried out using various configurations of amplified recirculating loops, and had the following characteristics:

a- 2x5 Gbit/s WDM transmission with fixed frequency-guiding filter of photosensitive fiber FP type;

b- 2x10 Gbit/s WDM transmission with sliding frequency-guiding filters of bulk FP-etalon type;

c- Large-effective area fiber (LEA) management in 5-10 Gbit/s single-channel transmission with frequency-guiding filters;

d- 20 Gbit/s synchronously-modulated transmission with electroabsorption intensity/phase modulator;

We finally conclude this section with a description of an experimental all-optical clock recovery device, based on a mode-locked ring-fiber laser, which has potential applications in OTDM soliton systems for modulation and demultiplexing.

It might be useful at first to review the characteristics of different types of semiconductor sources that were used for soliton generation in the experiments described here. On one hand, mode-locked erbium-doped fiber-lasers (EDFL) can provide short and ultrashort pulse generation over a wide tunable range; while realistic field applications of EDFLs are remote, they could likely concern dense OTDM systems (100-160 Gbit/s). On the other hand, three other main types of soliton sources are based on integrated

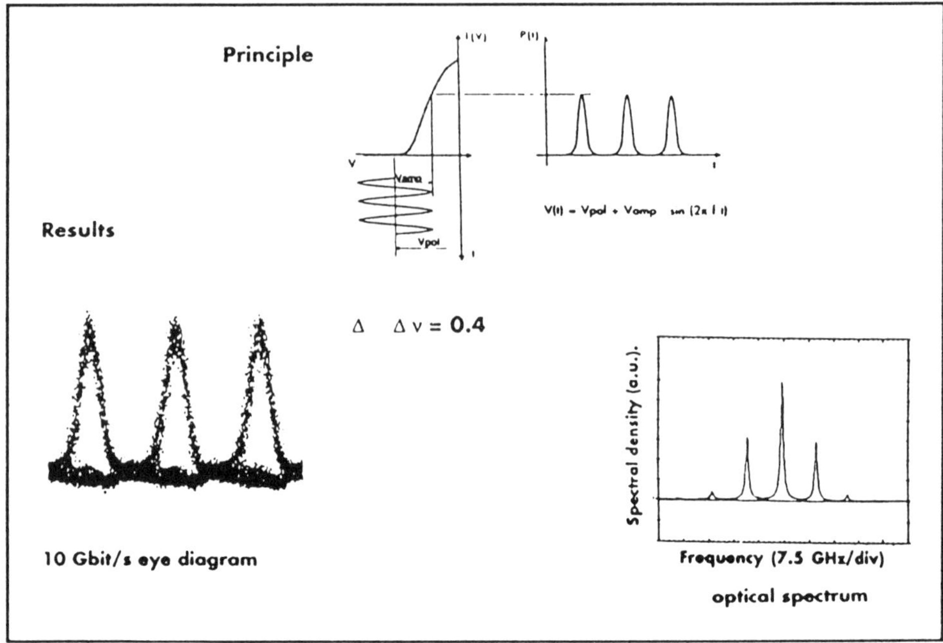

Figure 2. Principle of soliton pulse generation through electroabsorption modulation, and experimental output characteristics at 10 GHz in the time and frequency domains, after 10 Gbit/s external modulation.

semiconductor-diode devices: gain-switched DFBs, three-section DBRs, and electroabsorption-laser-modulator (ILM). For encoding the pulse train with data, all three types require external modulation. The advantage of gain-switched DFBs is to provide the shortest pulses (duty cycle $\Delta T/T < 0.1$, typically), although a compression fiber of high negative dispersion is required to suppress the chirp and produce a Fourier-transform-limited (FTL) signal [21]. They also provide the highest output power (0 dBm average, typically) and can be directly modulated. Three-section DBRs offer the advantage of tunability, which is important to adjust the dispersion value in soliton propagation; FTL pulses can be obtained by simultaneous modulation of gain and phase sections of the laser, which produces direct chirp compensation [22]. The modulation bandwidth of DBRs, however, is currently limited to 5 GHz or so. The combination of a DFB laser and an electroabsorption (EA) modulator, or under its integrated form, the ILM, can produce FTL pulses at high repetition rates (10-20 GHz), although with relatively limited output powers (-20 dBm bulk, -10 dBm integrated) [23]. The principle of operation of the EA modulator, are shown in Figure 2. It

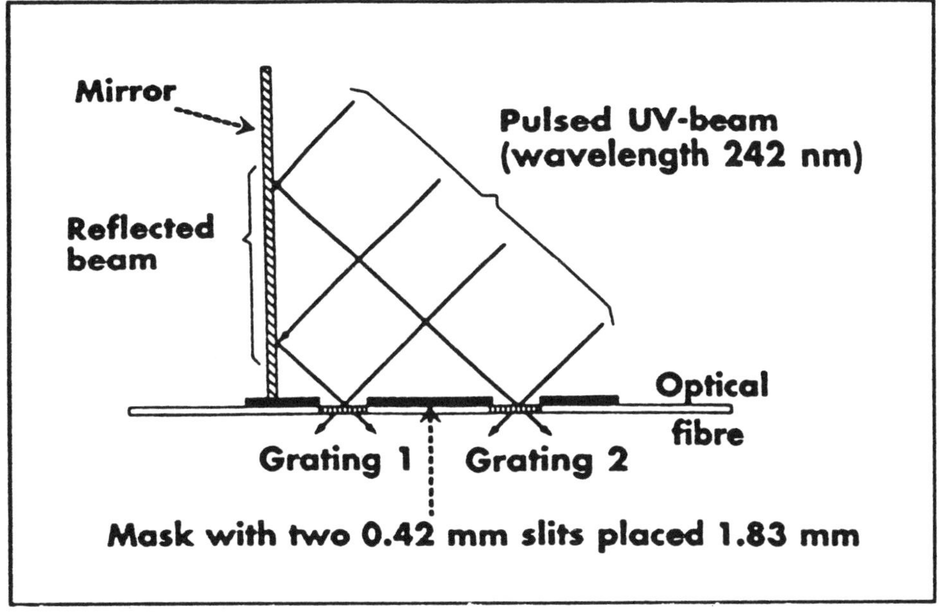

Figure 3. Fabrication principle of photosensitive fiber Fabry-Perot filters.

is seen that a sinusoidal voltage modulation applied to the appropriate region of the modulator's response T(V) produces low-duty-cycle pulses with high extinction ratio. The experimental time/spectral output characteristics under 10 GHz direct and 10 Gbit/s external modulations, which are also shown in the figure, correspond to a time-bandwidth product of TBP $= \Delta T \Delta \nu = 0.40$ (note that TBP = 0.44 and 0.315 for Gaussian pulses and sech^2 pulses, respectively). The DFB/MEA or ILM sources were primarily used throughout our experiments.

3.1. 2X5 GBIT/S WDM TRANSMISSION WITH FIXED PHOTOSENSITIVE FREQUENCY-GUIDING FILTER

We consider first the 2x5 Gbit/s WDM loop transmission experiment, which is based on the principle of fixed frequency-guiding filter control [16]. The goal of this experiment was to test a Fabry-Perot filter of the photosensitive fiber grating type, which is a promising candidate for realistic system implementation. The results obtained with the fiber filter are also compared to that obtained with a commercial bulk FP filter. The fabrication principle of the photosensitive fiber FP filter is illustrated in Figure 3.

A pulsed UV-beam produced by a $\lambda = 242$ nm excimer laser is made to interfere onto the Ge-core fiber through a mask. The mask has two 0.42 mm

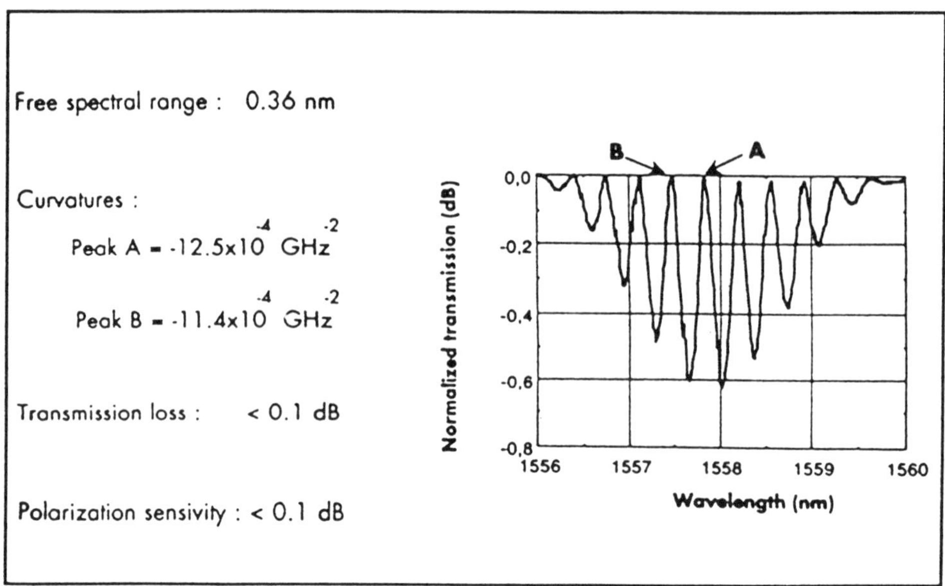

Figure 4.   Transmission characteristics of fiber FP filter.

slits spaced 2.2 mm apart (center of Bragg gratings). The Bragg gratings resulting from this interference pattern form a FP interferometer of whose transmission characteristics vs. signal wavelength are shown in Figure 4. The 0.36 nm free-spectral range is adapted to 5 Gbit/s WDM signals, at points shown by A and B, for instance, which can be made to match the desired wavelengths by mechanical stretching of the filter. Note that by accurately controlling the peak transmission wavelength locations during fabrication, it is also possible to use such a device for sliding-filter, straight-line soliton systems. The transmission loss and polarization sensitivity of this filter were measured to be less than 0.1 dB. In order to suppress the EDFA gain peak ($\lambda = 1,532$ nm) which falls outside the bandwidth of interest, an additional interference filter of 2nd-order Butterworth type (flat peak transmission) and of 2.4 nm bandwidth was included. Figure 5 shows the IF transmission, as well as that of a commercial bulk glass FP filter (FSR = 0.32 nm) having very similar peak curvature and also used for comparison. The experimental setup for 2x5 Gbit/s WDM transmission is shown in Figure 6. The transmitter is made of two DFB/EA sources producing 40 ps pulses at 0.36 nm separation, corresponding to points A and B in Figure 4, followed by two separate $LiNbO_3$ modulators driven by 5 Gbit/s, PRBS signals. The recirculating loop consists of two segments of 30 km, 50 μm^2-area dispersion-shifted (DSF) fibers, and includes three EDFAs, an acousto-optic (A/O) switch, the FP and the IF filters. Note that the A/O

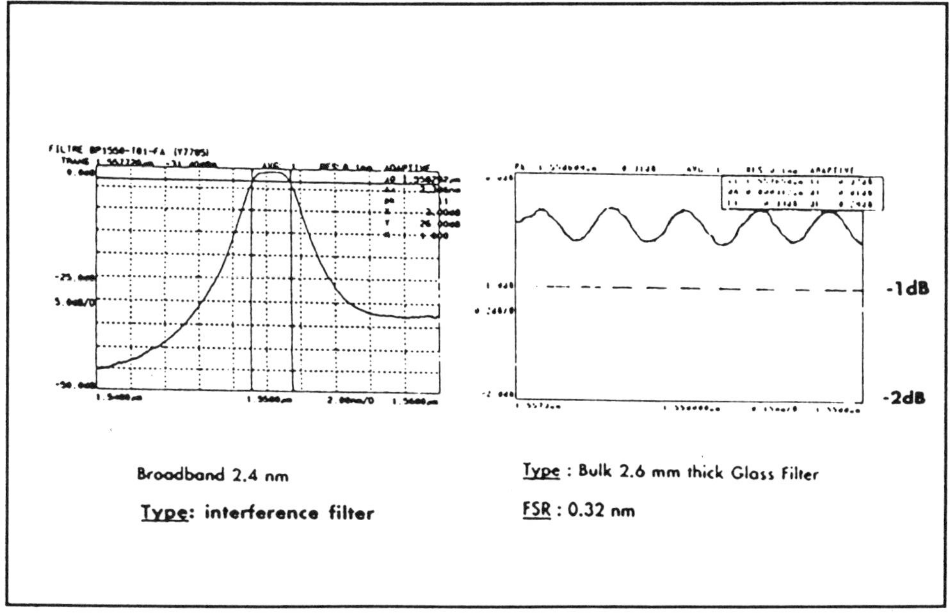

Figure 5. Transmission characteristics of interference filter (left) and bulk Fabry-Perot filter (right) used in the experiment.

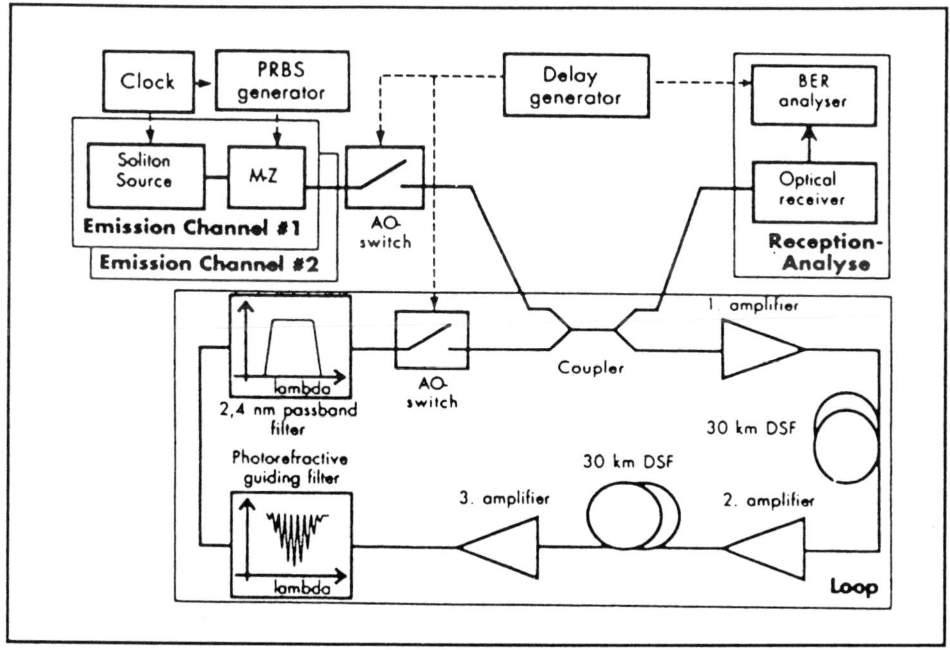

Figure 6. Experimental loop setup for 5 Gbit/s WDM transmission.

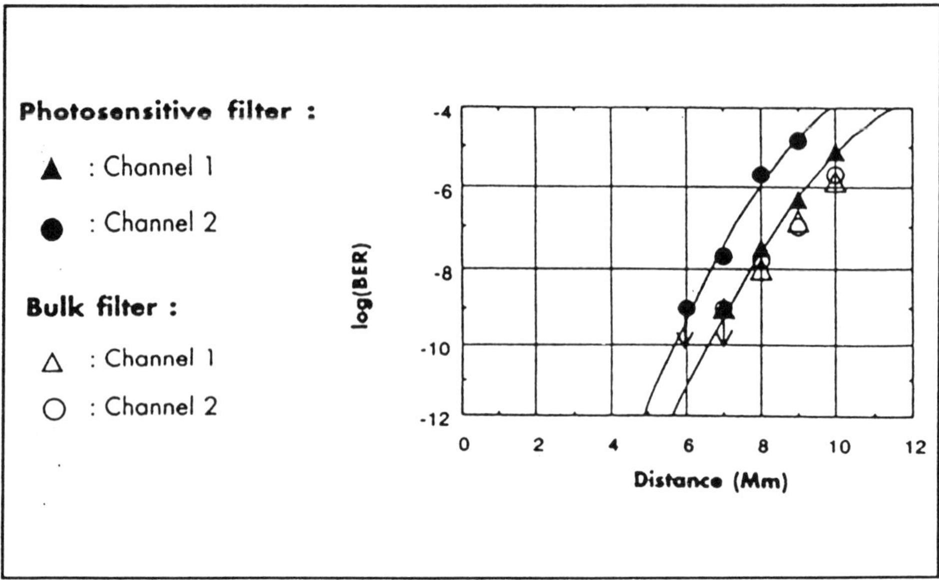

Figure 7.   BER measurements vs. distance in 5 Gbit/s.WDM transmission obtained with both types of FP frequency-guiding filters.

switch is operating in the zero-order configuration, so as to prevent unwanted signal frequency shift at each loop recirculation. The bit-error-rate (BER) measurements as function of the transmission distance, as obtained in both channels and for both FP filter types, are shown in Figure 7.

As seen from the figure, error-free transmission (BER$< 10^{-9}$) is achieved up to 6,000 km and 7,000 km for channels 2 and 1, respectively (the lower transmission performance, in comparison to results of [24], is attributed to the fact that the loop had three amplifiers for only two spans, thus generating a higher noise level). The difference in error-free distance between the two channels is attributed to the fact that they do not experience the same filter curvature (channel 1 having the highest), as seen in Figure 4. This is confirmed by the measurements obtained with the bulk filter (Figure 5), which show no meaningful difference. Finally, it is seen that bulk and fiber FP filters provide identical transmission performance. With improvements to include the IF passband function, photosensitive fiber FP filters are promising components for sliding-filter soliton system implementation.

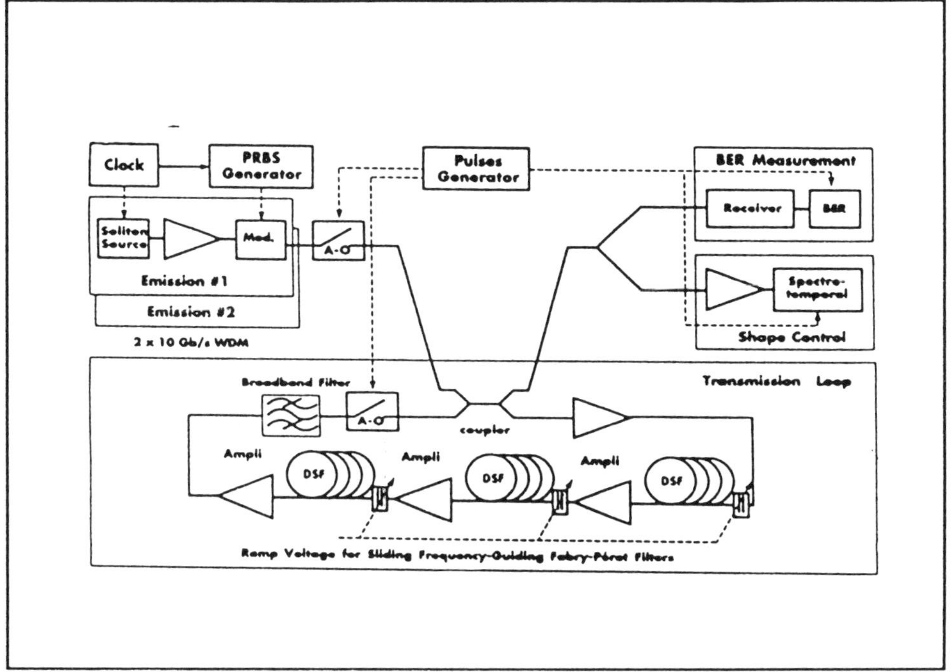

Figure 8.   Experimental setup for 2x10 Gbit/s WDM sliding-filter loop transmission.

3.2. 2X10 GBIT/S WDM TRANSMISSION WITH SLIDING FREQUENCY-GUIDING FILTERS

We consider next a 2x10 Gbit/s WDM transmission with sliding frequency-guiding filters. The goal of this experiment is to optimize the filter sliding rate (in particular to compare up- vs. down-sliding), as well as to increase the amplifier (Za) and/or filter (Zf) spacing. The loop experimental setup is shown in Figure 8. The setup now includes a bulk air-gap FP-etalon filter (1.5 mm thick, FSR = 0.8 nm) controlled by a voltage ramp, which is synchronized to the switches and provides variable sliding rates from -15 GHz/Mm to +15 GHz/Mm. The DSF spool lengths are 30 km, and the power of the amplifiers is adjusted for either 30 km or 60 km loss compensation. The measurements relevant first to single-channel propagation with a sliding rate of -11 GHz/Mm (Za = Zf = 30 km, D = 0.4 ps/nm/km) are summarized in Figure 9. After optimization of the soliton power, the initial 20 ps pulse are seen to rapidly evolve with distance into narrower (10 ps) or reshaped solitons. Error-free transmission is achieved up to 16,000 km. Systematic BER measurements also show a decrease of the timing decision tolerance to tw = 25 ps - 12 ps (z = 9,000 - 15,000 km) from the receiver acceptance window of tw = 55 ps, which is explained by the accumulation of Gordon-Haus (GH) jitter. A comparison of down-sliding and up-sliding

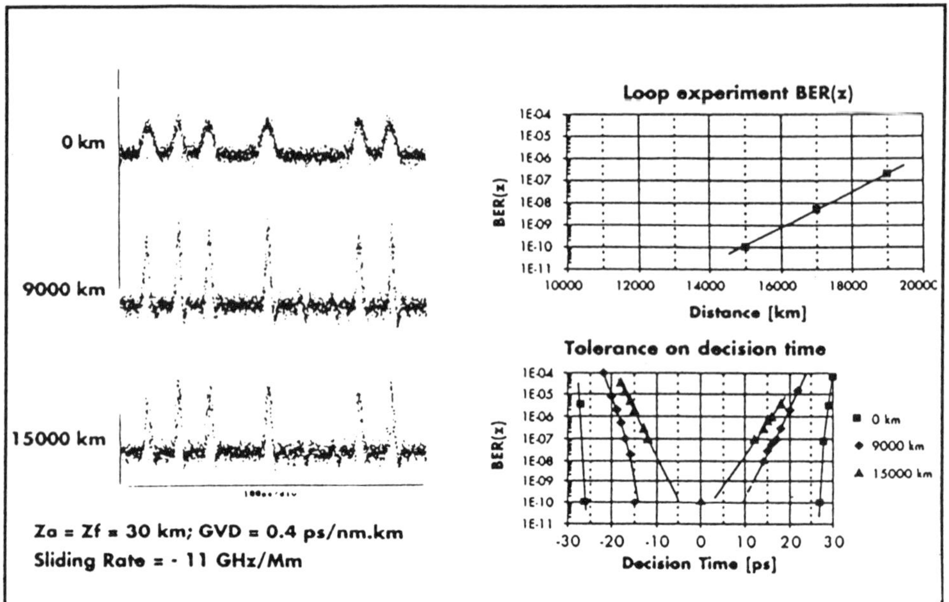

Figure 9. Temporal, BER and decision window measurements concerning 10 Gbit/s, single-channel transmission up to 19,000 km with down-sliding rate of -11 GHz/Mm (Za = Zf = 30 km and D = 0.4 ps/nm/km)

configurations (± 11 GHz/Mm, Za = Zf = 30 km, D = 0.4 ps/nm/km) where the same average dispersion was kept by switching start and stop wavelengths, is provided in Figure 10. The top oscilloscope traces correspond to the recirculating signal over a 19,000 km span with its background of amplified spontaneous emission (ASE). Error-free transmission distances corresponding to down-sliding and up-sliding are 15,000 km and 18,000 km, respectively. This difference can be attributed to the phase response of the filter, which introduces an asymmetry between the two configurations (the frequency lag experienced by the soliton is lower in the up-sliding case, which corresponds to lower filter loss and consequently, lower excess amplification noise and jitter [25]). This feature is well observed in the corresponding ASE background buildup (Figure 10), which is more important in the down-sliding case; while the signal-to-noise ratios at 19,000 km are similar, the GH jitter accumulated throughout the link is greater indeed in this last case.

We consider next the effect of sliding rate in the up-sliding configuration. Figure 11 shows the BER measurements corresponding to the three sliding rates: +5 GH/Mm, +11 GHz/Mm and +15 GHz/Mm. In each configura-

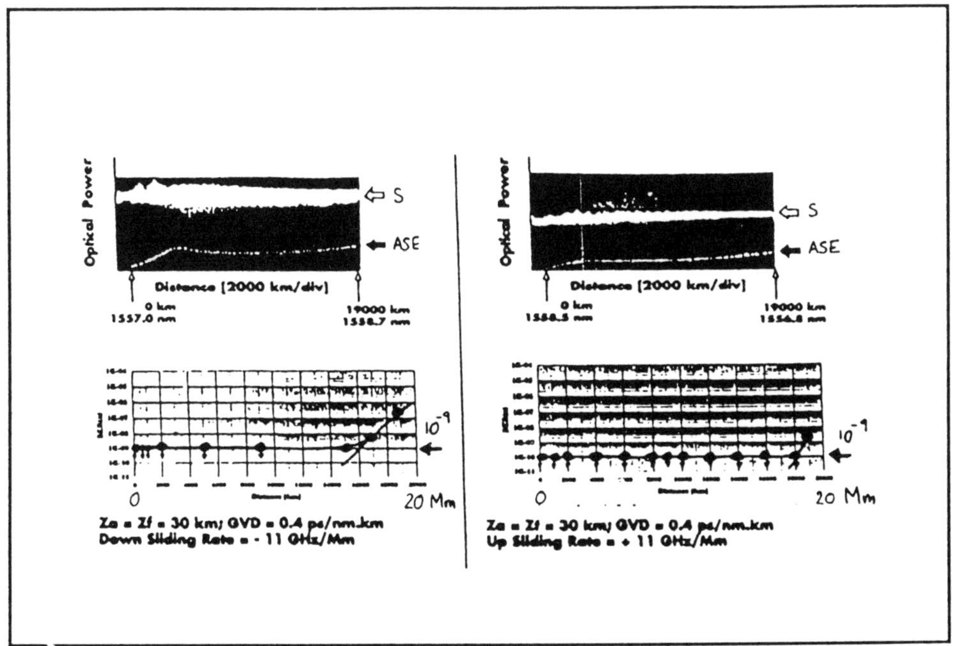

Figure 10. Comparison between down- and up-sliding (± 11 GHz/Mm) in 19 Mm transmission (Za = Zf = 30 km, D = 0.4 ps/nm/km).

tion, the soliton power is adjusted to minimize the error count. At rates below +5 GHz/Mm, transmission was observed to fail between 1,000-3,000 km, due to strong ASE noise buildup and excess continuum generated by soliton reshaping, both of which being unsufficiently removed by the filtering action. On the other hand, transmission degradation is also observed at rates above +11 GHz/Mm. This is attributed to the fact that in this case, the signal wavelength shifts down too far away from the peak transmission region of the wideband filter, which causes excess loss. Having found the optimum sliding rate (i.e. +11 GHz/Mm, Z = 18,000 km) for the system configuration (Za = Zf = 30 km, D = 0.4 ps/nm/km), we study now the effect of amplifier and filter spacings, when the filter strength is kept constant and the sliding rate is near +11 GHz/Mm. The results obtained for various combinations of Za, Zf and number of spans are summarized in Figure 12. It is seen first (configurations #1 and #2) that a small degradation is caused by placing the three filters together at Zf = 90km instead of distributing them every 30 km. The observed degradation can attributed to the fact that for this soliton with period Zo = 250km, a filtering perturbation at Zf = 90km is too strong to ensure adiabatic frequency control. Reducing the number of spans from 3 to 1 (configurations #1 and #3)

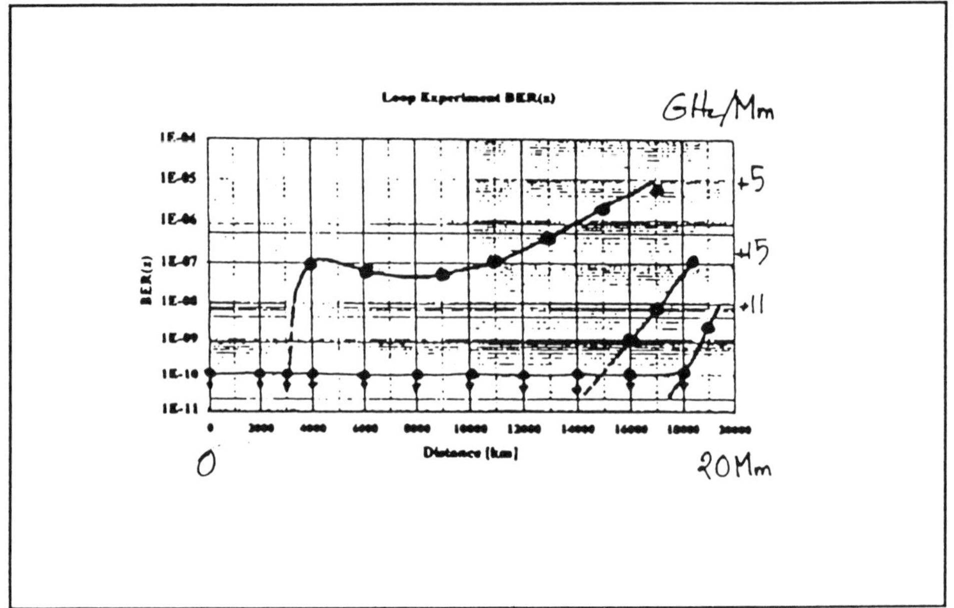

Figure 11. Optimization of up-sliding rate in 10 Gbit/s, single-channel soliton transmission (Za = Zf = 30 km, D = 0.4 ps/nm/km).

causes a substantial penalty (19,000 km to 15,000 km error-free distance), which is attributed to the excess noise of the extra amplifier used in the loop for component loss compensation. Finally, increasing the amplifier spacing to Za = 60 km with two concatenated filters (Zf = 60 km) in the one-span configuration (#4) introduces some extra penalty (12,000 km distance) as expected, even after sliding rate optimization to +13 GHz/Mm. We have thus identified three independent causes of performance degradation in our experimental loop apparatus. Finally, we consider 2x10 Gbit/s WDM transmission in the configuration Za = Zf = 30 km (+11 GHz/Mm sliding rate). The signal outputs monitored at z=9,000km and BER measurements up to z = 16,000 km are shown in Figure 13. As seen in the figure, the short-wavelength channel (#2) could be transmitted error-free up to 10,000 km, to compare with 13,000 km for channel #1. These results are similar to that reported in Reference [26], where two 10 Gbit/s channels could be transmitted simultaneously over 13 Mm, with an amplifier spacing of Za = 26 km. Our transmission performance has been considerably improved ever since, as will be reported in the near future [27]. This is due to the fact that, countrarily to the case of straight-line systems, transmission in recirculating loops is very sensitive to polarization-dependent loss, and best results are achieved when the proper polarization eigenmodes of the loop, different for each WDM channel, are excited.

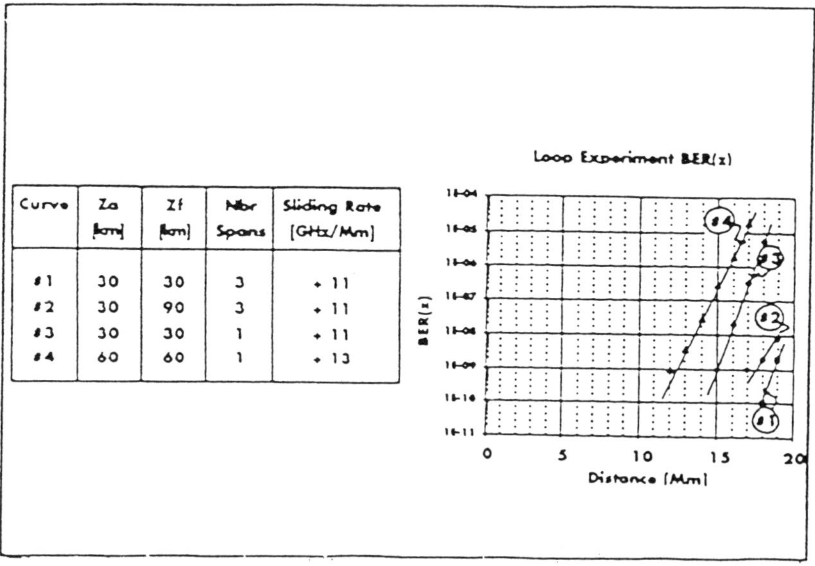

Figure 12.   Effect of amplifier spacing (Za), filter spacing (Zf) and number of spans in 10 Gbit/s sliding-filter soliton transmission.

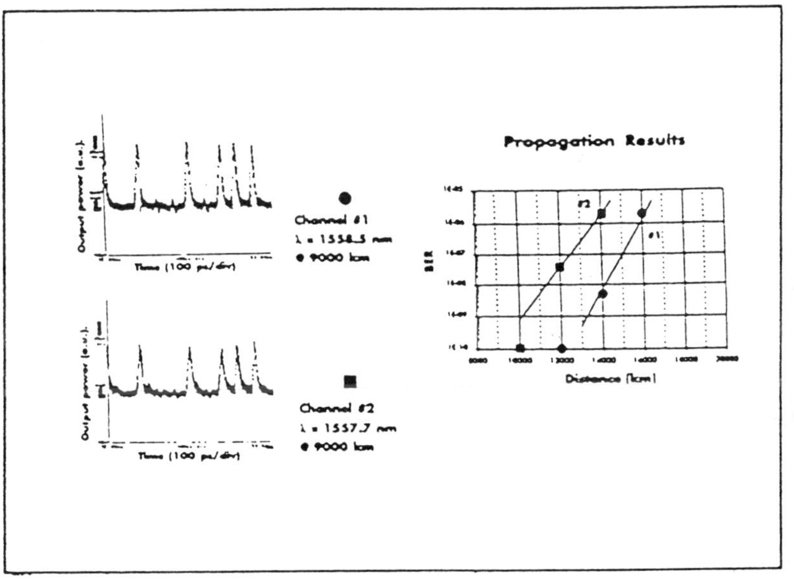

Figure 13.   BER measurements in 2x10 Gbit/s sliding-filter soliton transmission.

3.3. LARGE-EFFECTIVE AREA FIBER (LEA) MANAGEMENT

The third part of this review concerns the use of large-effective area fibers (LEA) in soliton transmission. Before describing results obtained in a 5/10

Gbit/s sliding-filter loop, it is useful to briefly recall the principle and interest of LEA fibers in amplified systems, i.e. whether soliton or NRZ. For both types of amplified systems, basically, the output signal-to-noise ratio (SNR) is proportional to the quantity $A_{eff}/F(G)$ where A_{eff} is the fiber effective area and F(G) a penalty function given in its most general form by [28]:

$$F(G) = \frac{\eta_{out}\left(\frac{G}{\eta_{in}\eta_{out}} - 1\right)(G - 1)}{G(logG)^2} \qquad (1)$$

where $\eta_{in}, \eta_{out} \leq 1$ represent the line amplifier's input and output coupling coefficients, respectively (Expression (1) reduces to that of Reference [29] when $\eta_{in} = \eta_{out} = 1$), and G = $\exp(\alpha Za)$ the fiber-to-fiber gain (α = fiber loss coefficient). On the other hand, the GH jitter variance in soliton systems is proportional to $F(G)/A_{eff}$ (Equation (7) of Reference [29]). Thus, increasing the effective area of the transmission fiber results in higher SNR (i.e. greater amplitude Q-factors) and lower GH jitter (i.e. greater timing Q-factors). The resulting improvement of BER can be exploited in terms of either longer transmission distances (or higher system margins), or longer amplifier spacings. This last point is shown in Figure 14, where we compare plots of $F(G)/A_{eff}$ vs. amplifier spacing Za as defined above, corresponding to a typical dispersion-shifted fiber (DSF) of area $A_{eff} = 50\mu m^2$ and an LEA fiber of area $A_{eff} = 70\mu m^2$, respectively ($\eta_{in} = 1$ dB, $\eta_{out} = 3$ dB, $\alpha = 0.22$ dB/km). It is first seen that plots predicts optimal transmission performance (SNR and GH jitter) for an amplifier spacing near Za = 30 km. Changing the transmission fiber from DSF to LEA leads then two possible improvements: either (1) a greater system margin for the same amplifier spacing (R→A in figure), or (2) identical performance for a longer amplifier spacing of Za = 50 km (R→B). Both improvements were experimentally demonstrated in a 5 Gbit/s, fixed-filter soliton loop [30], the BER measurements being shown in Figure 15. As predicted by the penalty factors in Figure 14, the transmission performance is identical for configurations R and B (Za = 30 km-DSF and Za = 50 km-LEA), while it is greater in configuration A (Za = 30 km-LEA). Note also the fair agreement with theoretical simulations, which accounts for both GH and electrostrictive timing jitter. The characteristics of both DSF and LEA fibers used in this experiment are shown in Figure 16. Among other features, the LEA attenuation (including splice loss) is 0.26 dB/km, and its index profile is trapezoidal with a surrounding step ring. This optimum profile provides a microbending loss similar to that of commercially-rated DSF. The fabrication of long lengths of LEA with substantially larger cross-sections (e.g. $A_{eff} = 90\mu m^2$), low attenuation (e.g. 0.22 dB/km) and other required characteristics in high production yield is certainly constraining from an industrial point of view.

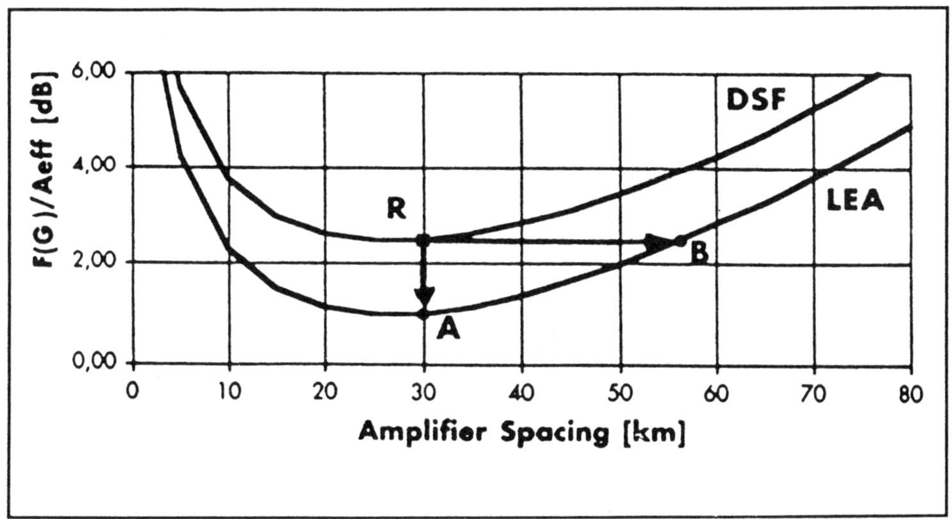

Figure 14. Penalty factor F(G)/A_{eff} as a function of amplifier spacing Za for dispersion-shifted fibers (DSF) and large-effective-area fiber (LEA), as plotted from Equation (1) with η_{in} = 1 dB, η_{out} = 3 dB, α = 0.22 dB/km, A_{eff}(DSF) = 50 μm^2 and A_{eff}(LEA) = 70 μm^2.

Therefore, it is preferable to use concatenation of LEA and DSF between amplifiers. Remarkably, only a small length of LEA followed by a longer strand of DSF is sufficient to increase the average effective area of the link: this is the principle of LEA management [9]. This property can be shown by considering a link of LEA (length L) followed by DSF (length (Za-L)), with input power Po. The corresponding path-averaged intensity $\langle P\rangle/A_{eq}$, is then:

$$\frac{\langle P\rangle}{A_{eq}} = \frac{P_0}{Z_a}\left\{\int_0^L \frac{e^{-\alpha_{LEA}z}}{A_{LEA}}dz + \int_L^{Z_a}\frac{e^{-\alpha_{DSF}z}}{A_{DSF}}dz\right\} \qquad (2)$$

On the other hand, the path-averaged intensity defined in Equation (2) is equal to that of a trunk of identical length, loss and input intensity, i.e.

$$\frac{\langle P\rangle}{A_{eq}} \equiv \frac{P_0}{Z_a}\int_0^{Z_a}\frac{e^{-\alpha_{eq}z}}{A_{eq}}dz = \frac{P_0}{A_{eq}}\frac{1-e^{-\alpha_{eq}Z_a}}{\alpha_{eq}Z_a} \qquad (3)$$

with the equivalent attenuation coefficient being:

$$\alpha_{eq} = \frac{\alpha_{LEA}L + \alpha_{DSF}(L-Z_a)}{Z_a} \qquad (4)$$

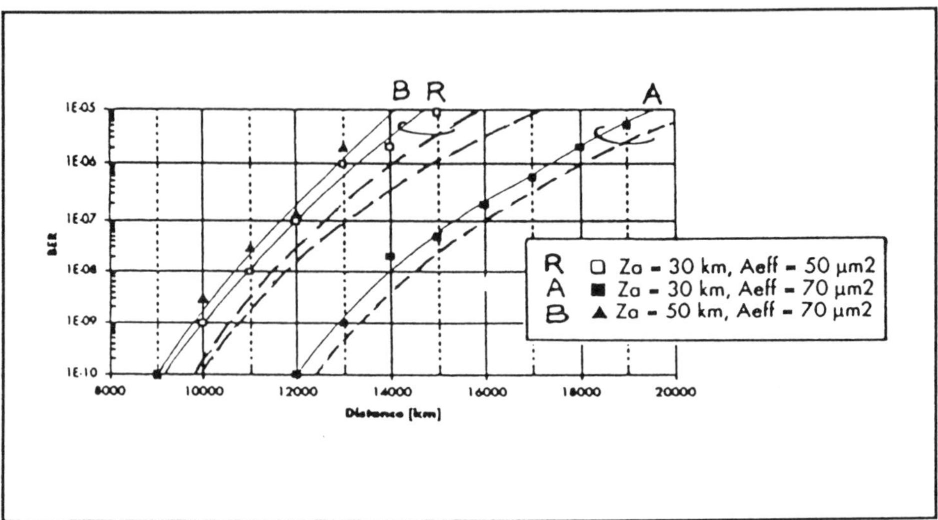

Figure 15. BER measurements obtained in 5 Gbit/s fixed-filter soliton transmission with DSF ($A_{eff} = 50\mu$m^2) for an amplifier spacing of Za = 30 km, and with LEA fiber ($A_{eff} = 70\mu$m^2) for amplifier spacings of Za = 30 km and 50 km, corresponding to penalty factors indicated by R, A, B in Figure 14, respectively. Dashed curves represent theoretical simulation.

Using Equations (2)-(4) one can determine analytically the equivalent effective area A_{eq} of the LEA/DSF hybrid link. The dependence of A_{eq} in the length L of LEA in the hybrid link is plotted in Figure 17 (Za = 63 km, $\alpha_{LEA} = \alpha_{DSF} = 0.24$ dB/km, $A_{LEA} = 70\mu$m^2 and $A_{DSF} = 50\mu$m^2). It is seen that in this example, the 50% − 50% concatenation of 31.5 km of LEA ($A_{LEA} = 70\mu$m^2) with 31.5 km of DSF ($A_{DSF} = 50\mu$m^2) provides a link having an equivalent effective area of $A_{eq} = 67\mu$m^2.

We implemented the principle of LEA management into a 10 Gbit/s sliding-filter soliton loop experiment with record amplifier spacing of Za = 63km [9] (note that the 63 km record value was first achieved in a 20 Gbit/s sliding-filter system [8]). The transmission fiber used in the experiment is the hybrid link corresponding to the previous example in Figure 17 ($A_{eq} = 67\mu$m^2); for comparison, transmission was also performed using Za = 63km of DSF ($A_{DSF} = 50\mu$m^2). The BER measurements made in these two configurations are shown in Figure 18. It is seen that with DSF, the maximum error-free distance is near 14 Mm, while with LEA/DSF hybrid the distance is increased to 19 Mm. No error was observed in the configuration Za = 32 km-DSF. This experiment clearly demonstrates the system margin enhancement introduced through LEA management.

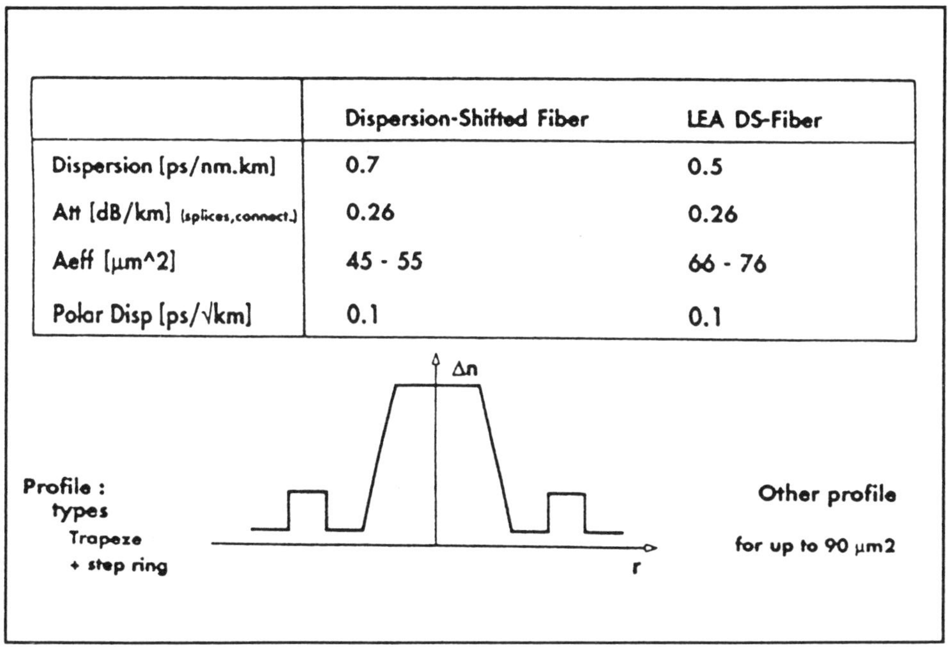

	Dispersion-Shifted Fiber	LEA DS-Fiber
Dispersion [ps/nm.km]	0.7	0.5
Att [dB/km] (splices,connect.)	0.26	0.26
Aeff [μm^2]	45 - 55	66 - 76
Polar Disp [ps/√km]	0.1	0.1

Figure 16. Characteristics of LEA fiber used in the 5 Gbit/s soliton transmission experiment.

3.4. 20 GBIT/S SYNCHRONOUSLY-MODULATED SOLITON TRANSMISSION

In this fourth part, we consider 20 Gbit/s soliton systems with active transmission control based on in-line synchronous modulation. The modulator used in the experiment is of the electro-absorption intensity/phase type [31], and is nearly polarization-independent (<2 dB); it is driven by an RF clock recovery circuit based on a high-finesse cavity filter. The 20 Gbit/s soliton signal is generated by time-multiplexing two independently-coded tributaries of a 10 GHz ILM source ($\lambda = 1,555$ nm, $\Delta T = 14$ ps). The receiver includes an EA optical demultiplexer, followed by a negative-dispersion fiber (DC) for pulse stretching (BER optimization requires careful adjustment of the compensating dispersion, here $D = -180$ ps/nm). The experimental loop setup is shown in Figure 19. The openig/closing of the loop is controlled by AO switches, the ones in the loop being in a -1/+1 order configuration to avoid signal frequency-sliding. The loop includes one to three amplified link sections of 30-140 km lengths. In some measurements, the average dispersion of the loop is optimized by splicing small sections of standard SMF ($D = +17$ ps/nm/km) into the links. Frequency-guiding filtering is provided by two IFs with 0.8 nm equivalent bandwidth. An

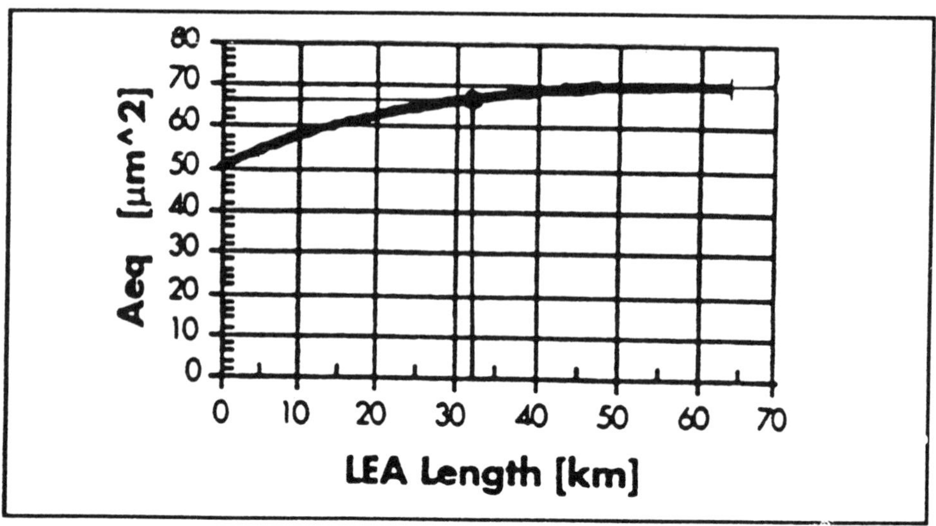

Figure 17. Plot of equivalent effective area A_{eq} of hybrid LEA/DSF link, as a function of LEA fiber section length L (Za = 63km, $\alpha_{LEA} = \alpha_{DSF} = 0.24$ dB/km, $A_{LEA} = 70\mu$m^2, and $A_{DSF} = 50\mu$m^2).

EDFA booster is used to compensate both modulator and AO switches insertion loss. The oscilloscope traces shown in Figure 20 correspond to the signal and ASE background evolution over 20,000 km (left) and the corresponding 20 GHz RF component (right). Note the complete stabilization of the ASE noise over distance, due to regeneration. A first optimization of the loop operating conditions (i.e. launched soliton power, polarization states, dispersion, receiver decision level and phase) provides the uniform, 20 GHz RF component shown in Figure 20. However, error-free transmission requires finer adjustments of these conditions. The results obtained for amplifier/modulator spacings of Za = Zr = 60 km and 140 km, with corresponding dispersions of D = 0.24 ps/nm/km and 0.37 ps/nm/km, respectively(which are found to be within optimal ranges), are shown in Figure 21. The figure shows that for the first configuration (Za = Zr = 60 km), no error is detected over a 20,000 km distance. In the second (Za = Zr = 140 km), transmission with $\leq 1.10^{-9}$ is achieved up to 8 Mm. The amplifier spacing of 140 km, which represents the current record for all types of soliton systems (i.e. active or passive), was previously reported for >150 Mm error-free transmission [11]. By increasing the fiber dispersion to D = 0.45 ps/nm/km and reducing further the excess loss, we have achieved ever

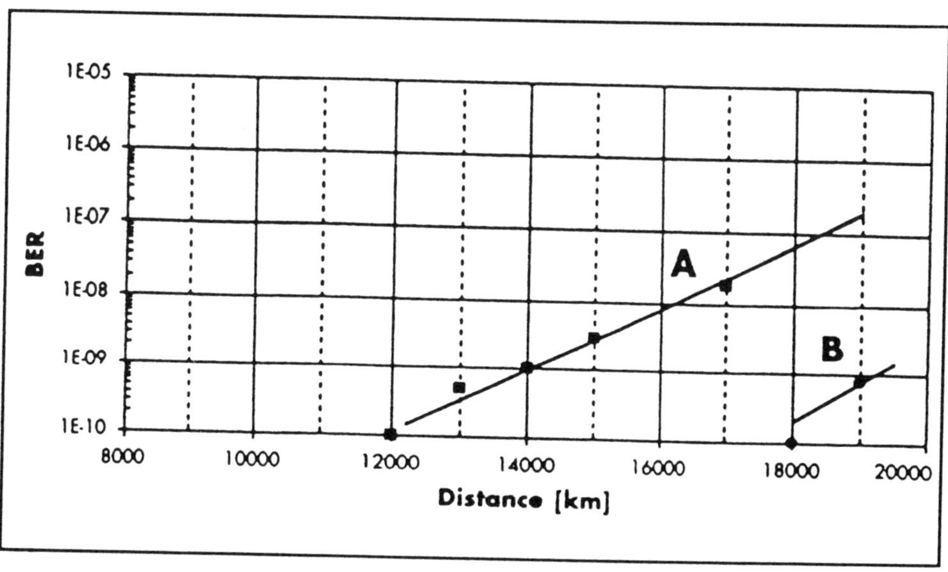

Figure 18.   BER measurements in 10 Gbit/s sliding-filter soliton transmission with amplifier spacing of Za = 63km, showing effect of LEA management: (A) transmission in DSF ($A_{DSF} = 50\mu\text{m}^2$), and (B) transmission in hybrid LEA/DSF ($A_{eq} = 67\mu\text{m}^2$).

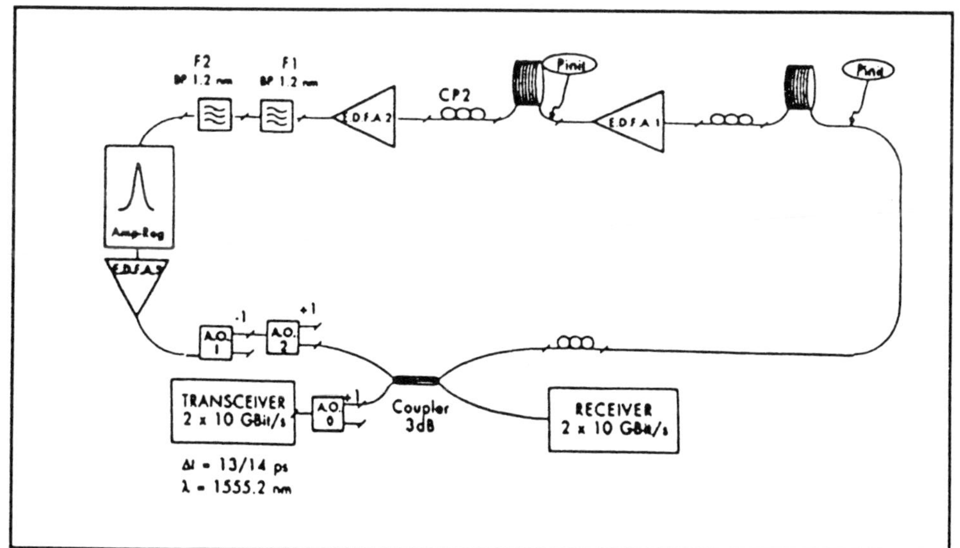

Figure 19.   Experimental loop setup for 20 Gbit/s regenerated soliton transmission.

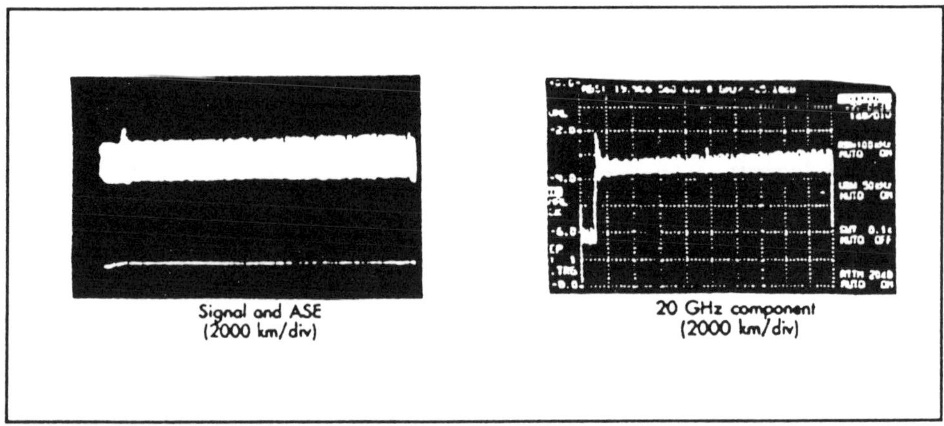

Figure 20. 20 Gbit/s recirculating soliton signal with ASE background evolution over 20,000 km (left) and corresponding 20 GHz RF component.

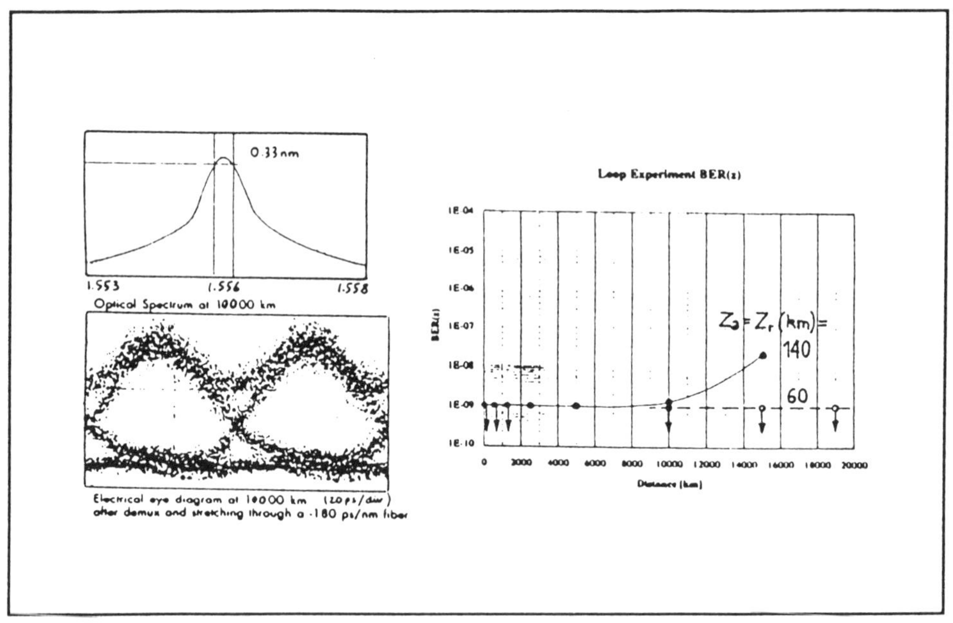

Figure 21. BERs obtained in 20 Gbit/s regenerated soliton transmission with amplifier/regenerator spacings of Za = Zr = 60 km and Za = Zr = 140 km. Optical spectrum and eye diagram after DC at z = 10,000 km are also shown.

since identical performance [32].

An improved experimental recirculating loop configuration was used to study the effect of regeneration period Zr. The corresponding setup is shown

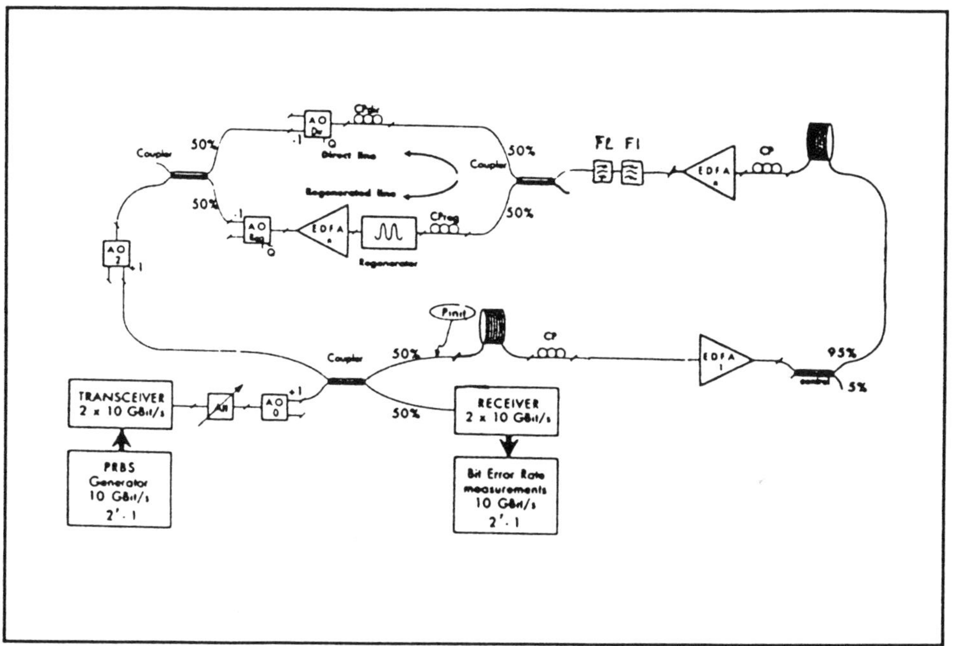

Figure 22.   Improved loop for 20 Gbit/s regenerated soliton transmission with regeneration period control.

in Figure 22. Compared to the previous configuration of Figure 19, the main modification is a controllable bypass circuit which enables signal recirculation in the loop without regeneration. The bypass function is controlled by two AO switches (-1 order) driven by complementary logic signals. A +1 order AO switch is placed at the output in order to null the signal frequency shift experienced in the bypass circuit. Synchronization of the bypass switches requires a very accurate division of the clock synchronizing the loop. Other major difficulties are (1) to balance the loss experienced in either direct or regenerated branches, (2) to obtain the same polarization state at the bypass output, regardless the branch followed by the signal, (3) to optimize this polarization state for minimum BER, and (4) to keep the optimum polarization conditions from fluctuating due to environment changes during BER measurements. The BER results achieved for three different values of Zr (120, 150 and 180 km) in a configuration Za = Zr/2 and with a dispersion D = 0.4 ps/nm/km are shown in Figure 23. The observed BER degradation can be attributed mainly to the effect of amplification period increase; a penalty is also introduced by the 6 dB extra loss of the bypass circuit. Finally, the filtering strength should be ideally optimized for each new regenerator spacing. The BER difference between Zr = 120 km (BER = 0.10^{-10}) and Zr = 150 km (BER = 0.10^{-9}) configurations can be explained by a polarization fading effect *during* BER measurements.

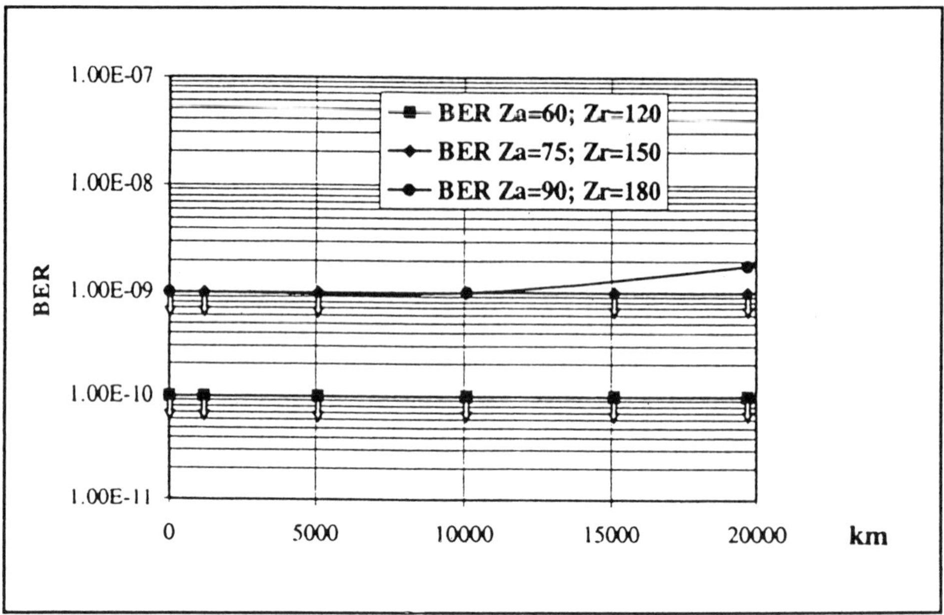

Figure 23.   BER measurements for 20 Gbit/s regenerated soliton transmission showing the effect of regenerator spacing in the configuration Za = Zr/2.

Indeed, measurements of BER $= 0.10^{-9}$ and 0.10^{-10} take 35s and 6 nm, respectively, which is the scale over which unwanted fading occurs. The higher Q margin achieved in the Zr = 120 km configuration actually hides such degradation. A more complete interpretation could be provided by actual Q-factor estimates, which can be obtained by varying the decision time or the decision level, such as those shown in Figure 9. However, such measurements require a high stability of the operating conditions once the optimum transmission regime is reached. These results show the difficulty of parametric experimental investigations in regenerated soliton transmission loops. Numerical simulations, on the other hand, are indicative of best performance in ideal, polarization-independent conditions. The results of a full numerical investigation of 20 Gbit/s regenerated transmission with optimized synchronous intensity/phase modulation will be reported in the near future.

The key component of regenerated soliton systems is the optical modulator and associate intensity/phase modulation characteristics. As discussed in the previous section, soliton regeneration at higher bit rates (40-160 Gbit/s) will require an all-optical implantation, which should include clock recovery (CR). Many approaches for all-optical regeneration have been proposed and demonstrated [17]-[20], but progress is still required to achieve

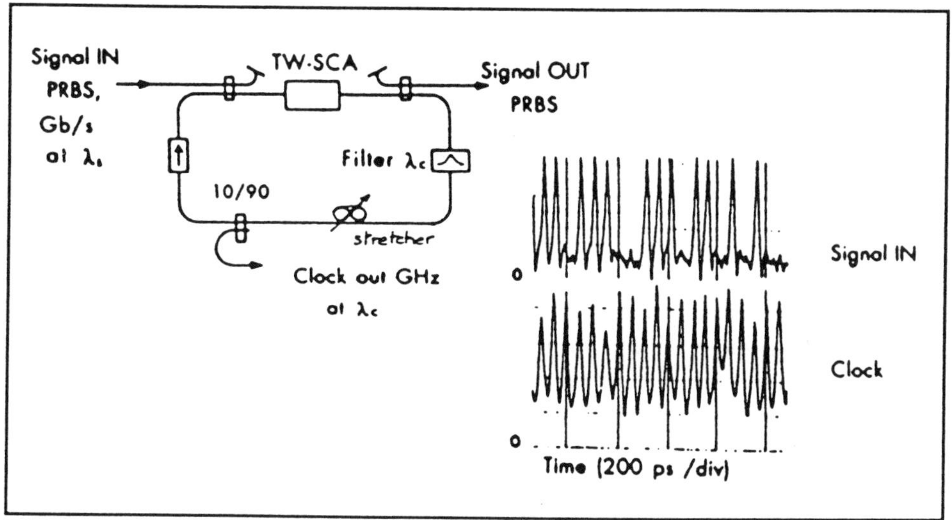

Figure 24.   All-optical clock recovery at 20 Gbit/s through mode-locked, ring fiber/semiconductor laser: experimental apparatus (left), and oscilloscope traces showing input pseudorandom signal sequence along with recovered output clock (right).

at once all the critical features previously mentioned. Concerning CR, one promising approach is based on a mode-locked ring fiber laser [33]. The basic principle is that a pseudo-random data sequence can be used as the control signal in harmonic active mode-locking. The laser output then represents the recovered clock. The occurrence of a limited number of "0s" in the control signal sequence does not affect the clock output, due to the averaging effect of pulse recirculation in the laser cavity. The key component remains the AM/FM modulator that is driven by the coded signal. FM modulation through optical Kerr effect is not desirable due to the long lengths required (10 km) and the resulting cavity instability. A 20 Gbit/s → 20 GHz implementation of this CR principle was demonstrated, using a saturated semiconductor amplifier as the FM modulator [33]. While in this experiment the intracavity gain was provided by an EDFA, we realized a simpler device where both intracavity gain and AM/FM modulation were provided by a polarization-independent semiconductor amplifier, as shown in Figure 24. Apart from simplicity and compactness, the advantage of this last CR configuration is twofold: (1) low control signal requirement (here -20 dBm), due to the amplifying function of the modulator, and (2) shorter cavity lengths (here 10 m), which provides greater laser stability and accuracy in RF frequency control. The recovered clock intensity dependence in control signal polarization was measured to be less than 1 dB, in spite of the fact that the fiber ring was non-polarization maintaining (PM). Ideally, this type of CR device should be implemented with all PM components, including the fiber and the intracavity filter. Highly stable ring cavities of lengths near 1 m can be technically realized. Precise matching the cavity FSR to a multiple of STM-1 communication frequencies can be achieved through accurate fiber splicing. To date, the remaining open question concerns the ultimate speed at which semiconductor-based CR devices could operate. While semiconductor-based, direct 40 Gbit/s $\rightarrow$ 40 GHz CR function remains yet to be demonstrated, solutions exist for prescaled CR at comparatively higher input rates and lower output rates. For instance, 400 Gbit/s $\rightarrow$ 6.33 GHz was recently demonstrated [34], which was based on ultrafast four-wave-mixing in a semiconductor amplifier. Should semiconductor nonlinearity dynamics prevent CR at rates higher than 20 GHz, optical multiplication to 40-80 GHz through passive OTDM is possible. Finally, further progress in ultrafast Kerr fiber devices is required in order to reach even higher clock frequencies.

4. Acknowledgements

The authors gratefully acknowledge J.-P. Hamaide, B. Biotteau, F. Pitel, E. Kolltveit, E. Brun, P. Brindel, S. Bigo and O.Leclerc for the experimental work reviewed in this paper.

References

1. Desurvire, E.: The golden age of optical fiber amplifiers, *Physics Today* **Vol.47, No.1** (1994), 20; id. *Erbium-doped fiber amplifiers: Principles and applications*, J.Wiley, New York, (1994).
2. Bergano, N. S. et al.: 40 Gbit/s WDM transmission of eight 5 Gbit/s data channels over transoceanic distances using the conventional NRZ modulation format, *Proc. OFC'95* **PD19-1**; id. 100 Gbit/s WDM transmission of twenty 5 Gbit/s NRZ data channels over transoceanic distances unsing gain-flattened amplifier chain, *Proc. ECOC'95* PD-paper **ThA3.1**
3. Amano, K. and Iwamoto, Y.: Optical fiber submarine cable systems, *IEEE J. Lightwave Technology* **Vol.8, No.4** (1990), 595.
4. Hasegawa, A. and Tappert, F.: Transmission of stationary nonlinear optical pulses in dispersive dielectric fibers. I. Anomalous dispersion, *Appl. Phys. Lett.* **Vol.23, No.3** (1973), 142.
5. Mollenauer, L. F. and Smith, K.: Demonstration of soliton transmission over more than 4,000 km in fiber compensated by Raman gain, *Optics Lett.* **Vol.13, No.8** (1988), 675; see also L. F.Mollenauser, L. F. et al.: Experimental study of soliton transmission over more than 10,000 km in dispersion-shifted fiber, *Optics Lett.* **Vol.15, No.1** (1990), 1203.
6. Mollenauer, L. F. et al.: The sliding frequency-guiding filter: an improved form of soliton jitter control, *Optics Lett.* **Vol.17, No.22** (1992), 1575.
7. Nakazawa, M. et al.: 10 Gbit/s soliton data transmission over one million kilometers, *Electron. Lett.* **Vol.27, No.14** (1991), 1270.

8. Favre, F. et al.: Robustness of 20 Gbit/s 63 km span 6 Mm sliding-filter controlled soliton transmission, *Electron. Lett.* **Vol.31, No.18** (1995), 1600.

9. Hamaide, J.-P. et al.: Experimental 10 Gbit/s sliding-filter guided soliton transmission up to 19 Mm with 63 km amplifier spacing using large effective-area fiber management, *Proc. ECOC'95* PD-paper **ThA3.7** (1995).

10. Suzuki, K. and Nakazawa, M.: Recent progress in optical soliton communications, *Optical Fiber Technology* **Vol.1, No.4** (1995), 289.

11. Aubin, G. et al.: 20 Gbit/s soliton transmission over transoceanic distances with a 105 km amplifier span, *Electron. Lett.* **Vol.31, No.13** (1995), 1079; see also Aubin, G. et al.: Electroabsorption modulator for a 20 Gbit/s soliton transmission experiment over 1 million km with a 140 km amplifier span, *Proc.IOOC'95* **PD2-5** (1995), 29.

12. Figure based from an original diagram kindly communicated by Suyama, M. et al., Fujitsu Limited, (1995).

13. Mollenauer, L. F. et al.: WDM with solitons in ultra-long distance transmission using lumped amplifiers, *IEEE J.Lightwave Technol.* **Vol.9, No.3** (1991), 362.

14. Uchiyama, K. et al.: Ultrafast polarisation-independent all-optical switchinh using a polarisation-diversity scheme in the nonlinear optical loop mirror, *Electron. Lett.* **Vol.28, No.20** (1992), 1864.

15. Clesca, B. et al.: 1.5 μm fluoride-based fiber amplifiers for wideband multichannel transport networks, *Optical Fiber Technology* **Vol.1, No.2** (1995), 135.

16. Kolltveit, E. et al.: Soliton frequency-guiding in a 2x5 Gbit/s WDM system using a UV-written fibre Fabry-Perot filter, *Proc.ECOC'95* **MoA3.6**.

17. Jinno, M. and Abe, M.: All-optical regenerator based on nonlinear fibre Sagnac interferometer, *Electron. Lett.* **Vol.28, No.14** (1992), 1350.

18. Widdowson, T. et al.: Soliton shepherding: all-optical active soliton control over global distances, *Electron. Lett.* **Vol.30, No.12** (1994), 990.

19. Lucek, J. K. and Smith, K.: All-optical signal regenerator, *Optics Lett.* **Vol.18, No.15** (1993), 1226; see also W. A. Pender, W. A. et al.: 10 Gbit/s all-optical regenerator, *Electron. Lett.* **Vol.31, No.18** (1995), 1587.

20. Bigo, S. and Desurvire, E.: 20 GHz all-optical clock recovery based on fibre laser mode-locking with fibre nonlinear loop mirror as variable intensity/phase modulator, *Electron. Lett.* **Vol.31, No.21** (1995), 1855; see also Bigo, S. et al.: Analysis of soliton in-line regeneration through two-wavelength nonlinear loop mirror as synchronous amplitude/phase modulator, *Electron. Lett.* **Vol.31, No.25** (1995).

21. Takada, A. et al.: Picosecond optical pulse compression from gain-switched 1.3 μm distributed-feedback laser diode through highly dispersive single-mode fibre, *Electron. Lett.* **Vol.21, No.21** (1985), 969.

22. Lourtioz, J.-M. et al.: Fourier-transform-limited pulses from gain-switched distributed-Bragg-reflector lasers using simultaneous modulation of gain and phase sections, *Electron. Lett.* **Vol.28, No.16** (1992), 1499.

23. Kataoka, T. et al.: 20 Gbit/s transmission experiments using an integrated MQW modulator/DFB laser module, *Electron. Lett.* **Vol.30, No.11** (1994), 872.

24. Mollenauer, L. F. et al.: Demonstration of error-free soliton transmission over more than 15,000 km at 5 Gbit/s, single-channel, and over more than 11,000 km at 10 Gbit/s in two-channel WDM, *Electron. Lett.* **Vol.28, No.8** (1992), 792.

25. Golovchenko, E. A. et al.: Soliton propagation with up- and down-sliding frequency guiding filters, *Optics Lett.* **Vol.20, No.6** (1995), 539.

26. Mollenauer, L. F. et al.: Demonstration, using sliding frequency-guiding filters, of error-free soliton transmission over more than 20 Mm at 10 Gbit/s, single-channel, and over more than 13 Mm at 20 Gbit/s in two-channel WDM, *Electron. Lett.* **Vol.29, No.10** (1993), 910.

27. Hamaide, J.-P. et al.: to be submitted to *Electron. Lett.*.

28. Hamaide, J.-P. et al.: Transoceanic optical communications with optical amplification, in *Proc. European Fiber and Optical Communications Conferences (EFOC),*

Paris, France, **54** (1992); see also O. Audouin and Hamaide, J.-P.: Enhancement of amplifier spacing in long-haul optical links through the use of large-effective-area transmission fiber, to be published in *IEEE Photonics Technology Letters*.

29. Gordon, J. P. and Mollenauer, L. F.: Effects of fiber nonlinearities and amplifier spacing on ultra-long distance transmission. *IEEE J. Lightwave Technology* **Vol.9, No.2** (1991), 170.

30. Biotteau, B. et al.: Enhancement of soliton system performance by use of new large effective area fibres, *Electron. Lett.* **Vol.31, No.23** (1995), 2026.

31. King, J. P. et al.: Polarisation-independent 20 Gbit/s soliton transmission over 12,500 km using amplitude and phase modulation soliton transmission control, *Electron. Lett.* **Vol.31, No.13** (1995), 1090.

32. Brun-Maunand, E.: Chromatic dispersion influence in 20 Gbit/s regenerated transoceanic soliton systems having up to 140 km amplifier spacing, submitted to *Electron. Lett.*.

33. Patrick, D. M. and Manning, R. J.: 20 Gbit/s all-optical clock recovery using semiconductor nonlinearity, *Electron. Lett.* **Vol.30, No.2** (1994), 151.

34. Kamatani, O. and Kawanishi, S.: Ultrahigh speed characteristics of a phase-locked loop based on four-wave mixing in a laser-diode amplifier, *Proc.IOOC'95* **WC1-4** (1995), 74.

SOLITON FAMILY AND SOLITONICS

YOICHI FUJII

Institute of Industrial Science, University of Tokyo
22-1 Roppongi-7, Minatoku, Tokyo 106, Japan

Abstract. Recect theoretical developments of the fiber-optic soliton are reviewed. New family members of the solitons are introduced, such as the coupled solitons, the paired solitons and the supersoliton. The practical application of the solitons are also reviewed. A new technology of the "solitonics" is expected.

1. Introduction

The soliton has given a serious impact on the optoelectronics. Historically, the concept of the soliton was first introduced by a scotch, Russel, for the solitary water wave about two hundred years ago. This concept includes essentially the adoption of the nonlinearity of the medium. From this reason, it was forgotten for long years in the nineteenth and the twentieth century, because these periods were the tremendous development of the "linear" science and technology, for example, the linear electromagnetism, the linear mechanics, etc. This situation did not change after the mathematical refining of the soliton by Korteweg and De Vlies.

This concept of the soliton is very recently recollected considering its application to the plasma. In this occasion, the terminology "soliton" is given. This trend gave a strong impetus to Prof. Hasegawa, who had sought an application of the soliton concept to the fiber-optics. He has introduced a nonlinear Schrödinger-type wave equation, so-called Hasegawa equation.

This fiber-optic soliton assumes the nonlinear effect of the optical fiber. This effect is called as the optical Kerr effect by which the permittivity of the fiber and thus the optical power concentrates into its pulse peak. This pulse compression effect is balanced at a certain waveform so-called fiber-optic soliton to the pulse-broadening by the dispersion.

237

A. Hasegawa (ed.), Physics and Applications of Optical Solitons in Fibres '95, 237–247.
© 1996 *Kluwer Academic Publishers. Printed in the Netherlands.*

The soliton is considered to be very promising because its waveform is unchanged while its propagation. Usually, this **FUNDAMENTAL SOLITON** is called as the soliton and it is used for the very long-distance optical communication. The technology which handles the soliton is now going to be well developed. [1],[2]

Besides, the other solitons can exist. For example, the DARK SOLITON is a propagation of the dark portion in the continuous bright light in the positive dispersive medium. The HIGHER-ORDER SOLITON which has more complex waveform but higher amplitudes also exists.

2. Soliton Dynamics and Edge Drift

The property of the nonlinear transmission in the fiber can be easily understood by the dynamics of the pulse edge.

The optical Kerr effect:

$$n = n_0 + n_2 E^2 \tag{1}$$

gives the index increase proportional to the light intensity. Thus the phase-shift after the transmission is given:

$$\phi = n\beta_0 z = n_0\beta_0 z + n_2 E^2 \beta_0 z. \tag{2}$$

The frequency of the transmitted light is given by the differentiation of the phase shift:

$$\frac{d\phi}{dt} = n_2 \beta_0 z \frac{dE^2}{dt}. \tag{3}$$

It is found that the frequency shift appears only when the pulse intensity changes to the time. It means the frequency chirps only at the pulse edge, not at the pulse flat top. When this medium is dispersive, the drifting velocity of the edge becomes different from the flat top of the pulse. The relative velocity difference of the pulse edge will be determined from the signs of the dispersion and of the slope of the edge.

As the result, the pulse broadens or sharpens according to the sign of the dispersion.

The sharpening of the pulse balances with the broadening due to the dispersion. This is the soliton condition.

This edge dynamics can be applied to the modulation instability, dark soliton phenomenon etc., because this concept can describe the local nature of the nonlinear wave propagation, while Hasegawa's equation describes the property of whole nonlinear pulse.[3]

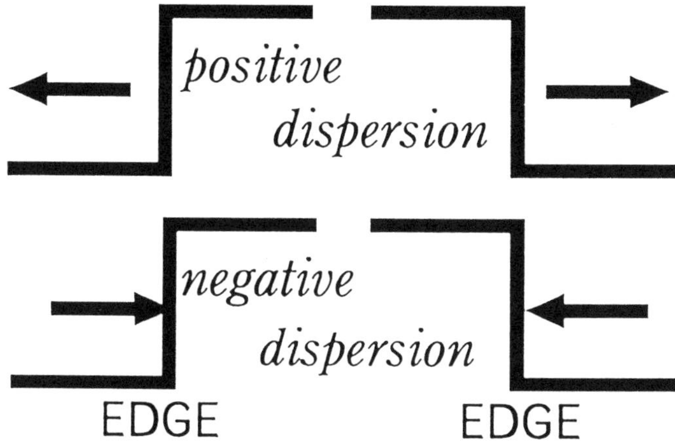

Figure 1.   Edge drifting in nonlinear fiber.

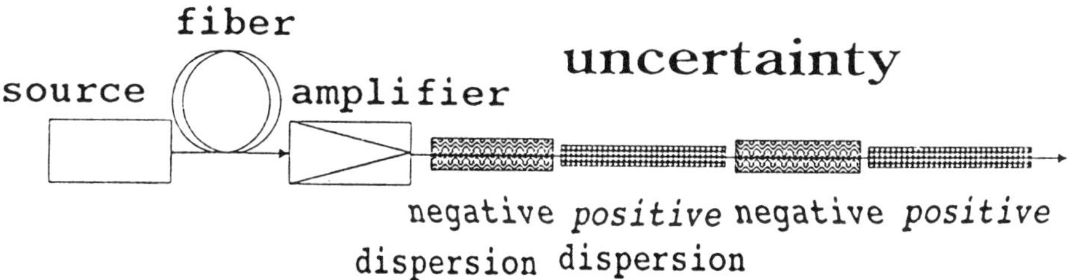

Figure 2.   Pulse separation and limitation due to finiteness of interacting distance.

3. Limitation of Soliton Interaction

The edge dynamics can be applied to the reformation of the optical pulse shape. A small gap which exists on the pulse can be reinforced and exagerated. By this way, the original pulses can be regenerated. This is a pulse separation technique. [4],[5]

In this process, the finite length of the interation is required for the fiber with negative dispersion. This finiteness of the interaction length limits the compensation due to the uncertainty in the spatial coordinate.

$$\Delta z \Delta k > \pi \tag{4}$$

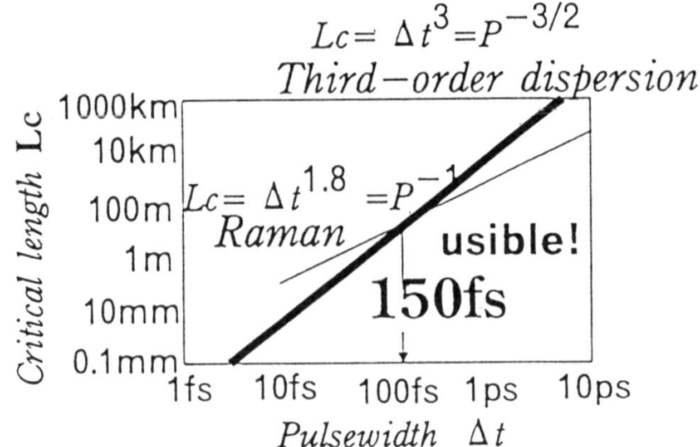

Limitation of soliton propagation
due to Raman effect and the third−order dispersion

Figure 3.   150 fs Raman 3rd-order dispersion

The nonlinear pulse compression increases by the length of the fiber but
the limitation due to the spatial uncertainty becomes significant. So there
is an optimum solution.

Another limitations to the soliton interaction are Raman effect and the
third-order dispersion. Raman effect makes the interaction dissipative and
thus the soliton condition is not held in the ultra-short pulse less than 150 fs.

4. Dynamic and WDM Soliton

Recently, various kinds of solitons are now developed. One is the **DY-
NAMIC SOLITON.** This is the soliton propagation in the practical lossy
fiber combined with the fiber-optic amplifier. The transmission length can
be optimized by introducing slightly higher-order solitons to compensate
thepulse broadening due to the attenuation in the fiber. The other possibility is the **WAVELENGTH-DIVISION-MULTIPLEX (WDM)
SOLITON.** The solitons with different wavelengths can be transmitted in
one single-mode optical fiber. This scheme, the **WDM** solitons, can extend
the transmission capacity of the fiber and also can be applied to the wavelength switching system but it has several problems to be solved.

One of such problems is the collision of the non-fundamental solitons of
the same direction. The solitons with different wavelengths have the different group velocity due to the dispersion of the fiber. Thus these solitons may

suffer collisions between the transmitting solitons even with same group velocity. In such collisions, the solitons are interacting with each other and as the result, one soliton may take off a part of the other soliton. Such interaction, or, a kidnapping of the soliton pulse, may cause the crosstalk between the transmission channels.[6]

5. Supersoliton

The new and strange member in this "soliton family" is the SUPERSOLITON. The Hasegawa equation, conventionally used in the analysis of the solitons, is essentially under an approximation for the slowly-varying envelope of the carrier sinusoidal wave. In this approximation, the dispersion is assumed to be linearized to the frequency (wavelength). So only a small dispersion for narrow bandwidth in the fiber can be considered. Instead, the generalized nonlinear wave equation is expressed by a generalized dispersion coefficient and the conventional nonlinear term.

$$(\Box + \alpha + \beta(d/dt)^2|E|^2)E = 0 \tag{5}$$

The eigen solution of this equation is given by a very narrow (about 14 fs) pulse with the fixed amplitude. This eigen solution is named as a "supersoliton". [7]

6. Switching and Logic Soliton

The fiber-optic coupler which has a nonlinear wave propagation but a linear coupling coefficient is now very much interested and many applications are proposed. Usually two fibers in this coupler are assumed to be symmetric.

The asymmetric nonlinear fiber coupler, the coupling coefficient of which between two lines is linear, has uneven nonllinear coefficients of each line.

The behavior of this coupler can be analyzed by using the coupled Hasegawa equation.

Asymmetric Coupled Normalized Hasegawa equation:

$$i(d/dz) + 1/2(d/dt)^2 + |u|^2 + i\Gamma u + \kappa v = 0 \tag{6}$$

$$i(d/dz) + 1/2(d/dt)^2 + m|v|^2 + i\Gamma v + \kappa u = 0 \tag{7}$$

where m is the coefficient of asymmetry and κ is the coupling coefficient between u and v modes. In this formulation, the linear loss Γ is included.

The asymmetric coupler has a property that the coupled light power is easily moving into the fiber with narrower core or with higher nonlinearity. This is utilized for the self-power switch.

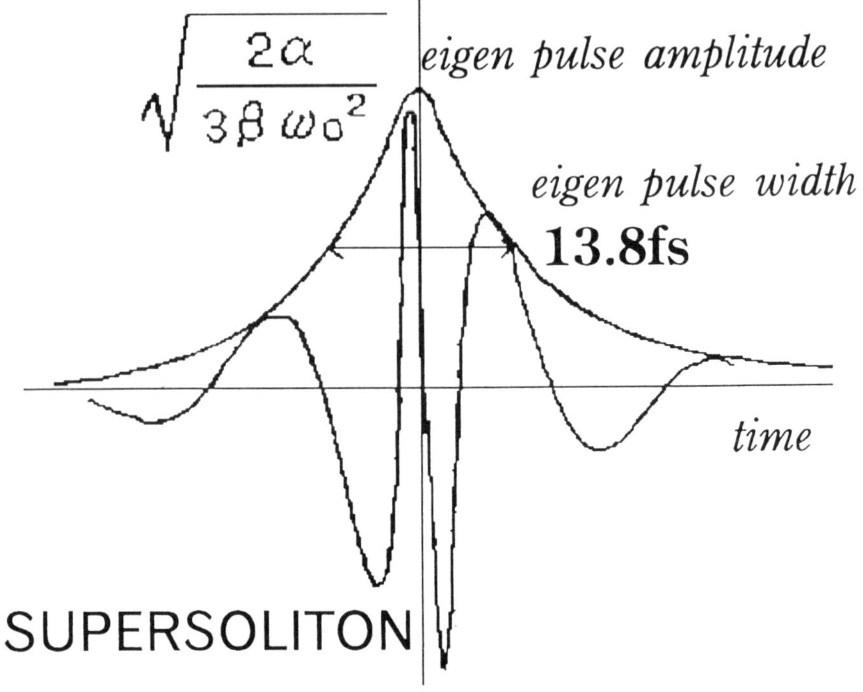

Figure 4.   Eigen waveform of supersoliton

Recently, a new device which can act as a logic element is devised. This coupler is designed to meet the certain conditions, gives two outputs; an AND-output and an OR-output. This may be called as "LOGIC SOLI-TON".

7. Coupled Soliton

The coupled Hasegawa equation for the nonlinear fiber-optic coupler gives, in some special cases, a special solution in which the soliton waveforms in each lines are transmitted unchanged or periodically changed.

In such cases, a new concept of the "COUPLED SOLITON" may be introduced.

In the coupled soliton, the unchanged transmission may be the fundamental coupled soliton and the periodically changed soliton is the higher-order coupled soliton. [8]

In the symmetrical two-core fibers, the fundamental coupled solitons are obtained for the in- and out-of-phase input solitons to each fiber. But in the asymmetrical two-core fiber coupler, it appears only in the case of

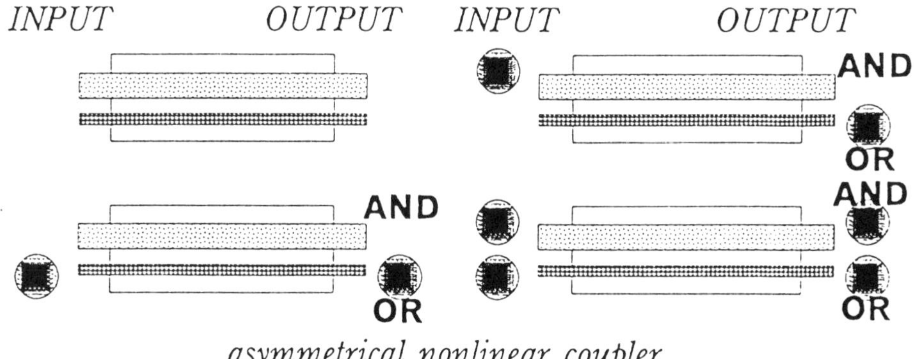

Figure 5. The logic soliton or the nonlinear coupler for the logic element.

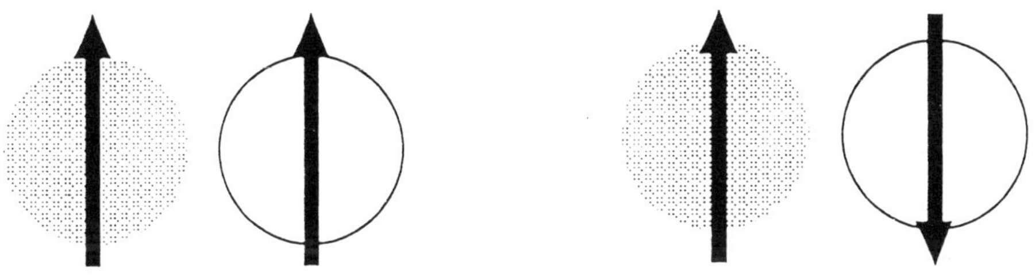

Figure 6. The coupled soliton in two-core fibers. The in- and out-of-phase cases.

the out-of-phase soliton inputs.

In the three-core fiber coupler, the same discussion will be held. The state of the coupled solitons is realized mainly for the cases in which the out-of-phase inputs. In the special case, a three-phase solitons can be obtained. This is a set of fibers triangularly placed can have a state of the coupled soliton when each phase of lines is of 120 degrees.

In this discussion, the spatial distribution of the soliton is not considered. The effect of spatial distribution is included in the concept of the eigen mode for the soliton transmission. In the nonlinear fiber coupler, the effect of the field distribution is included in the linear coupling coefficient. So the field distribution can be included in the coupled Hasegawa equation when the coupling is weak and the transmission has eigenmodes.

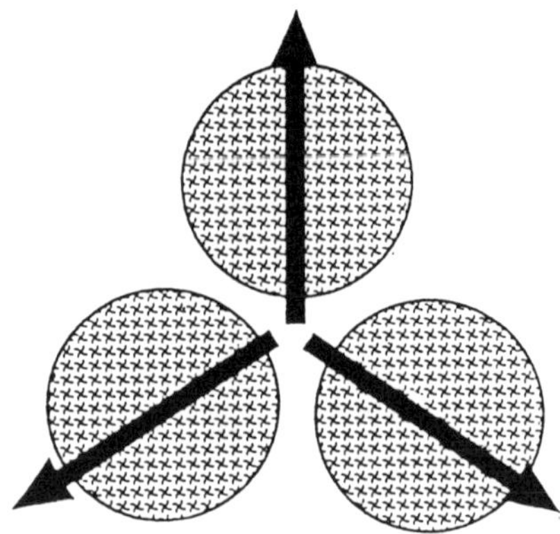

Figure 7. The coupled soliton in three-core fibers. The three-phase soliton.

The coupling power flow between the coupled lines is expressed by the crossterm between the fields from both lines. In the case of out-of-phase, the amplitude of the field should be zero at a certain boundary surface in between. So the power coupling does not exist beyond this boundary surface. In such case, if the soliton could be propagate, it should be a pair of the coupled soliton.

8. Paired Soliton and Phase-Conjugation

Another new member, such as the **PAIRED SOLITON**, may be included in the soliton family.

In the paired soliton system, two nonlinear lines transmit two solitons. These solitons travel for a certain distance in the fiber transmission system. As a result, these solitons suffer a certain timing jitter and the amplitude noise due to the Gordon-Haus jitter and the noise due to the spontaneous emission from the amplifying fiber. After traveling and being suffered by these noises are coupled at a coupler. After the coupling, two solitons with timing jitter etc. fuses into one soliton due to the nonlinear self-focusing effect. Thus it can reduce random timing jitter. This self-correcting effect of the random timing jitter is limited within a certain extent of timing difference.

This effect is also only for the two solitons in phase. When two solitons are out of phase, the coupled output disappear and the in-phase output appears from the other port.

This system can be considered as a diversity transmission system.

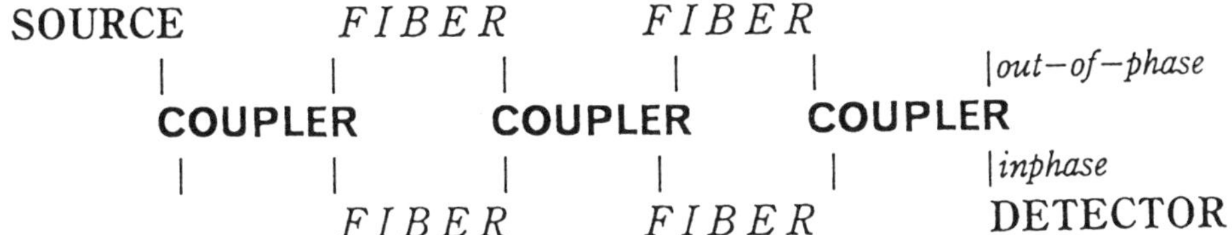

Figure 8.   Paired transmission

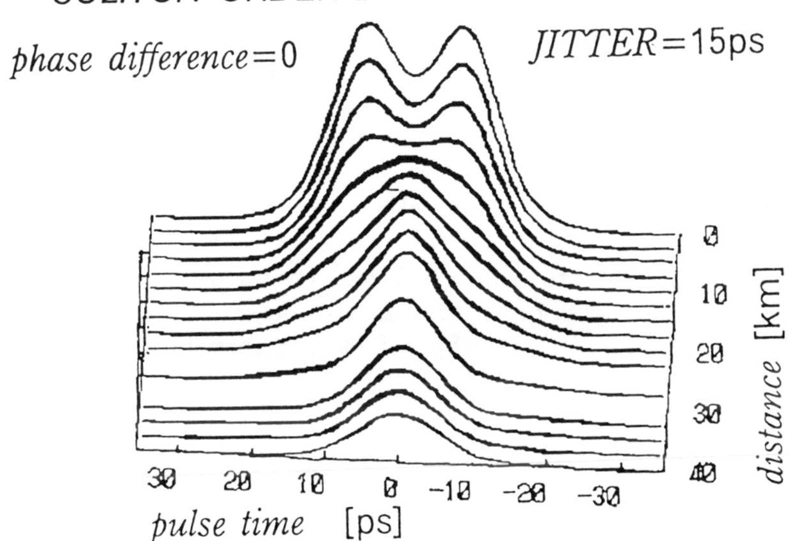

Figure 9.   Cancellation of random timing jitter.

Another example is a solitonic PHASE-CONJUGATION. The phase-conjugation invert the original spectrum of the input soliton. So, the soliton is transmitted through two channels with different dispersion. Thus it can reduce the random jitter and timing jitter due to the self-phase modulation.

SOURCE −*fiber*−> **AMP** −*fiber*−>...........**AMP** −*fiber*−> **AMP** −*fiber*−>

PHASE_CONJUGATOR −*fiber*−> **AMP** −*fiber*−>...............

AMP −*fiber*−> **AMP** −*fiber*−>...................*DETECTOR*

Figure 10.   Solitonic phase-conjugation system.

This system can improve the repeater distance.[9]

9. Solitonic OTDR

The application of soliton to the fiber-optic sensor system will be a solitonic OTDR system. The utilization of the soliton to the OTDR system will be beneficient because the solitons have high peak power and narrow bandwidth thus this system has a possibility to improve the spatial resolution and the sensitivity. In this system, it should be considered that the fact that the transmitting pulse is a soliton but the backscattered light is a conventional nonsoliton pulse. When a higher-order soliton is incident, the backscattered pulse has still frequency chirping due to the nonlinear propagation in the fiber. Thus a certain amount of the pulse narrowing may be realized even in the linear scattered pulse.

In the solitonic OTDR system, not only one-pulse measurement, but also a burst of the soliton can be utilized. The incident of burst of solitons, especially of the higher-order solitons, will increase the input power without deforming the soliton waveform. So it may contribute the sensitivity of the solitonic OTDR system.[10],[11]

10. Conclusion - Evolution of Solitonics

Now the technology of the solitons is the very fundamental element in the field of the optical technology. So now a new concept of the "SOLITONICS" should be introduced to cover all the technology of soliton. The solitonics corresponds to soliton as the electronics to electron or the photonics to photon.

The solitonic technology will contribute not only for the communication, but also for the sensing, the computing, information processing, machining, etc. The solitonics will become an indispensable element of the optoelectronics.

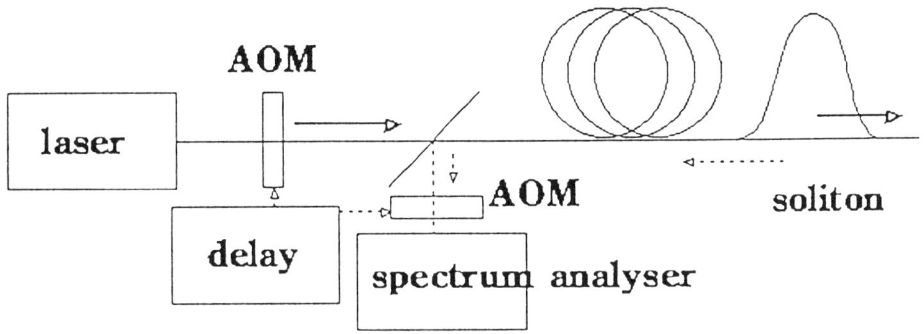

Figure 11.   Solitonic OTDR system

References

1. Agrawal, G. P.: *Nonlinear Fiber Optics*, (Academic Press, 1995).
2. Stolen, R. H., Mollenauer, L. F. and Tomlinson, W. J.: *Opt.Lett.* **8** (1983), 186; Mollenauer, L. F., Stolen, R. H., Gordon, J. P. and Tomlinson, W. J.: *Opt. Lett.* **8** (1983), 289-291.
3. Shiojiri, E. and Fujii, Y.: Transmission capability of an optical fiber communication system using index nonlinearity, *Applied Optics* **Vol.24, No.3** (1995), 358-360.
4. Fujii, Y.: Information-maintaining separation of optical pulses employing nonlinearity of silica fiber, *Applied Optics* **Vol.29, No.6** (1990), 864-869.
5. Fujii, Y. and Shimosaka, N.: Compression and Separation of Optical Pulses by Using Nonlinearity of Optical Fiber and Gires-Tournois Interferometer, *The transactions of the IECE of Japan* **Vol.E.69, No.4** (1986), 420-422.
6. Shimizu, K. and Fujii, Y.: Nonlinear Interactions of Soliton and the Maximum Transmission Capacity, *J. IECEJ.* **vol.J74-C1, No.11** (1991), 440-448, in Japanese.
7. Osawa, Y. and Fujii, Y.: Stationary Solutions of Nonlinear Propagation of Femtsecond Optical Pulse by Cuicuit Model, *Journal of the Physical Society of Japan* **Vol. 61. No.11** (1992), 3977-3983.
8. Sotobayashi, S. and Fujii, Y.: Soliton switching and coupled soliton transmission by three-core nonlinear fiber coupler, *J.IECEJ.* **vol.J78-C1, No.3**; (1995), 157-165, in Japanese.
9. Yoshioka, H. and Fujii, Y.: Soliton Transmission Using Optical Phase-Conjugators-Compensation for Timing Jitter due to Soliton-Soliton Interaction and SSFS-, *IOOC'95* (1995), Hongkong.
10. Levanon, A., Friberg, S. R. and Fujii, Y.: Solitons for High Resolution Optical Time Domain Reflectometry, submitted to *J. Opt. Soc. Am. B.*
11. Levanon, A. and Fujii, Y.: Solitons for Optical Time Domain Reflectometry, *IOOC'95* (1995), Hongkong.

OPTICAL CAVITONS

STEFAN WABNITZ

Fondazione Ugo Bordoni
Via B. Castiglione 59,00142 Rome, Italy

Abstract. We present an analysis of soliton creation in a frequency-shifted feedback fiber ring laser with an injected continuous wave light acting as a seed. We compare perturbative estimates and numerical results for the temporal and spectral characteristics of the generated soliton trains. The birth of fiber optic solitons is analogous to the formation of plasma density cavitons through the resonance absorption mechanism.

1. Introduction

The generation of a stable pulsed output from continuously frequency-shifted feedback (FSF) and bandwidth limited lasers was discussed and demonstrated by several Authors [1]-[5]. In these lasers, frequency shifting is achieved by means of an acousto-optic modulator (AOM).CW and pulsed operation of fiber-based FSF lasers was recently demonstrated using Nd^{3+} [6, 7] or $Er^{3+}-$doped [8] fibers. In particular, in References [6, 7] the role of spectral broadening by self-phase modulation (SPM) in the fiber was pointed out as a mechanism for pulse shortening. In the presence of anomalous group-velocity dispersion (GVD), a FSF laser is a self-starting soliton generator [9]-[13]. Whereas in the case of normal average GVD short chirped pulses are obtained [14]. The principle of operation of FSF soliton lasers is analogous to the sliding-frequency filter method of Mollenauer *et al.* [15, 16] for the stabilization of soliton transmission systems. Indeed, the use of AOM frequency-shifting elements combined with fixed-frequency filters for the control of soliton transmission, as proposed in Reference [9], was recently demonstrated in [17]- [19].

In this work I study the dynamics of a FSF soliton laser that is seeded by a weak cw signal. As we shall see, an interesting analogy exists between

249

A. Hasegawa (ed.), Physics and Applications of Optical Solitons in Fibres '95, 249–261.

the generation of optical solitons in a cw-seeded FSF laser, and the resonant absorption of a laser beam in a plasma with a linear transverse variation of its density [23]-[25]. Here, resonant absorption of the laser energy occurs at that position where the plasma density has a certain critical value, which corresponds to phase-matching between the laser and the electrostatic field in the plasma. The absorbed energy leads to a growing spatially localized (about the critical density position) electrostatic field pulse. The pulse energy grows larger until the combined action of plasma nonlinearity, diffraction, and the plasma density gradient displace this pulse away from the critical position so that another pulse may be formed. These pulses represent plasma density holes and are known as *cavitons*. In the bandwidth-limited FSF laser case, I will show that the process of caviton build-up may be described by the simple linear superposition of amplified frequency-shifted waves. Moreover, after reaching a certain critical amplitude or time width, the nonlinear dynamics of optical cavitons and the temporal characteristics of the FSF laser emission may be predicted by means of a reduced system of perturbative nonlinear differential equations for the pulse parameters.

2. Self-Starting

The build-up of a periodic pulse train in the caviton laser as depicted in Figure 1 may be simply understood in terms of a linear superposition of the circulating waves [2]. The cw seed $E = A_0 \exp\{-i\omega_0 t\}$ is injected into the ring through the zeroth-order diffraction peak of the AOM, with amplitude transmission and reflection coefficients τ and ϱ, respectively. On each transit, the AOM shifts the circulating light by $+\Delta\omega = 2\pi f$. The ring of length $L = L_a + L_n$ is composed of a bandpass filter (BF), an erbium-doped fiber amplifier (EDFA) of length L_a, and a fiber of length L_n.

Let $E_m(z,t) = A_m(z,t)\exp\{-i\omega_0 t\}$ be the field after m ($m \geq 1$) circulations through the cavity. Neglecting at first nonlinearity and dispersion, the envelope $A_m(z = 0, t)$ obeys

$$A_m(z = 0, t) = \tau A_0 + \varrho' e^{i\phi_m - i\Delta\omega(t+t_s)} A_{m-1}(z = 0, t) \qquad (1)$$

where $z = 0$ is at the AOM, t_s is an arbitrary time, $\phi_m = k_0 L + (m - 1)k_0'\Delta\omega L \equiv \phi_0 + \bar{\phi}_m$ is the linear cavity detuning, $\varrho' = \varrho\sqrt{G_0}\exp\{-\alpha L_n\} > 1$, α is the fiber loss coefficient and $k_0' = \partial k_0/\partial\omega|_{(\omega-\omega_0)}$. We assumed that the filter does not influence the initial spectral development of the field around ω_0. Whenever the resonance condition

$$k_0'\Delta\omega L \equiv \omega t_r = 2N\pi \qquad (2)$$

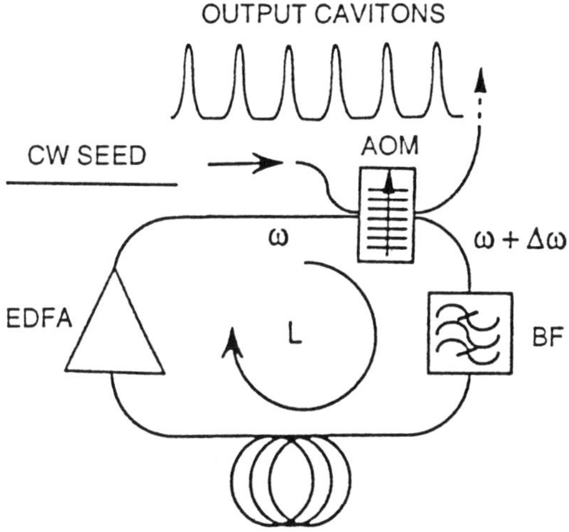

Figure 1. Schematic of caviton laser.

(with N integer) is verified, one obtains that $\phi_m \equiv \phi_0(mod(2\pi))$, and Equation (1) yields

$$A_m(z = 0, t) = \tau A_0(1 + \varrho' e^{-i\Delta\omega t} + \varrho'^2 e^{-i2\Delta\omega t} + ... + \varrho'^m e^{-im\Delta\omega t}) =$$
$$\tau A_0 \frac{\varrho'^{(m+1)} e^{-i(m+1)\Delta\omega t} - 1}{\varrho' e^{-i\Delta\omega t} - 1} \quad (3)$$

where we have set $t_s = k_0 L/\Delta\omega$. Note that for a passive cavity (i.e., with $G_0 = 1$) Equation (3) converges to the Airy function for the intensity [2]

$$|A(t)|^2 = \frac{\tau^2 |A_0|^2}{(1 - \varrho')^2 + 4\varrho' \sin^2(\Delta\omega t/2)} \quad (4)$$

With an active cavity, the number of circulations, say, M, that are necessary for building-up a unit amplitude pulse read

$$M = \frac{ln[1 + (\rho' - 1)/|\tau A_0|]}{ln\rho'} - 1. \quad (5)$$

After M round trips, nonlinear and dispersive effects should be included in the analysis.

3. Nonlinear Dynamics

The nonlinear dynamics of the cavity field $A_m(z,t)$ is described by the nonlinear Schrödinger (NLS) equation

$$\frac{\partial A_m}{\partial z} + k_0' \frac{\partial A_m}{\partial t} + \frac{ik_0''}{2}\frac{\partial^2 A_m}{\partial t^2} = i\gamma |A_m|^2 A_m \tag{6}$$

where $\gamma = \omega_0 n_2/cA_{eff}$, n_2 is the nonlinear refractive index and A_{eff} is the effective core area. In order to understand the role of nonlinearity and dispersion on the dynamics of pulse generation in the cw-pumped FSF laser, one may first rewrite Equations (1,6) in terms of dimensionless coordinates and variables, and then average the map(1,6) over a circulation through the ring [21]. This leads to the perturbed NLS equation for the cavity mean field, say, v,

$$i\frac{\partial v}{\partial Z} + \frac{1}{2}\frac{\partial^2 v}{\partial T^2} + |v|^2 v = iR \equiv$$
$$(\alpha_0(T + T_s) + i\delta)v + iV_f\frac{\partial v}{\partial T} + i\sigma\frac{\partial^2 v}{\partial T^2} - iSe^{i\Lambda Z}. \tag{7}$$

where $\alpha_0 = \Delta\omega'/Z_l, \Lambda = \phi_0/Z_l$ is the linear cavity detuning, $\delta = ln(\varrho')/Z_l > 0$ is an excess gain, and S represents the injected cw beam. We also defined the filter-induced inverse group velocity shift $V_f = -2/(Bt_0 Z_l)$, and $\sigma = -g''/2|k_0''| \simeq 2/(|k_0''|B^2 Z_l)$, where B is the filter bandwidth in the Lorentzian approximation.

Without the filter, we may set $\delta = \sigma = V_f = 0$, $\alpha_0 T_s = -\Lambda$, and define $Q = v\alpha_0^{2/3}2^{-1/3}/S$, $T' = (2\alpha_0)^{1/3}T$, $Z' = \alpha_0^{2/3}2^{-1/3}Z$. Then Equation (7) may be rewritten as

$$iQ_Z + Q_{TT} + (P|Q|^2 Q - T)Q = 1, \tag{8}$$

where for simplicity we dropped the primes, and $P \equiv 2S^2/\alpha_0^2$. Equation (8)is well-known to model the onset of Langmuir-wave turbulence in a plasma [23, 24, 25]. Whenever the proper balance exists between the plasma and ponderomotive pressure, one obtains Equation (8) from the Zakharov equations describing the transfer of energy from an electromagnetic laser field into the plasma in the form of an electrostatic field of amplitude q. The energy coupling is phase-matched through the inhomogeneous plasma density. The forcing term in Equation (8) describes the external injection of energy from the laser beam, whereas the time-dependent phase-shift in Equation (8) is associated with the linear density variation across the plasma: resonant absorption of the pump laser occurs at the critical density, where $T = T_{cr} = 0$. At this point, energy is coupled into the electrostatic field in the form of localized density cavities or *cavitons*. Right after their formation, the cavitons propagate down the linear potential that is created by the density gradient. Therefore, the process of energy coupling from the cw field into short optical pulses in the FSF laser is the optical analog of the caviton-mediated resonance absorption mechanism in inhomogeneous plasmas.

No analytical solution of Equation (8)is available. Nevertheless, numerical simulations show that, if $P \geq P_{cr} \simeq 0.6$, cavitons are periodically generated near T_{cr}, and then accelerate away from T_{cr}. For $P \gg P_{cr}$, spatio-temporal chaotic behavior is observed in the energy absorption by the plasma [23, 24]. Caviton dynamics in the presence of frequency-shifting alone is well described by the NLS soliton ansatz [22]

$$v(T, Z) = \eta \mathrm{sech}[\eta(T - \xi(Z))]\exp[-i\kappa(Z)T + i\theta(Z)] \qquad (9)$$

where $\xi = -(\kappa_0 + V_f)Z - \alpha_0 Z^2/2$, $\kappa = -d\xi/dZ$, and $d\theta/dZ = (\eta^2 - \kappa^2)/2$. For an active cavity with cw pumping and filtering, one obtains from first order soliton perturbation theory [26] the adiabatic variation with Z of the soliton parameters

$$\frac{d\eta}{dZ} = 2\delta\eta - 2\sigma\eta(\eta^2/3 + \kappa^2) + S\pi\mathrm{sech}(\beta)\sin(\chi),$$

$$\frac{d\kappa}{dZ} = \alpha_0 - 4\sigma\eta^2\kappa/3 - \frac{\pi\kappa S}{\eta}\mathrm{sech}(\beta)\sin(\chi),$$

$$\frac{d\xi}{dZ} = -\kappa - V_f + \frac{\pi^2 S}{2\eta^2}\mathrm{sech}(\beta)\mathrm{tgh}(\beta)\cos(\chi), \qquad (10)$$

$$\frac{d\theta}{dZ} = \frac{\eta^2 - \kappa^2}{2} + \xi\dot{\kappa} - \frac{\pi^2 \kappa S}{2\eta^2}\mathrm{sech}(\beta)\mathrm{tgh}(\beta)\cos(\chi).$$

where $\chi \equiv \Lambda Z + \kappa\xi - theta - \pi/2$, and $\beta \equiv \pi\kappa/(2\eta)$. Provided that $|\alpha_0| \leq \alpha_c \simeq (2/3)^{3/2}\sigma$, $V_f \simeq 0$ and $\Lambda = 1/2$, the above equations have a stable eigensolution with $\eta = \eta^* = 1$, $\kappa = \kappa^* = 3\alpha_0/(4\sigma + 3\pi S\sin(\chi^*))$, $\xi = \xi^* = 0$ and $\chi = \chi^* = \arcsin\{[2\delta - 2\sigma(k^{*2} + 1/3)]/[\pi S]\}$. In the next section, we compare the predictions of Equations (10) with the behavior of the numerically generated cavitons.

4. Numerical Results

In this section, we present numerical simulations of caviton build-up and propagation in a coherently seeded FSF cavity. The fiber length is $L_n \simeq L = 650$ m, the average GVD is $k_0'' = -19.2\mathrm{ps}^2/\mathrm{km}$ at the seed wavelength $\lambda_0 = 1550$ nm. The BF bandwidth is equal to 1.32 nm, or 164 GHz, which

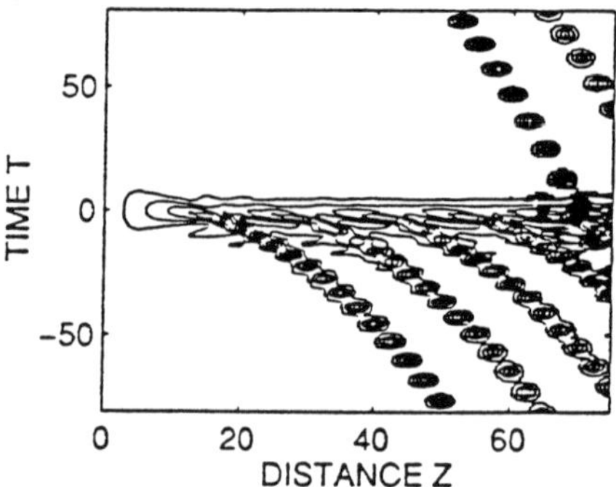

Figure 2. Caviton generation without filtering.

leads to the distributed filtering coefficient $\sigma = 0.15$, where

$$\sigma = \frac{2}{B^2 L |k_0''|}. \tag{11}$$

Here $B/(2\pi)$, and we express the BF bandwidth in THz, L in km, k_0'' in ps^2/km. The averaged sliding rate reads

$$\alpha_0 = \frac{10^{-6} 2\pi f t_0^3}{L |k_0''|}, \tag{12}$$

where f is measured in MHz, t_0 in ps, and L in km. In the examples below, we chose the AOM frequency shift $f = 1.23$ GHz, which permits to reduce the computation time, without affecting the generality of the conclusions. In fact, Equation (12) shows that one may change both f and L, as long as their ratio remains constant.

Perturbation theory permits to estimate the temporal duration of the generated cavitons as follows. In fact, for a weak seed one may separate the action of the cw from that of the filter. In other words, we may assume that the cw influences the build-up of the cavitons only, whereas their dynamics is ruled by Equations (10) with $S = 0$. In this case, the cavitons remain trapped by the filters for $|\alpha_0| < \alpha_c$ [15, 20]: the caviton time width is then $t_{fwhm} = 1.763 t_0$, where $t_0 \simeq t_c$, and t_c is given by Equation (12) with $\alpha_0 = \alpha_c$. With $\sigma = 0.15$, one obtains $\alpha_c = 0.08$, and $t_c \simeq 5$ ps. We further take $n_2 = 3.2 \times 10^{-20} \mathrm{m}^2/\mathrm{W}$, and $A_{eff} = 80\,\mu\mathrm{m}^2$, which leads to $\gamma = 1.6 \times 10^{-3} (\mathrm{mW})^{-1}$.

Figure 2 shows the contour plot of the cavity field amplitude $|q(Z, T)|$ in a filterless laser ($\delta = \sigma = 0$) with an injected cw power of 240 μW, and a resonant cavity (see Equation (2)). The time window in Figure 2 is

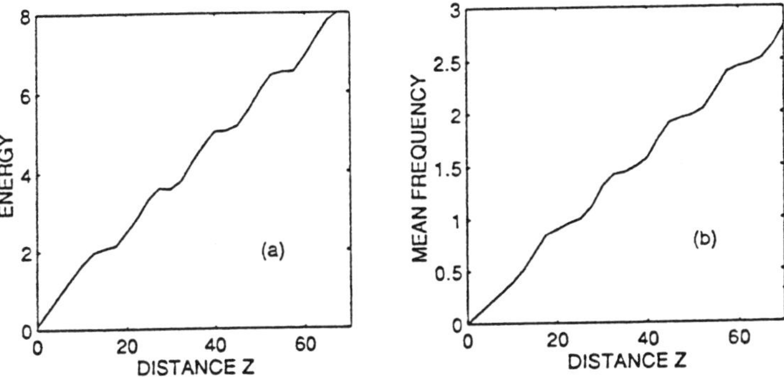

Figure 3.   Cavity energy and mean frequency vs. distance for the case of Figure 2.

$t_{\Delta\omega'} = 1/f$, and the time unit $t_0 = 5$ ps. As can be seen, the injected seed builds up a field at $T \simeq 0$, where the detuning between the pump and the cavity field vanishes at low powers. Note that the distance Z measures the number of circulations in the cavity $M = Z/Z_l$.

Figure 3(a) displays the growth of energy $E(Z)$ in the laser, where

$$E(Z) \equiv \int_{-T_{\Delta\omega'}/2}^{T_{\Delta\omega'}/2} |q(Z,S)|^2 dS. \tag{13}$$

The staircase growth of E is due to the periodic caviton emission of Figure 2. After its creation, the caviton dynamics is ruled by Equations (10). Without a filter, the soliton frequency is untrapped and it gets continuously up-shifted by the AOM. This is shown by Figure 3(b), where the average frequency $< \kappa >$ is defined as

$$< \kappa > \equiv \frac{\int_{-T_{\Delta\omega'}/2}^{T_{\Delta\omega'}/2} \kappa |q(Z,\kappa)|^2 d\omega}{E}. \tag{14}$$

Here $q(Z,\kappa) \equiv q(Z, 2\pi F)$ denotes the Fourier transform of $q(Z,T)$.

Figures 4(a-b) show contour plots of the cavity field amplitude $|q(Z,T)|$ in a bandwidth-limited active cavity with $\sigma = 0.15$, $|\alpha_0| = 0.078$, and $\delta = 0.073$ (so that Equations (10) with $S = 0$ predict $\eta^* = 1$). As can be seen, caviton generation strongly depends on the sign of the AOM frequency shift. Figure 4(a) was obtained for frequency up-shifting (i.e., with $\alpha_0 = +0.078$), whereas Figure 4(b) corresponds to down-shifting. Up-sliding produces a train of solitons that move with a certain relative velocity, say V_s, in the (Z,T) plane. This velocity may be estimated from Equations (9,10) with $S \simeq 0$. With the present parameters, one obtains $\kappa^* = 0.39$, and

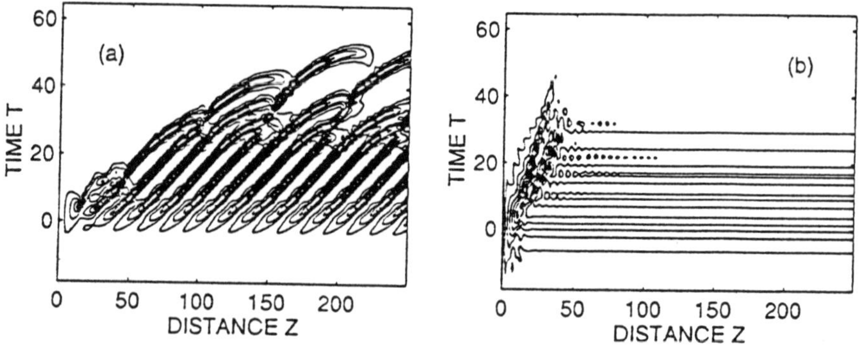

Figure 4. Caviton generation with filtering and (a) AOM up-shifting or (b) down-shifting.

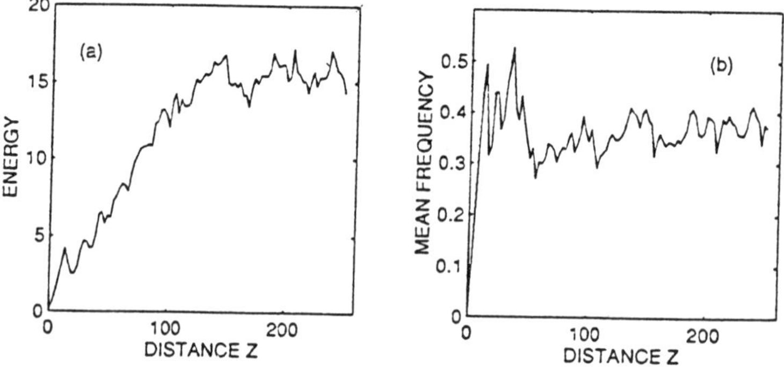

Figure 5. Cavity energy and mean frequency vs. distance for the case of Figure 4(a).

$V_f \simeq -0.78 = -2\kappa^*$, so that $V_s = V_f + \kappa^* = -0.39$. Whereas from Figure 4(a) one estimates $V_s \simeq -0.44$. On the other hand, Figure 4(b) reveals that caviton generation is inhibited with down-sliding, where $\kappa^* = -0.39$ and $V_s \simeq -1.2$. Figure 4(a) also shows that after a certain propagation distance the solitons are absorbed and eventually disappear.

Figure 5(a) shows that, after an initial growth, the total energy in the AOM period $T_{\Delta\omega'}$ remains clamped to a constant value. The saturation of the cavity energy corresponds to the caviton absorption of Figure 4(a). Figure 5(b) illustrates the evolution of the average frequency $< \kappa >$ of the soliton train of Figure 4(a). As can be seen, the mean frequency quickly relaxes to its asymptotic value $\kappa^* = 0.39$ after a few tens of circulations.

The caviton absorption in Figure 4(a) may be ascribed to the destructive interference between the cavitons and the injected cw seed. In fact, Equa-

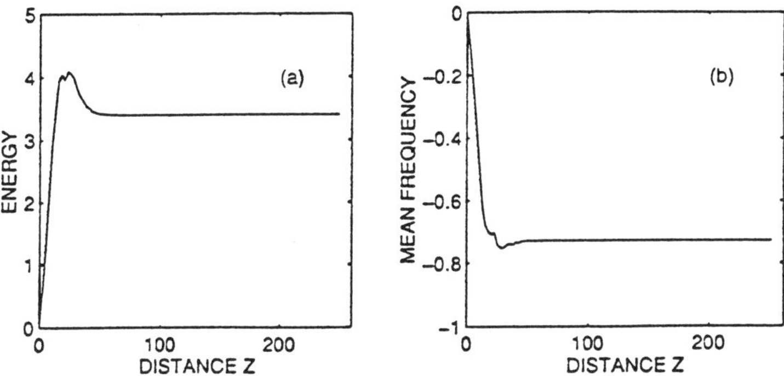

Figure 6.   Cavity energy and mean frequency vs. distance for the case of Figure 4(b).

tions (10) yield that, for $S \simeq 0$, $\theta^*(Z) = (\eta^{*2} - \kappa^{*2})Z/2 = (1 - \kappa^{*2})Z/2$. On the other hand, whenever the soliton is centered at the time $\xi(Z) = (\kappa^* + V_f)Z$, the AOM induces a phase-shift (or detuning) between the caviton and the injected seed equal to $\Lambda(Z) = \alpha_0\xi(Z)$. The caviton was initially in-phase with the cw seed when it was formed near $T_{cr} = 0$, and a relative phase slip of π (which leads to destructive interference) with respect to the cw occurs after a distance Z_d such that

$$\left| \int_0^{Z_d} \Lambda(Z)dZ - \theta^*(Z_d) \right| = \pi, \tag{15}$$

or

$$Z_d = \frac{1 - \kappa^{*2} + \sqrt{(1 - \kappa^{*2})^2 + 8\pi\alpha_0|\kappa^* + V_f|}}{2\alpha_0|\kappa^* + V_f|}. \tag{16}$$

Equation (16) yields $Z_d \simeq 33$, which agrees with the numerical result of Figure 4(a). As a result, the individual caviton lifetime in the FSF laser is equal to 0.2 ms.

Figures 6(a) and 6(b) display the evolution of the energy and the mean frequency of the field that is generated with down-sliding as in Figure 4(b). In this case, the field reorganizes itself into a single, time-asymmetric, broad pulse after just a few circulations in the cavity. The mean negative frequency shift of this pulse from the center of the filter passband is nearly twice the value that is obtained with up-sliding.

Details on the caviton formation process with up-sliding as in Figure 4(a) are presented in the temporal and spectral profiles of Figures 7-8. Here we plot the field amplitude and its corresponding Fourier spectrum after $M = 5, 25, 35$, and 250 circulations through the ring, respectively. Figures 7(a) and 8(a) illustrate the field that is formed in the active filtered cavity

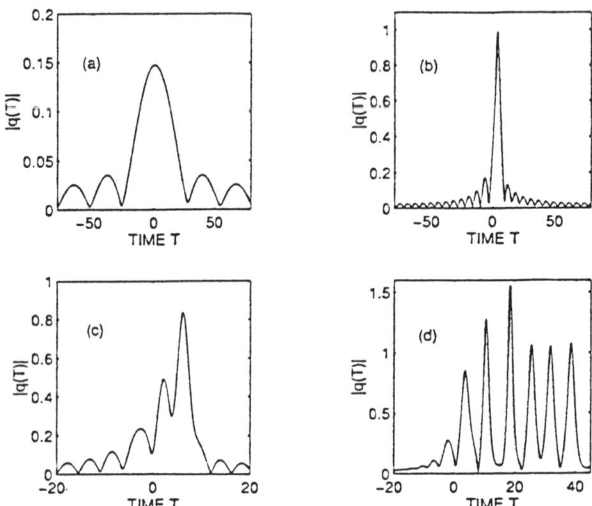

Figure 7. Temporal profiles of the cavity amplitude for the case of Figure 3(a), after (a) 5, (b) 25,(c) 35, and (d) 250 circulations.

after $M = 5$ circulations. As can be seen, the field builds-up near $T_{cr} = 0$ from the constructive interference of frequency-shifted waves.

Figures 7(b) and 8(b) show that after $M = 25$ circulations the peak amplitude of the pulse reaches the unit value, in agreement with the prediction of Equation (5). In the present dimensionless units, the unit amplitude corresponds precisely to the amplitude of a soliton whose time width $t_{fwhm} = 1.763t_0$ yields the critical sliding rate α_c through the relation (12). The corresponding logarithmic plot of the spectrum in Figure 8(b) shows the development from the square shape, that is typical of the linear regime, to the triangular shape of a hyperbolic secant pulse. The caviton build-up time, say, $T_b = k_0' M L$, that one estimates from Equation (5), permits to obtain the requirement on the minimum coherency time T_c of the injected seed laser. In fact, the coherency between all the frequency comb components inside the laser up to 25 circulations requires a seed coherency time $T_c \geq T_b = 81$ μs, or a seed laser line width less than 15 kHz.

The plots in Figure 7(c) (and the corresponding spectrum in Figure 8(c)) illustrate the initial stage of the splitting between the first soliton (after its formation, the soliton is delayed as we discussed with reference to Figure 4(a)) and the background near $T_{cr} = 0$, where another caviton will subsequently grow up. The result of this multiple pulse generation process is shown by Figures 7(d) and 8(d), that show the field after $M = 250$ circulations in the cavity. The irregular shape of the spectral profile results from

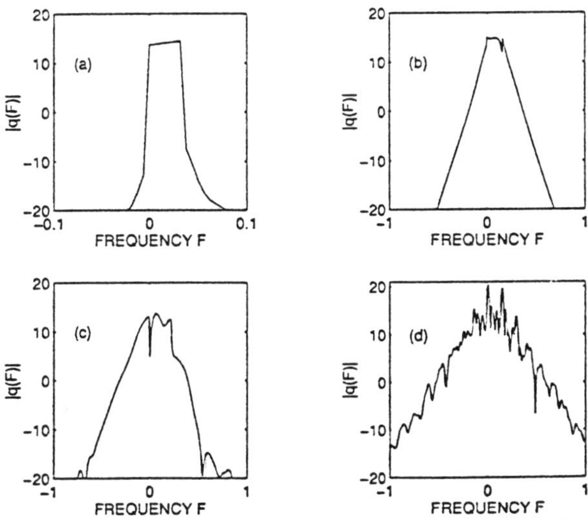

Figure 8. Spectral intensity of the cavity amplitudes of Figure 7.

the interference between the subsequent pulses (with different phases) of the train. The time separation, say, T_s, between caviton pulses(or repetition rate of the train) is roughly determined by the caviton inverse velocity V_s and Equation (5), which estimates the number of round-trips that occur between the splitting-off of a caviton from T_{cr} and the creation of the next pulse. One obtains $T_s = M Z_l V_s$: in the present case $T_s \simeq 5$, while from the numerical results in Figure 7(d) one obtains $T_s \simeq 6.5$. In real units, these values yield a caviton train repetition rate of about 30 GHz.

5. Conclusions

We have shown that energy may be coupled from a cw seed into a frequency-shifted cavity in the form of short pulses, in analogy with the coupling of energy from a laser into the cavitons that propagate in a plasma with a density gradient. The soliton build-up time may be estimated by a linear superposition formula. Soliton perturbation theory may be used to predict the stability properties and the characteristics of the generated high-repetition rate pulse trains. This work was carried out in the framework of the agreement between the Bordoni Foundation and the Italian Post and Telecommunications Administration.

References

1. Streifer, W. and Whinnery, J. R.: Analysis of a dye laser tuned by acousto-optic filter, *Appl. Phys. Lett.* **17** (1970), 335-337.

2. Kowalski, F. V., Squier, J. A. and Pinckney, J. T.: Pulse generation with an acousto-optic frequency shifter in a passive cavity, *Appl. Phys. Lett.* **50** (1987), 711-713.

3. Kowalski, F. V., Shattil, S. J. and Hale, P. D.: Optical pulse generation with a frequency shifted feedback laser, *Appl. Phys. Lett.* **53** (1988), 734-736.

4. Hale, P. D., and Kowalski, F. V.: Output characterization of a frequency shifted feedback laser: theory and experiment, *IEEE J. of Quantum Electron.* **26** (1990), 1845-1851.

5. Phillips, M. W., Liang, G. Y. and Barr, J. R. M.: Frequency comb generation and pulsed operation in a Nd:YLF laser with frequency-shifted feedback, *Optics Communic.* **100** (1993), 473-478.

6. Sabert, H. and Brinkmeyer, E.: Stable fundamental and higher order pulses in a fibre laser with frequency shifted feedback, *Electron. Lett.* **29** (1993), 2122-2124.

7. Sabert, H. and Brinkmeyer, E.: Pulse generation in fiber lasers with frequency shifted feedback, *J. Lightwave Technol.* **LT-12** (1994), 1360-1368.

8. Perry, I. R., Wang, R. L. and Barr, J. R. M.: Frequency shifted feedback and frequency comb generation in an Er^{3+}-doped fibre laser, *Optics Commun.* **109** (1994), 187-194.

9. Kodama, Y., Romagnoli, M. and Wabnitz, S.: Stabilization of optical solitons by an acousto-optic modulator and filter, *Electron. Lett.*, **30** (1994), 261-262.

10. Fontana, F., Bossalini, L., Franco, P., Midrio, M., Romagnoli, M. and Wabnitz, S.: Self-starting sliding-frequency fibre soliton laser, *Electron. Lett.* **30** (1994), 321-322.

11. Romagnoli, M., Wabnitz, S., Kodama, Y., Fontana, F., Bossalini, L., Franco, P. and Midrio, M.: Acousto-optic filter soliton control, *Pure Appl. Opt.* **4** (1995), 441-449.

12. Romagnoli, M., Wabnitz, S., Franco, P., Midrio, M., Fontana, F. and Town, G. E.: Tunable erbium-ytterbium fiber sliding-frequency soliton laser, *J. of the Opt. Soc. of Am. B* **12** (1995), 72-76.

13. Town, G., Chow, J. and Romagnoli, M.: Sliding-frequency figure-eight optical fibre laser, *Electron. Lett.* **31** (1995), 1452-1453.

14. Romagnoli, M., Wabnitz, S., Franco, P., Midrio, M., Bossalini, L. and Fontana, F.: Role of dispersion in pulse emission from a sliding-frequency fiber laser, *J. of the Opt. Soc. of Am. B* **12** (1995), 938-944.

15. Mollenauer, L. F., Gordon, J. P. and Evangelides, S. J.: The sliding-frequency guiding filter: an improved form of soliton jitter control, *Opt. Lett.* **17** (1992), 1575-1577.

16. Mollenauer, L. F., Lichtman, E., Neubelt, M. J. and Harvey, G. T.: Demonstration, using sliding-frequency guiding filters, of error-free soliton transmission over more than 20 Mm at 10 Gbit/s single channel, and over more than 13 Mm at 20 Gbit/s in two-channel WDM, *Electron. Lett.* **29** (1993), 910-911.

17. Aubin, G., Montalant, T., Moulu, J., Nortier, B., Pirio, F. and Thomine, J.-B.: Demonstration of soliton transmission at 10 Gbit/s up to 27 Mm using signal frequency sliding technique, *Electron. Lett.* **31** (1995), 52-54.

18. Toda, H., Yamagishi, H. and Hasegawa, A.: 10-GHz optical soliton transmission experiment in a sliding-frequency recirculating fiber loop, *Opt. Lett.* **20** (1995), 1002-1004.

19. Kawai, S., Iwatsuki, K. and Nishi, S.: Demonstration of error free optical soliton transmission over 30000 km at 10 Gbit/s with signal frequency sliding technique, *Electron. Lett.* **31** (1995), 1463-1464.

20. Kodama, Y. and Wabnitz, S.: Analysis of soliton stability and interactions with sliding filters, *Opt. Lett.* **19** (1994), 162-164.

21. Haelterman, M., Trillo, S. and Wabnitz, S.: Hopf sideband bifurcation and chaos in fiber lasers with injected signal, *Phys. Rev. A* **47** (1993), 2344-2353.

22. Chen, H. H. and Liu, C. S.: Solitons in nonuniform media, *Phys. Rev. Lett.* **37**

(1976), 693-697.

23. Adam, J. C., Gourdin Serveniere, A. and Laval, G.: Efficiency of resonant absorption
     of electromagnetic waves in an inhomogeneous plasma, *Phys. Fluids* **25** (1982), 376-
     383.
24. Larroche, O. and Pesme, D.: Source-driven dissipative nonlinear Schrödinger model
     of resonance absorption, *Phys. Fluids B* **2** (1990), 1751-1767.
25. Bussac, M. N., Lochak, P., Meunier, C. and Heron-Gourdin, A.: Soliton generation
     in the forced nonlinear Schrödinger equation, *Physica D* **17** (1985), 313-322.
26. Karpman, V. I. and Maslov, E. M.: Perturbation theory for solitons, *Zh. Eksp. Teor.
     Fiz.*, **73** (1977), 537-559 [*Sov. Phys. JETP* **46** (1977), 281-291].

LINEAR PERSPECTIVE OF SOLITONS

A.W. SNYDER, D.J. MITCHELL AND Y.S. KIVSHAR
Optical Sciences Centre, The Australian National University
ACT 0200 Canberra, Australia

Abstract. Soliton propagation can be approached using linear physics only. This applies generally to both stationary and nonstationary waves. Examples include bright and dark solitons, vector solitons, dynamic solitons, cascading solitons in $\chi^{(2)}$ medium, interacting solitons and solitons in the presence of gain and loss.

1. Introduction

The conventional approach to solitons is highly mathematical, often involving sophisticated studies of nonlinear differential equations. It may then come as a surprise that soliton propagation can be understood intuitively by using elementary concepts of linear guided-wave optics. This linear perspective has enabled significant advances. For example, through the linear perspective, the very notion of the soliton itself was generalized to include solitons with internal dynamics. Such dynamic solitons, among other things, opened the way to new phenomena including non-interacting solitons, vector solitons with rotating polarization, and domain wall solitons. The linear perspective also demonstrates that solitons in a dispersive (or diffractive) medium near the second-harmonic resonance can be understood precisely in the same manner as Kerr-type (cubic or $\chi^{(3)}$) solitons. Furthermore, analytical solutions can be borrowed directly from the literature of linear waveguide physics. Perhaps most importantly, the linear perspective imparts deep insight. Among other things, it provides a physical explanation for why certain solitons have radiation free collisions and why higher order solitons result from the process of scaling up the fundamental soliton.

Now, the linear perspective to optical solitons lends itself most naturally to the *spatial domain* where we can appeal to fundamental concepts of

A. Hasegawa (ed.), Physics and Applications of Optical Solitons in Fibres '95, 263–275.

classical optics. It is possible, in principle, to develop the equivalent linear physics for *temporal solitons*, but the models are somewhat artificial in that they involve time-varying refractive indices.

Accordingly, we find it easier to conceptualize in the spatial domain, particularly because we can use the well developed *linear physics of guided waves*, and then to convert our results into the temporal domain whenever this is possible. For example, one of the most beautiful, yet intuitive, concepts of spatial solitons is that a soliton of one colour (or polarization) can induce a waveguide which, in turn, can guide a weak beam of a different colour (or polarization), e.g., [1, 2, 3]. The weak beam would normally diffract. We anticipate an analogous phenomenon in the temporal domain, e.g. *temporal simultons* [4].

2. Links between Spatial and Temporal Solitons

History has demonstrated that the conceptual leap for advances in optical solitons has come mainly from the spatial domain, including the (stationary) optical soliton itself first discussed in 1964 by Chiao *et al.* [1]. Only in 1973, did Hasegawa and Tappert [5] give life to *temporal (or pulse) solitons*. They predicted that temporal dispersion could be balanced by nonlinear self-induced phase modulation for pulses travelling along optical fibres. The era of temporal solitons was launched.

In the main, solitons of space are much richer than those of time. First, spatial solitons have two transverse directions thus imbuing them with an additional richness compared to temporal solitons. This extra dimension leads to novel phenomena that *have no temporal analogue*. Examples, in which the linear perspective has led to rich insight, include solitons that spiral around one another [7]. While such results are relevant to futuristic photonic systems, where light itself guides and manipulates light [3], they are not of concern in this paper.

Secondly, spatial solitons have also been studied in a medium where the linear part of the refractive index can be inhomogeneous as well as in non-Kerr materials, neither of whose temporal analogue has been fully explored. On the other hand, not all temporal phenomena appear to have spatial analogues. The concept of zero dispersion is one example. When an analogue exists, then a simple mapping takes us from space to time [8]. In particular, the transverse spatial dimension gets mapped into time, and the longitudional coordinate becomes the propagation coordinate along a fiber. Furthermore, we recognize that dispersion plays the role of diffraction, while self-induced phase modulation plays the role of induced refraction.

Finally, we emphasize that temporal solitons for practical applications are comparatively a 'weak' nonlinear phenomena. They are guided by pre-existing waveguides, whereas spatial solitons induce their own waveguides. Our intent here is to concentrate on those spatial, one-dimensional soliton phenomena which, in most of the cases, have a clear temporal analogue.

3. Linear Perspective and Self-consistency

Our thesis is that both the physical understanding and the mathematical description of nonlinear stationary and even nonstationary self-guided waves follows naturally from linear guided waves.

Conceptually speaking, nonlinear beams interact with matter to create their own linear optical waveguides of *arbitrary shape and form*. These beams then propagate along their induced waveguides according to the familiar physics of linear optics. This elementary concept of linear physics can be used for a systematic approach to nonlinear guided wave optics.

To set the stage, recall the physics of an optical waveguide. Optical beams have an innate tendency to spread as they propagate in a homogeneous medium. However, this beam *diffraction* can be compensated for by beam *refraction* if the refractive index is increased in the region of the beam. The resulting optical waveguide provides *an exact balance* between diffraction and refraction if the medium is uniform in the direction of propagation. But, waveguides are not generally axially uniform, whereupon the balance between beam refraction and beam diffraction is not necessarily maintained. This gives rise to radiation 'leakage'.

The propagation of optical beams is described by Maxwell's equations. For material with a slow response or for waves that are monochromatic, the refractive index n of a nonlinear medium depends on the time-averaged intensity $|\mathbf{E}(x, z)|^2$ of the electric field for the usual reasons elaborated upon in [9, 10]. Accordingly, as a beam propagates in a nonlinear medium, it creates its own (linear) optical waveguide (or modifies existing ones) and then travels along the induced waveguide, according to the familiar physics of linear guided waves optics.

This induced optical waveguide is characterized by the refractive index profile $n(x, z) = n(|\mathbf{E}(x, z)|^2)$. In general, the waveguide is of rather arbitrary shape in x, z, and it is generally anisotropic. We have used the isotropic notation for simplicity. Of course, the unique aspect of nonlinear propagation is that the particular induced (linear) waveguide depends upon the initial condition of the beam.

In theory, the linear perspective can be used to solve beam propagation in any specified nonlinear medium. To see this, suppose we know how an initial beam propagates along every possible inhomogeneous linear optical waveguide, $n(x, z)$. Call this solution $\mathbf{E}(x, z)$. Now, out of this infinity of linear waveguides, one is exactly that induced by the beam propagating in a specified nonlinear medium. This equivalent waveguide is identified by *the self-consistency relation*

$$n(x, z) = n(|\mathbf{E}|^2).\tag{1}$$

where we have suppressed the explicit spatial dependence $n(|\mathbf{E}|^2, x, z)$ which for inhomogeneous material. More generally, n is a tensor to describe anisotropy. While exact, the above procedure is impractical. We do not have the solutions of all possible linear problems and, if we did, it would be time-consuming to find the one which is self-consistent. Nevertheless, *the fact that every nonlinear problem has a linear equivalent provides a powerful conceptual tool as we now demonstrate.*

4. 'Stationary Solitons'-Intensity Unchanged with Propagation

Of the myriad of possibilities, those induced waveguides which are *uniform* in the direction of propagation comprise the simplest class. If *a linear waveguide* is axially uniform with a refractive index profile $n(x, y)$, then its electric field vector $\mathbf{E}$ can be represented as a sum of modes

$$\mathbf{E}(x, z) = \mathbf{e}_a(x)e^{i\beta_a z} + \mathbf{e}_b(x)e^{i\beta_b z} + \ldots,\tag{2}$$

where $\mathbf{e}_a, \mathbf{e}_b$ and β_a, β_b are the modal eigenfunctions and propagation constants, respectively.

The vector field $\mathbf{E}$ is a solution to the Maxwell's equations. However, waveguides of practical interest have a maximum and minimum refractive index that is nearly equal. It is then well known that [11, 12]

$$\frac{\partial^2 \mathbf{E}}{\partial z^2} + \frac{\partial^2 \mathbf{E}}{\partial x^2} + k^2 \mathbf{n}^2 \cdot \mathbf{E} = 0\tag{3}$$

which holds exactly for TE waves only, where $k = 2\pi/\lambda$ and $\mathbf{n}$ is the tensor refractive index.

Invoking only the standard weak guidance limit of linear waveguides [11, 12], when the maximum and minimum refractive index are approximately equal, this equation holds approximately for all wave types and it further reduces to the nonlinear Schrödinger (NLS) equation,

$$2i\frac{\partial \mathbf{\Psi}}{\partial Z} + \frac{\partial^2 \mathbf{\Psi}}{\partial X^2} + (\rho k)^2[\mathbf{n}^2 \cdot \mathbf{\Psi} - \bar{n}^2 \mathbf{\Psi}] = 0,\tag{4}$$

where $\mathbf{E} = \mathbf{\Psi}e^{ik\bar{n}z}$, with $\bar{n}$ the minimum refractive index, $X = x/\rho$, $Z = z/\rho$, and ρ a characteristic beam width. We note that the slowly varying envelope condition [9] follows as a consequence of weak guidance, no additional assumption is necessary.

Now, if Equation (2) is to describe *nonlinear waves*, then the self-consistency relation Equation (1), demands tha their induced waveguides, $n = n(|\mathbf{E}(x,z)|^2)$ be independent of the direction of propagation z. Thus, nonlinear beams can only be represented by one mode, by two orthogonal modes obeying $\hat{\mathbf{e}}_a \cdot \hat{\mathbf{e}}_b^* = 0$ or even by two like polarized modes of different wavelength when four wave mixing is neglected.

Equation (2) specifies the complete class of beams whose induced waveguides are axial uniform. No other representation will induce an axially uniform waveguide. Despite its simplicity, Equations (3), (4) describe a multitude of nonlinear waves including classes of self-guided beams of a homogenous medium, surface waves at a nonlinear interface and modes of nonlinear waveguides.

We classify beams based upon whether they are one or two modes of the induced waveguide. If the beam is one mode of its induced waveguide, its polarization is maintained with propagation. These are the stationary solitons, e.g. classical sech-type soliton [1] and vector solitons [13]. If the beam is two modes of its induced waveguide, then novel polarization dynamics arise as exemplified by 'dynamic' spatial solitons [14].

4.1. BRIGHT SOLITONS

The simplest bright soliton is one mode of the **isotropic** linear waveguide it induces. From Equation (2), we observe that the beam polarization is arbitrary and independent of spatial position. Then $\mathbf{E}(x,y,z) = \hat{\mathbf{e}}e(x,y)e^{i\beta z}$, with $\hat{\mathbf{e}}$ a unit vector and $e(x,y)$, β found from the eigenvalue equation

$$\left[\frac{d^2}{dx^2} + k^2 n^2(|e|^2) - \beta^2\right] e = 0 \tag{5}$$

In particular, solutions of Equation (5) include the 'classical' bright soliton of a cubic (Kerr) homogeneous nonlinear medium [1]. Other examples can be borrowed directly from the literature of linear waveguides applying *the concept of inversion* [15]: we know that a nonlinear guided wave is the mode (or modes) of the axially uniform waveguide it induces. Reversing the argument, a mode of a linear waveguide must be a nonlinear guided wave of some nonlinear medium [15]. The particular medium is found by inverting the self-consistency relation Equation (1) as shown in Figure 1.

The simplest illustration of our approach may be easily understood for the solitons of the power-law nonlinearity. Indeed, the familiar [11] sech^2 profile **linear** waveguide has the form $n^2(x) = n_\infty^2 + (n_0^2 - n_\infty^2)\text{sech}^2(x/\rho)$, where $n_0^2 = n^2(0)$, $n_\infty^2 = n^2(\infty)$ and ρ is a characteristic half-width. Taking the modal field amplitude $e(x,z) = e(x)e^{i\beta z}$, where $\{d^2/dx^2 + k^2 n^2(x) - \beta^2\}e(x) = 0$, leads to $e(x) = e_0\text{sech}^s(x/\rho)$ with $s = (1/2)[\sqrt{1 + 4V^2} - 1]$.

Here V is the waveguide parameter $V = k\rho\sqrt{n_0^2 - n_\infty^2}$, and the modal propagation constant is, $\beta^2 = (kn_\infty)^2 + (s/\rho)^2$. The sech^2 waveguide is single moded provided that $V \leq \sqrt{2}$ or, equivalently, $s \leq 1$.

To find the nonlinearity that supports solitons that have a sech$^s(x/\rho)$ modal field, we use the self consistency relation (Figure 1) that $n_I^2[I(x)] = n^2(x)$, where n_I^2 is the material nonlinearity. It is necessary *to invert* $I(x)$ to obtain an explicit relation for the dependence of the refractive index on intensity, i.e. $n_I^2(I) = n^2[x(I)]$, where $x(I) = \rho\,\mathrm{sech}^{-1}[I(x)/I_0]^{1/2s}$. This leads to a power-law nonlinearity of the form $n_I^2(I) = n_\infty^2 + (\alpha I)^q$ with $q = 1/s$. To obtain a solution of the nonlinear problem we replace s by $1/q$ and, for example, this specifies the parameter V of the soliton-induced waveguide in terms of the nonlinearity power q, $V = \sqrt{1+q}/q$. The soliton intensity $[I = e^2(x)]$ is $I(x) = I_0\mathrm{sech}^{2/q}(x/\rho(I_0))$. Further details are in [15]. Solitons of more general nonlinearities can be treated similarly [2].

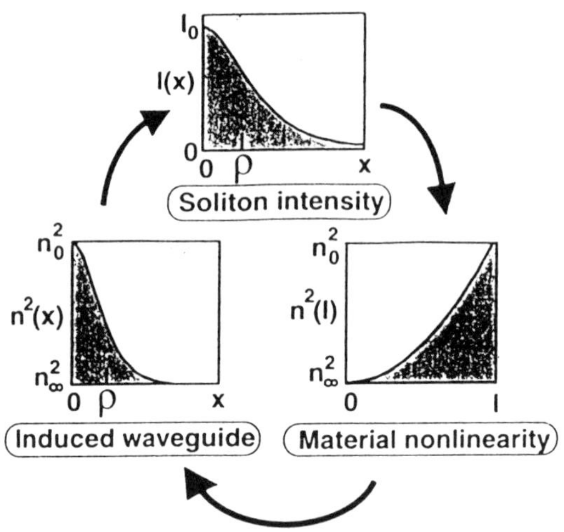

Figure 1. The electric field intensity $I(x)$ of a mode of a (linear) refractive index profile $n^2(x)$, is a self guided beam of a homogeneous medium whose material nonlinearity is $n_I^2(I)$. Self-consistency demands that $n_I^2[I(x)] = n^2(x)$.

4.2. DARK SOLITONS

Optical waveguides also have unbounded modes which comprise the continuous mode spectrum [11]. These correspond to plane wave scattering from the waveguide and, in general, consist of an incident and scattered wave, except at special angles of incidence [16]. However, the sech^2 waveguide (with $V = \sqrt{2}$) is unusual, in that it is reflectionless to every angle of plane wave incidence. Consequently, arbitrary beams will pass through this linear waveguide without reflection. Upon inverting the self consistency relation Equation (1), we find [16] that the fields of the unbound (reflectionless) modes of the linear sech$^2(x/\rho)$ waveguide,

$$e(x) = e_\infty e^{iqx}[\tanh(x/\rho) - iQ]/(1 - iQ),$$

where $\beta^2 = (kn_\infty)^2 - q^2$ and $Q = \rho q$, with $V \equiv (k\rho)\sqrt{n_0^2 - n_\infty^2} = \sqrt{2}$ are the fields of the dark fundamental solitons propagating in a (negative) Kerr medium with $n^2 = n_0^2 - \alpha|\mathbf{E}|^2$ In sum, the familiar sech^2 profile of linear optical waveguides (with $V = \sqrt{2}$) gives us the family of bright and fundamental dark solitons of a Kerr medium as well as the knowledge that their induced waveguides are reflectionless. Non-Kerr dark solitons can be treated similarly (see, e.g., [16]).

4.3. VECTOR SOLITONS

When the beam is one mode of the *birefringent* linear waveguide it induces, then the polarization as well as its dependence on positions x are set by the characteristics of the birefringence. In general, the direction of the optical axis of **n** depends on spatial position x as is known from linear waveguides [11]. However, the field can be expressed as $\mathbf{E}(x, y, z) = \{\hat{\mathbf{e}}_a e_a(x, y) + \hat{\mathbf{e}}_b e_b(x, y)\}e^{i\beta z}$ when the directions of the optical axis $\hat{\mathbf{e}}_a, \hat{\mathbf{e}}_a$ are independent of position. From Equation (3), this leads to

$$\left[\frac{d^2}{dx^2} + k^2 n_a^2(|e_a|^2, |e_b|^2) - \beta^2\right] e_a = 0, \tag{6}$$

$$\left[\frac{d^2}{dx^2} + k^2 n_b^2(|e_b|^2, |e_a|^2) - \beta^2\right] e_b = 0. \tag{7}$$

The simplest class of beams are polarized along one optical axis 'a' or 'b', e.g. those that are linearly or circularly polarized in a birefringent Kerr medium [17]. These induce an effectively isotropic waveguide.

Solitons requiring both 'a'and 'b' components are called *vector solitons* [13]. The simplest conceptual example is for a homogeneous medium where the birefringence is due exclusively to the nonlinearity, although examples including linear birefringence are often discussed. Note that vector solitons result due to fortuitous degeneracies, $\beta_a = \beta_b$. This 'linear' view of vector solitons is physically insightful.

4.4. DYNAMIC SOLITONS

It is remarkable, following the linear perspective introduced at the outset of Section 4, that we can generalize the notion of nonlinear waves from their traditional stationary character (one mode of the waveguide they induce) to the novel *dynamic soliton* which is composed of *two orthogonal modes* of the linear waveguide it induces [27]. It is clear from Equation (2), that when $\beta_a \neq \beta_b$ the polarization state of this wave *changes* as the wave propagates, due to modal beating. Unlike vector solitons, dynamic solitons do not necessitate a birefringent medium unless mode 'a' and 'b' are degenerate. Indeed, the simplest conceptual examples are the so-called dynamic spatial soliton of an isotropic homogeneous medium [14] where $n_a^2 = n_b^2 = n^2(|e_a|^2 + |e_b|^2)$. Dynamic spatial solitons embrace a large class of waves, depending on the choice of modes in (2) and on the material properties as dictated by $\mathbf{n}$ in Equation (3). The motivation for 'dynamic' soliton arises from the linear perspective in th spatial domain. But, once the concept is appreciated, temporal analogies become apparent, e.g., [18, 19].

5. Solitons in a $\chi^{(2)}$ Medium - cascading

All previous approaches to soliton propagation in a medium which exhibits harmonic generation are based exclusively on the 'new' physics and mathematics of the phase-involved parametric interactions rather than the familiar concept of refractive index change discussed above. But, the elementary concepts of induced linear waveguides applies elegantly to such parametric solitons, leading to an intuitive description. In other words, both soliton classes, for cubic and quadratic nonlinearities, can be treated identically. Again, we emphasize that the conceptual leap for advances in parametric optical solitons of quadratic nonlinearities comes mainly from the spatial domain [20]-[23] whereas their temporal analog, have been only recently discussed [24].

From the linear perspective, harmonic generation arises when light at frequency ω causes the molecules to vibrate sinusoidally in time. Temporal propagation in such a medium is thus analogous to spatial propagation through a periodic medium.

The familiar equations [9] for propagation in a medium exhibiting harmonic generation can be cast into the linear framework by simply identifying certain quantities as an effective refractive index. Taking the fields to be linearly polarized along the optical axes, this leads to $\mathbf{E} = \mathbf{E}(\omega)e^{-i\omega t} + \mathbf{E}(2\omega)e^{-2i\omega t}$ +c.c., where c.c. is the complex conjugate, and where $[d^2/dz^2 + d^2/dx^2 + k^2 n^2(\omega)]E(\omega) = 0$ and $[d^2/dz^2 + d^2/dx^2 + 4k^2 n^2(2\omega)]E(2\omega) = 0$. Here $E(\omega)$ and $E(2\omega)$ see two different nonuniform waveguides $n(\omega)$ and $n(2\omega)$ respectively with $n^2(\omega) = \bar{n}^2(\omega) + 2\chi^{(2)}E(2\omega)E^*(\omega)/E(\omega)$ and

$n^2(2\omega) = \bar{n}^2(2\omega) + \chi^{(2)}E^2(\omega)/E(2\omega)$ with $\bar{n}$ denotes the linear part of the refractive index and $\chi^{(2)}$ is the second order susceptibility.

In general, the refractive index is complex which corresponds to loss and gain. *Only when both $n(\omega)$ and $n(2\omega)$ are real can stationary waves exist. This demands that $E(2\omega)$ and $E^2(\omega)$ are in phase or π out-of-phase.*

For plane wave propagation, $E(\omega) = e(\omega)e^{i\beta z}$ and $E(2\omega) = e(2\omega)e^{2i\beta z}$. With the consistency relation that $\beta = kn(\omega)$ and $2\beta = 2kn(2\omega)$, then $\bar{n}^2(\omega) \pm 2\chi^{(2)}|e(2\omega)| = \bar{n}^2(2\omega) \pm \chi^{(2)}|e(\omega)|^2/|e(2\omega)|$ (where $\pm$ corresponds to $E(\omega)$ in or π out-of-phase with each other).

Similarly for solitons, the envelopes $e(\omega)$ and $e(2\omega)$'see' a *linear* waveguide characterized by $n(\omega)$ and $n(2\omega)$, respectively. Thus, from Equation (3) the envelopes $e(\omega)$ and $e(2\omega)$ satisfy the modal equations

$$\left[\frac{d^2}{dx^2} + k^2n^2(\omega)\right]e(\omega) = \beta^2 e(\omega), \tag{8}$$

$$\left[\frac{d^2}{dx^2} + 4k^2n^2(2\omega)\right]e(2\omega) = 4\beta^2 e(2\omega), \tag{9}$$

where the effective refractive indices $n(\omega)$ and $n(2\omega)$ are respectively

$$n^2(\omega) = \bar{n}^2(\omega) \pm 2\chi^{(2)}|e(2\omega)|, \tag{10}$$

$$n^2(2\omega) = \bar{n}^2(2\omega) \pm \chi^{(2)}|e(\omega)|^2/|e(2\omega)|. \tag{11}$$

These equations are the same as for vector solitons. Thus, soliton propagation in a $\chi^{(2)}$ medium is like a mode of a birefringent linear waveguide.

Suppose $E(\omega)$ sees a sech^2 waveguide, i.e. $n^2 = \bar{n}^2+(V^2/k^2\rho^2)\text{sech}^2(x/\rho)$ whose mode is $e = e_0\text{sech}^s(x/\rho)$ with $\beta^2 = k^2\bar{n}^2 + s^2/\rho^2$ where s is given by $V^2 = s(s+1)$. Self-consistency ahows that $E(\omega)$ and $E(2\omega)$ both see a sech^2 waveguide with $V = \sqrt{6}$, and

$$e(2\omega) = \sqrt{2}e(\omega) = \frac{3}{k^2\rho^2\chi^{(2)}}\text{sech}^2(x/\rho), \tag{12}$$

where β and ρ are given by: $\beta^2 = k^2n^2(\omega) + 4/\rho^2 = k^2n^2(2\omega) + 1/\rho^2$ so that $k^2\rho^2[n^2(2\omega) - n^2(\omega)] = 3$ and $\beta^2 = k^2[4\bar{n}^2(2\omega) - \bar{n}^2(\omega)]/3$. The soliton Equation (12) is the only solution known analytically [20, 22].

In the general case, no analytical form exists for the solitons in $\chi^{(2)}$ medium, and the stationary solutions of Equations (8), (9) are known only numerically [22]. Finally, cascading is a particular limit of the stationary soliton, when the d^2/dx^2 term in Equation (9) can be neglected.

6. Solitons Whose Intensity Changes with Propagation

We now consider self guided waves (spatial solitons) which induce optical waveguides that are no longer axially uniform. The linear perspective provides deep insight into such phenomena, including beams that collide or propagate in parallel; that have periodic changes in intensity (higher order solitons); that attract, repel or spiral around each other [6]; that self contract or expand; and that exchange their power on nonlinear couplers [25]. Not all of these phenomena find their analog in the temporal domain.

In general, beams induce *linear* waveguides that are not axially uniform. These beams obey the self-consistency relation given by Equation (1). The linear perspective leads again to the wave equation Equation (3). Unfortunately, there are few analytical solutions to this equation or to its NLS-type approximation, and usually approximate methods as the soliton perturbation theory or variational approach are used. Nevertheless, the linear perspective provides physical insight and approximate analytical forms for a number of important examples. Unlike the inverse scattering technique, the approach is not restricted to one-dimensional beams of an isotropic Kerr medium.

6.1. SOLITONS IN A MEDIUM WITH LOSS OR GAIN

Linear waveguides that change adiabatically slowly, but otherwise arbitrarily in the direction of propagation, have an elegant description [11]. The modes are then locally those of the axially uniform waveguide. Therefore, *the adiabatic approximation* applies when the wavelength of spatial change is long compared to relevant electromagnetic lengths. This result can be borrowed directly for nonlinear waves [26].

For example, self-guided beams (spatial solitons) of a self-focusing material, with either loss or gain, will self-taper. The beam power $P(z)$ changes according to the standard relation [11] of linear waveguides where

$$dP = 2k\{|\mathbf{E}|^2 n^i dA\} P(z) dz / \int_A |\mathbf{E}|^2 dA.$$

When the imaginary part n_i of the *linear* index n is not zero, the medium possesses loss or gain, then the power changes exponentially with distance z. Because the characteristic soliton width $\rho \propto 1/P$, we have $P = P_0 \exp(-z/z_0)$. When the *nonlinear part* of the refractive index is lossy as in the case of two photon absorption, then (for a Kerr material) n^i is proportional to $|\mathbf{E}|^2$. This leads to a power-law dependence of P vs. z.

6.2. SOLITONS WITH PERIODIC MODULATION

In Sections 4.3 and 4.4 we have mentioned how a spatial soliton can be two modes of its induced linear waveguide, but that the modes must be orthogonally polarized if the waveguide is to remain axially uniform. Next, suppose the modes are identically polarized. The soliton intensity changes periodically with propagation and so does the waveguide it induces.

This picture of periodic beams [27], suggests that higher-order solitons are described by *modal beating* and that such solitons result when the induced waveguide is *multimoded*. This is consistent with the axially varying intensity profiles of higher-order solitons [28], e.g. the second-order soliton appears qualitatively like the beating of the first two even modes and with the fact that periodic solitons of a Kerr medium occur only when the fundamental ($N = 1$) soliton has been sufficiently scaled up [28].

This linear approach to periodic solitons is not peculiar to Kerr material. But, periodic (higher-order) solitons are presumed to be a by-product of integrability and thus, in optics, applicable only to one-dimensional beams, propagating in a Kerr (cubic) nonlinear medium. For example an idealized saturating medium, the threshold nonlinearity, differs radically from a Kerr material. Theory shows that a beam in this medium induces a step profile waveguide where the width changes periodically. This result has been demonstrated in [27] for one and two dimensions.

7. Interacting and Colliding Solitons

When self-guiding beams of a homogeneous isotropic medium collide, they induce a complicated criss-crossed linear waveguide. Light travelling along one linear waveguide will refract through the other and in general leak due to reflection and scattering. This qualitative observation of linear physics anticipates a phase shift (refraction) as well as radiation upon collision. But, the inverse scattering technique [8] proves that it is generally possible to contrive an X-junction linear waveguide to be lossless.

First, recall that radiation free soliton collisions do not require the solitons to be equal in size or in polarization. So, we assume that one beam spreads out to approximate a plane wave, while the other beam is arbitrary. Further, we assume the beams are either orthogonally polarized or of different wavelength. Under these conditions, the quasi-plane wave does not affect the induced waveguide to first-order and so sees the linear waveguide induced by the other beam. The V value of the waveguide that is seen by the plane wave is exactly that seen by a weak signal propagating along the strong 'pump' induced waveguide. We can show that the plane wave will in general 'see' a waveguide which differs from $V = \sqrt{2}$. But we know (Section 4.2) that only the $V = \sqrt{2}$ waveguide is reflectionless. Accordingly, linear physics predicts that spatial solitons will radiate if they differ in wavelength or in polarization.

Solitons of approximately the same intensity are known to attract or repel each other when they are like polarized. But, elementary physics of parallel linear waveguides and the self-consistency suggest that the soliton can remain *parallel* if the even and odd modes of their induced two linear waveguide system are *orthogonally polarized* [30]. One (of several) temporal analogs of parallel solitons can be found in the phenomenon of suppression of the mutual interaction between solitons of different wavelengths and polarizations [31].

8. Conclusions

The conventional approach to solitons is highly mathematical. Furthermore, while linear physics forms the foundations for our intuition, it nonetheless has appeared to be of little use for understanding nonlinear waves. We have demonstrated that the knowledge of linear physics of guided waves provides a natural approach to understanding and to advancing the subject of spatial solitons. It is easier to conceptualize in the spatial domain and then to translate into the temporal domain. This concept is expanded upon elsewhere [32].

References

1. Chiao, R. Y., Garmire E. and Townes, C. H.: *Phys. Rev. Lett.* **15** (1964), 1005.
2. Snyder, A. W., Mitchell, D. J., Poladian, L. and Ladouceur, F.: *Opt. Lett.* **16** (1991), 21.
3. Thwaites, T.: *New Scientist,* **12** (1991), 14; Luther-Davies, B. and Yang, X.: *Opt. Lett.* **17**, (1992), 496, 1755; Snyder, A. W. and Sheppard, A.: *Opt. Lett.* **18** (1993), 482.
4. Konopnicki, M. J. and J. H. Eberly, J. H.: *Phys. Rev. A* **24** (1981), 2567.
5. Hasegawa, A. and Tappert, F.: *Appl. Phys. Lett.* **23** (1973), 142.
6. Poladian, L., Snyder, A. W. and Mitchell, D. J.: *Opt. Commun.* **85** (1991), 59.
7. Snyder, A. W., Poladian, L. and Mitchell, D. J.: *Opt. Lett.* **17** (1992), 789.
8. Zakharov, V. E. and Shabat, A. B.: *Sov. Phys. JETP* **34** (1972), 62.
9. Y. R. Shen, Y. R.: *The Principles of Nonlinear Optics* (John Wiley, New York, 1984); *Phys. Lett.* **20** (1966), 378.
10. Bloembergen, N.: *Nonlinear Optics,* (Benhamin, NY, 1965).
11. Snyder, A. W. and Love, J. D.: *Optial Waveguide Theory* (Chapman and Hall, London, 1983).
12. Snyder, A. W. and Young, W. R.: *J. Opt. Soc. Am.* **68** (1978), 279.
13. Christodoulides, D. N. and Joseph, R. I.:*Opt. Lett.* **13** (1988), 53; Tratnik, M. V. and Sipe, J. E.:*Phys. Rev. A* **38** (1988), 2011.
14. Snyder, A. W., Hewlett, S. J. and Mitchell, D. J.: *Phys. Rev. Lett.* **72** (1994), 1012.
15. Snyder, A. W. and Mitchell, D. J.: *Opt. Lett.* **18** (1993), 101.
16. Snyder, A. W., Mitchell, D. J. and Luther-Davies, B.: *J. Opt. Soc. Am. B* **10** (1993), 2341.
17. Maker, P. D., Terhune, R. W. and Savage, C. M.: *Phys. Rev. Lett.* **12** (1964), 507.

18. Silberberg, Y. and Barad, Y.: *Opt. Lett.* **20** (1995), 246.
19. Haelterman, M., Sheppard, A. P. and Snyder, A. W.: *Opt. Lett.* **18** (1993), 1406.
20. Karamzin, Yu. N. and Sukhorukov, A. P.: *Sov. Phys. JETP* **41** (1976), 414.
21. Schiek, R.: *J. Opt. Soc. Am. B* **10** (1993), 1848.
22. Buryak, A. V. and Kivshar, Yu. S.: *Opt. Lett.* **19** (1994), 1612.
23. Torner, L., Menyuk, C. R. and Stegeman, G. I.: *Opt. Lett.* **19** (1994), 1615.
24. Werner, M. J. and Drummond, P. D.: *J. Opt. Soc. Am. B* **10** (1993), 2390.
25. Snyder, A. W., Mitchell, D. J., Poladian, L., Rowland, D. and Chen, Y.: *J. Opt. Soc. Am. B* **8** (1991), 2102.
26. Snyder, A. W., Poladian, L. and Mitchell, D. J.: *Opt. Lett.* **17** (1992), 118, 267.
27. Snyder, A. W., Hewlett, S. and Mitchell, D. J.: *Phys. Rev. E* **51** (1995), 6297.
28. Satsuma, J. and Yajima, N. *Supp. Prog. Theor. Phys.* **55** (1974), 284; Haus, H. A. and Islam, M. N.: *IEEE J. Quant. Elect.* **QE-21** (1985), 1172.
29. Snyder, A. W. and Sheppard, A. P.: *Opt. Lett.* **18** (1993), 499.
30. Snyder, A. W., Mitchell, D. J. and Haelterman, M.: *Opt. Commun.* **116** (1995), 365.
31. Evangelides, S. G., Mollenauer, L. F., Gordon, J. P. and Bergano, N. S.: *J. Lightwave Technol.* **10** (1992), 28.
32. Snyder, A. W., Mitchell, D. J. and Kivshar, Yu. S.: *Mod. Phys. Lett.* **9** (1995), 1479.

DISPERSION DECREASING FIBRES
FOR SOLITON GENERATION
AND TRANSMISSION LINE LOSS COMPENSATION

D. J. RICHARDSON, R. P. CHAMBERLAIN
L. DONG AND D. N. PAYNE
Optoelectronics Research Center
Southampton University
Southampton, S017 1BJ, United Kingdom

Abstract. We describe recent results concerning the fabrication and applications of optical fibres with accurately controlled and defined axial dispersion variation. We focus on two applications. Firstly, we describe the development of a 40 Gbit/s soliton transmitter based on beat-signal to soliton train conversion in a dispersion decreasing fibre. 40 GHz, < 5ps pulses with <300 fs timing jitter are first generated. 40 Gbit/s data is then successfully encoded via direct, all-optical modulation in a Kerr gate. Secondly, we demonstrate true soliton loss compensation in a fibre with a dispersion matched to the power loss profile within the fibre. Low distortion propagation of 2.0 ps pulses over 20 km (44 soliton periods) and of 3.5 ps pulses over 38 km (18 soliton periods) is obtained in both the temporal and spectral domains illustrating the quality of the loss compensation achieved.

1. Introduction

The idea of varying the axial distribution of dispersion along a length of optical fibre as a means of manipulating and controlling the soliton supporting nature of the fibre and thereby the characteristics of soliton pulses propagating through the medium has been around for somewhile [1,2]. A number of specific applications have been suggested in particular techniques for bright and dark soliton generation [3,4], pulse compression [5] and most notably techniques for high frequency soliton transmission [1]. The experimental realisation of most of these techniques however has been hindered

A. Hasegawa (ed.), Physics and Applications of Optical Solitons in Fibres '95, 277–291.
© *1996 Kluwer Academic Publishers. Printed in the Netherlands.*

by difficulties in the reliable fabrication of dispersion varying fibres. A technique for the fabrication of such fibres was first developed by workers at General Physics Institute, Moscow [6]. Control of the waveguide dispersion was achieved by active control of fibre diameter during the pull. Fibre lengths of up to 2 km were fabricated and successfully used in the first experimental demonstrations of high frequency (> 60 GHz) bright soliton generation [3,7] and pulse compression [5,8]. Subsequent to these first experimental results a number of other groups have commenced fabrication programs on such fibres, extending the techniques to fibre lengths of 40 km. In this presentation we describe our latest achievements in Dispersion Decreasing Fibre (DDF) fabrication and report on two applications of the technology. Firstly, we describe a robust, diode-pumped, 40 GHz bright soliton source [9,10], and secondly we demonstrate loss compensation in a 38 km loss-matched dispersion varying fibre [11].

2. Dispersion Decreasing Fibre Fabrication

As is well known the total chromatic dispersion of a fibre D_{Tot} can be closely approximated as the sum of two components $D_{Tot} = D_M + D_W$, where D_M is the material dispersion and $D_W = (\Delta n/\lambda)(V d^2(BV)/dV^2)$, the waveguide dispersion. Δn is the index difference and B and V are the normalised propagation constant and frequency respectively [12]. Controlled variation of the wave guide characteristics along the fibre length can therefore result in controlled dispersion variation. In order to continuously vary the waveguide dispersion along a section of fibre one needs to change one of the waveguide parameters along its length. By far the most convenient parameter to vary is the core diameter- and hence the effective V-value since this can be done during the fibre draw and does not require special preform design. What is needed however, is accurate fibre diameter control during the pulling process and the production of highly length-homogenous preforms.

Our fibres were pulled on a commercial 7 m fibre pulling tower on which the fibre diameter was stabilised by active control of the fibre pulling speed. Computer software was developed to continuously and controllably vary the fibre diameter to follow any required diameter variation. The control loop feedback parameters were optimised over the full range of required pull speeds 30-80 m/min, enabling RMS diameter errors $< 0.2\mu$m over the diameter range of 75-150 μm, with negligible diameter offset error($< 0.1\mu$m). A (typical) plot of set diameter and experimentally determined fibre difference from the set value are shown in Figure 1 illustrating the quality of the diameter control obtained for the 20 km loss compensating DDF (LCDDF1) described in Section 4.

In order to fabricate useful/interesting fibres at 1550 nm one needs to

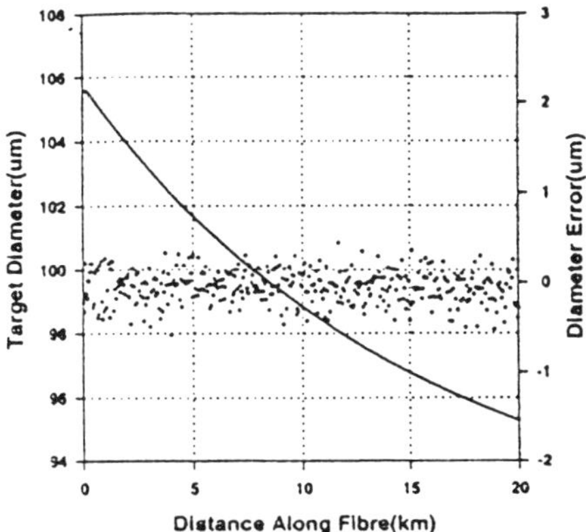

Figure 1. Diameter variation and measured diameter offset from set diameter along the fibre length for a 20 km loss compensating dispersion decreasing fibre (LCDDF1)

be able to make the waveguide dispersion opposite in sign, and equal, or greater to in magnitude, to the material dispersion ($\approx$ 20ps/nm/km at 1550 nm). This can be achieved either byusing a simple step index design with an NA>0.15, or by going to a more complex refractive index profile e.g. a dispersion shifted design. In Figure 3 we plot the experimentally determined variation in total dispersion as a function of fibre diameter obtained using the step refractive index profile shown in Figure 4 illustrating that both anomalous and normally dispersive, dispersion decreasing fibres are possible for diameter variations of order 30%. Note however that care has to be taken in the normal case to avoid bend loss effects at 1550 nm due to the low cut-off wavelength. The effective gradient at $D_{Tot} = 0$ of $d(D_{Tot})/dr = 0.3$ ps/(nm.km.μm) combined with our RMS diameter error = 0.2 μm means that the fundamental limit of our effective local RMS variation in dispersion is <0.1 ps/(nm.km). Even lower effective gradients can be obtained by using more complex R.I. profiles and ultimately a dispersion accuracy of <0.05 ps/(nm.km) should be possible. In addition, we have experimentally examined the absolute accuracy of dispersion varying fibre technology over long lengths of fibre by sampling sections of fibre (1km) from a 40 km DDF pulled in two sections (beginning, middle and end) and by measuring the average dispersion along two 20 km sections of the fibre. Our measurements were within 0.15 ps/(nm.km) of the expected value in each instance confirming the excellent pulling control and preform homogeneity.

To date we have pulled a wide range of dispersion varying fibres: fibres for bright soliton generation (1-8 km $\approx$ 120 $-$ 30GHz), fibres for dark soli-

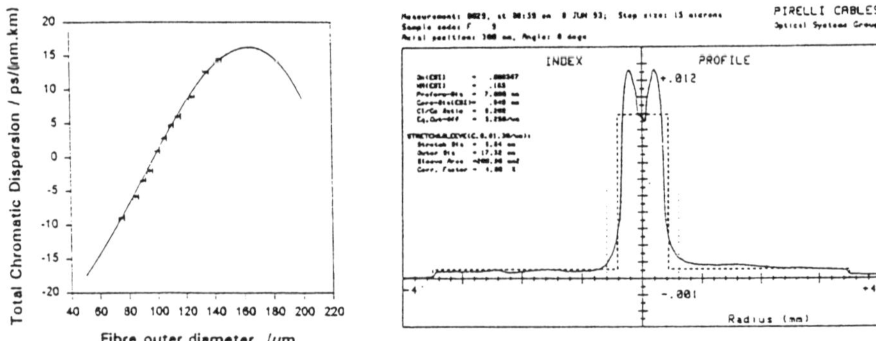

Figure 2. (a) Experimentally determined variation of total dispersion with fibre o/d for fibres fabricated from the preform with refractive index profile shown in (b).

ton generation (1.5 km $\approx$ 100GHz), fibres for soliton compression 50 m - 8 km, and for soliton loss compensation (38 km). We now go on to describe specific application demonstrations incorporating such fibres.

3. A Stabilised 40 Gbit/s Bright Soliton Transmitter [9,10]

There has been considerable recent interest in the development of optical techniques for the generation of high frequency soliton trains based on nonlinear beat signal to soliton train conversion in dispersion profiled fibre circuits (see e.g. References [3,7,13,14]). These techniques offer advantages of ultra-high repetition rates, high pulse quality, and broad wavelength and repetition rate tunability. Moreover, it is a non-resonant technique offering increased environmental stability relative to other more conventional fibre based short pulse generation schemes. However, although impressive source demonstrations have been made, their practical applications to date have been limited, due primarily to issues relating to timing jitter [7,13,14], Brillouin scattering [3,7,13] and difficulties in applying the techniques to repetition rate ranges <40 GHz to allow compatibility with state of the art, high data rate electronic signals [3]. In this section we report on the development and performance of a diode-driven, ultra-low jitter, 30-40 GHz soliton source with potential for telecommunication applications. We report the results of pulse propagation measurements of 35 GHz repetition rate, 5 ps pulses over a 205 km dispersion shifted fibre (DSF) transmission line [15] demonstrating for the first time that pulse trains from such sources can be transmitted with low distortion over terrestrial distances. Finally, we de-

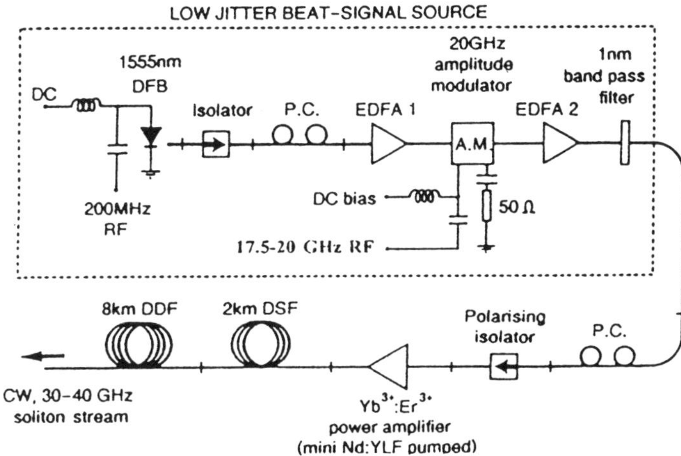

Figure 3. Schematic of diode-driven, stabilised 40 GHz bright soliton source.

scribe results on the all-optical modulation of pulses from such a source which illustrate that synchronisation, modulation and electrical detection issues specific to the use of such high frequency pulse sources can be overcome to give error free operation.

The source configuration (Figure 3) consists of three principal components: an optical beat-signal source, an Er^{3+}/Yb^{3+} optical power amplifier and a dispersion varying fibre section. In order to obtain a low timing jitter beat signal we used a 20 GHz amplitude modulator tuned to a transmission null and driven at 17.5 GHz to obtain 35 GHz sinusoidal modulation of the output from a DFB laser diode. Two equal amplitude frequency components separated by 35 GHz are obtained with almost no component at the carrier wavelength. The dispersion decreasing fibre section consisted of 2 km of DSF to spectrally enrich the input beat signal [7], and an 8 km DDF. The DDF dispersion followed a hyperbolic profile at 1550 nm tapering along the 8 km length from 13.75 to 2.75 ps/(nm.km). The profile and output dispersion were chosen so as to reduce the absolute physical length of fibre required to obtain high quality, adiabatic 40 GHz pulse generation at an MSR of 5:1, whilst maintaining a practical optical power requirement on the input beat signal (80 mW).

The source was tested under a wide range of input powers and beat frequencies in the range 32-40 GHz. Transform-limited, soliton pulses of durations 4.5-6.5 ps were obtained for input beat signal powers <20 dBm within the wavelength range of the available diodes 1547-1563 nm. A typical autocorrelation function (ACF) and spectrum of a 35 GHz pulse train at the source output are shown in Figure 4(a).

Pulse propagation experiments using the source were performed over a transmission line incorporating four $\approx$ 50 km spans of dispersion shifted fibre and 3 EDFAs [15]. The dispersion in the first and last two spans were D = 0.07 ps/(nm.km) and D = 0.21 ps/(nm.km), respectively. The ratio of soliton period to amplifier spacing was 3 over the first two DSF sections and 1 in the final two DSF spans. We therefore anticipate a strong departure from average soliton dynamics in the second half of the link resulting in the generation of significant dispersive wave radiation [16,17,18]. As expected the pulses were found to propagate with negligible distortion either temporally, or spectrally, over the first 100 km (see Figure 4(b)). Marked changes were however observed during propagation in the final two higher dispersion spans (Figure 4(c)). In the time domain the pulses are found to propagate with reasonable stability, with little distortion other than a slight temporal narrowing and the appearance of a low-level, flat background ($<$ 5%) on the ACF at 205 km. However a modulation of the train spectral envelope is observed due to the constructive interference of dispersive wave radiation shed from the pulses within the final two DSF spans [18]. The depth and position of the spectral modulation and the appearance of the low-level ACF background component at 205 km is in good agreement with our numerical simulations of the system.

Due to the low jitter beat signal seed source we were able to make direct electrical domain measurements on the pulse trains for the first time. The output pulse timing jitter was determined to be defined entirely by the phase noise of the modulator drive synthesizer ($<$300 fs). Furthermore, using electrical clock recovery circuits we have recently been able to synchronize a 6.4 ps, 40 Gbit/s optical data stream to the output of the soliton source. (The 6.4 ps pulses were generated using a gain-switched diode operating at 1545 nm. The diode output was externally modulated at 10 Gbit/s, and the resulting signal multiplexed to 40 Gbit/s). We could then employ an all optical modulator (Kerr Gate) to encode data all-optically directly onto the beat signal soliton stream. Demultiplexing using a second clock recovery unit and EA modulator could then be performed enabling BER measurements on the four 10 Gbit/s soliton data channels to be made. Error free operation, without a noise floor was readily obtainable with a power penalty of $<$1.5 dB (See Figure 5) [10].

In conclusion, we have demonstrated a practical low-timing jitter, diode-driven, soliton source capable of operating in the range 30-40 GHz. Furthermore, we have performed the first long distance, multi-amplifier stage pulse transmission experiments using a beat-signal source, demonstrating high quality, transmission of 35 GHz, 5.2 ps pulses trains over a total dis-

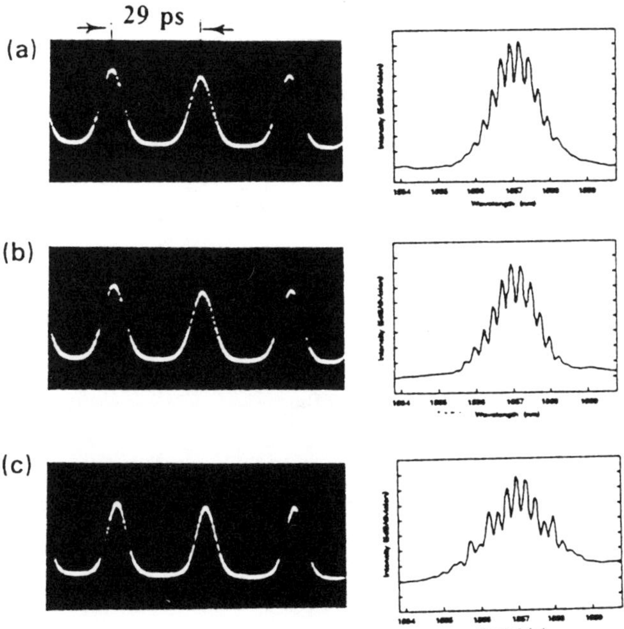

Figure 4. Autocorrelation and optical spectrum of 35 GHz pulse trains; (a) at the system input, (b) after 100 km, (c) after 200 km.

tance of 205 km. Good pulse propagation characteristics were obtained despite the large amplifier separation of 50 km and strong violation of average soliton dynamics in the latter half of the link. Transmission over considerably greater distances can be expected from an optimised system, with reduced dispersion variation and/or amplifier spacing. In addition, we have now demonstrated 40 Gbit/s, error free all optical data encoding onto the output of such sources illustrating for the first time that timing jitter and synchronisation issues associated with such sources can readily be overcome. Note, also that 40 GHz modulators and chip sets are currently under development and in future should permit direct electro-optic modulation at 40GHz [19]. The transmitter should enable a true assessment of the suitability of such sources for future high speed communication applications.

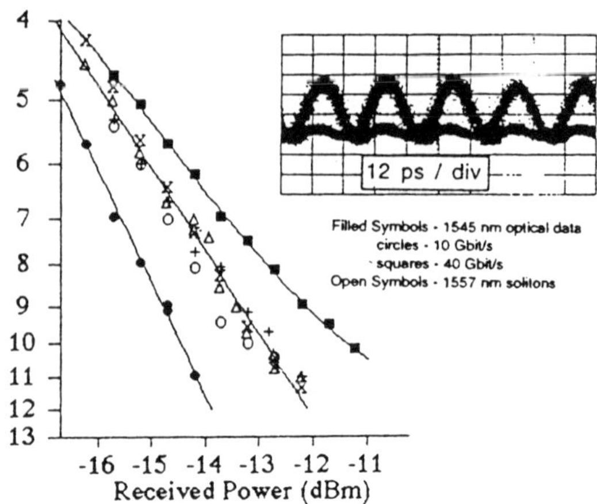

Figure 5. Eye diagram and BER curves for data encoded 40 Gbit/s data stream demultiplexed down to 10 Gbit/s illustrating error free all-optical modulation.

4. Soliton Loss Compensation [11]

Soliton communication systems based on EDFAs offer enormous potential for high data rate transmission. However, a number of fundamental limitations associated with the periodic amplification process restrict the ultimate capacity of such systems. The first limitation arises due to the interaction of the solitons with the EDFA noise and manifests itself as timing jitter at the system output [20]. Fortunately, a number of techniques have been successfully developed to eliminate/reduce these particular effects[21,22]. The second limitation results directly from the periodic amplification process itself and arises once the soliton period (z_0) of the pulses approaches the amplifier spacing (L_a) [23]. In this regime the perturbation to the pulses due to the fibre loss becomes too severe resulting in a violation of 'average', or 'guiding centre', soliton dynamics and eventual decay of the pulses [16,17]. The perturbation can only be eliminated by ensuring that the dispersive and nonlinear effects balance at all points along the fibre despite the presence of background optical loss. This can be achieved either by using a distributed amplifier to cancel the local fibre loss [24], or by the use of a Loss Compensating Dispersion Decreasing Fibre (LCDDF) [1]. In the latter case the dispersion of the transmission fibre is made to fall exponentially along its length so as to exactly follow the decrease in optical intensity and hence the strength of the nonlinear interaction. An exact balance between the two effects at all points along the fibre is thus ensured. To date, most experimental effort has centred on the use of distributed amplifiers [24]. However, recently improved transmission of 1-2 ps pulses over a 40 km dispersion varying fibre span has been reported renewing interest in the LCDDF option[25]. Unfortunately, the dispersion variation of the fibre in these earlier experiments did not follow the loss profile so that although improved pulse transmission was indeed obtained, true loss compensation was not demonstrated. In this section we report the fabrication of a 38 km LCDDF with a dispersion profile closely matched to the fibre loss and demonstrate high quality soliton loss compensation over distances corresponding to $> 40z_0$ (2 ps pulses over a physical length of 20 km). (Note, we have also recently become aware that the authors of Reference [9] have also now demonstrated true soliton loss compensation [26])

The LCDDF was fabricated from a step-index preform of NA $= 0.165$, designed to give a 125 μm single-mode fibre cut-off wavelength of 1310 nm. The preform had excellent homogeneity along its length. The required dispersion variation was obtained by tapering the fibre during the pulling process following a calibration procedure in which the dispersion, loss and mode-area variation with fibre diameter were determined. The exact fibre diameter variation required to achieve complete loss compensation over 38 km of fibre was evaluated for an input dispersion of 6.0 ps/(nm.km). A second order correction to compensate for the $\sim$ 5% mode-area variation was included. The fibre was pulled in two sections (LCDDF1 and LCDDF2) that could either be used individually, or spliced together to give the full 38 km span. The first section had a length of 20 km (dispersion variation based on fibre loss $= 0.27$ dB/km) and the second section a length of 18 km (dispersion based on fibre loss $= 0.26$ dB/km). An OTDR plot of the actual fibre losses with the superposed design loss profiles are illustrated in Figure 6(a) and (b).

The two glitches (≈ 0.3 dB additional loss over lengths of 200 m) observed in the measured loss profiles were due to localised imperfections within the preform cladding and resulted in slight deviations from the desired form. The total deviation at any point within the fibres was however small ($< +/- 0.2$ dB) with the average loss matching almost exactly in the case of LCDDF1 and within 0.2 dB within LCDDF2. The mismatch over the full 38 km was therefore 0.2 dB in 10.1 dB. The dispersion of LCDDF1 was set to vary exponentially from 6 ps/(nm.km) at the input to 1.7 ps/(nm.km) at the output (corresponding to 5.4 dB loss) and from 1.7 ps/(nm.km) to 0.55 ps/(nm.km) (loss $= 4.7$ dB) in LCDDF2. The effective mode area varied from 47 μm^2 at the input to 50 μm^2 at the output. The required external fibre diameter variation was from 105-96 μm in LCDDF1 and from 96 μm to 92 μm in LCDDF2. The local RMS dispersion variation along the fibre length was <0.1 ps/(nm.km) as determined from

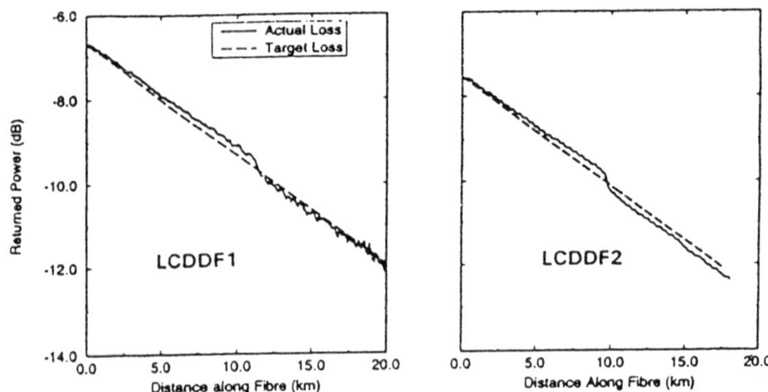

Figure 6. Experimental OTDR plot illustrating loss within LCDDF1 and LCDDF2 with superposed design loss profiles.

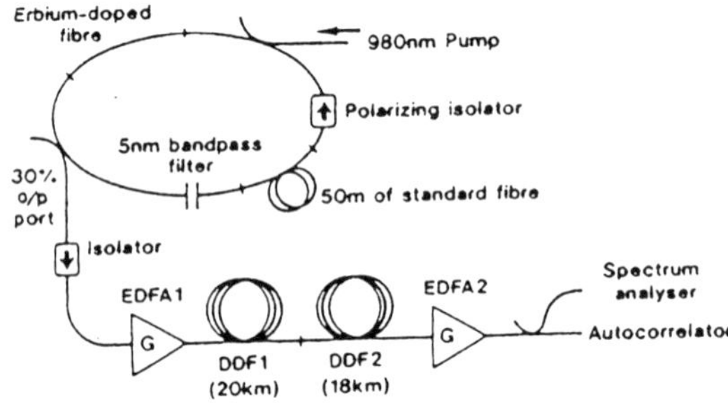

Figure 7. Experimental configuration used to examine soliton transmission characteristics of LCDDFs.

measurements of the RMS diameter error (0.18 μm). Measurements of the dispersion at the beginning and end of the two fibres along with average measurements along the two spans were in excellent agreement (+/- 0.15 ps/nm.km) with our target values. The third order dispersion within the fibre was measured to be 0.053 ps/(nm^2.km).

The experimental setup used to examine the pulse transmission characteristics of the LCDDF fibres is shown in Figure 7. Transform-limited pulses from a polarisation switch soliton laser with durations in the range 2.0-16 ps and tunable central wavelength in the range 1550-1565 nm were input into EDFA1. By varying the gain of EDFA1 we could then vary the signal power launched into the LCDDF in order to match it to the fundamental soliton power for the fibre. EDFA2 was used to boost the signal level prior to autocorrelation and spectral measurements of the pulses at the system output. Initial experiments were performed using just the 20 km LCDDF1 with input pulses of 3.6 ps and 2.0 ps duration and central wavelengths around 1555 nm. In Figure 8(a) we present the input and output spectra and autocorrelations for the 3.6 ps pulse case where the launched pulse corresponds to that of the fundamental (N=1) soliton. The transmission is seen to be excellent, the pulses are effectively indistinguishable in both the temporal and spectral domains indicating an excellent balance between dispersion and nonlinearity at all points along the LCDDF. The 20 km fibre corresponded to an effective 14 z_0 in this instance. Similar results are presented for the 2.0 ps case in Figure 8(b). In this instance the pulses were observed to broaden slightly in the time domain from 2.0-2.2 ps and to undergo slight spectral narrowing and deformation. In this instance the fibre corresponds to 44 z_0. The loss compensation for this case is worse for the shorter duration pulses due to increased sensitivity to localised imbalances in nonlinearity and dispersion e.g. the glitch in Figure 6(a) as the soliton period is reduced at all points along the fibre.

The pulse transmission was also investigated as a function of input pulse power. At low powers the pulse simply broadened linearly with no spectral distortion, however as the pulse power was increased significant spectral distortion in the form of spectral narrowing and side-band generation was obtained along with a narrowing of the pulse in the time domain. The spectral distortion, reminiscent in form to the side band generation observed in simple soliton lasers, arises due to the imperfect balance between dispersion and nonlinearity at powers away from the fundamental and was anticipated from our numerical simulations of the system. As the fundamental power is approached the depth of spectral modulation reduces as a more complete balance of the nonlinear and dispersive effects is obtained along the entire fibre length. At input powers significantly higher than the fundamental the excess nonlinearity results in a compression of the pulses within the LCDDF. The soliton self-frequency shift then comes into play and leads to a spectral walk off of the pulses towards longer wavelengths.

The two LCDDF fibres were then spliced together to create a single 38 km span. The total fibre loss was 10.4 dB (splice loss = 0.1 dB). 3.5 ps pulse transmission over the full fibre span was then examined. The fibre span constituted $\approx$ 18 soliton periods in this instance. The autocorrelations and spectra of the pulses at the input, 20 km and 38 km are illustrated in Figure 9 for the input fundamental soliton case. It is seen that the pulses are successfully transmitted over the fibre length with minimal temporal distortion. A degree of spectral lobing is apparent at both 20 and 38 km due to the slight mismatch in nonlinearity and dispersion over the fibre

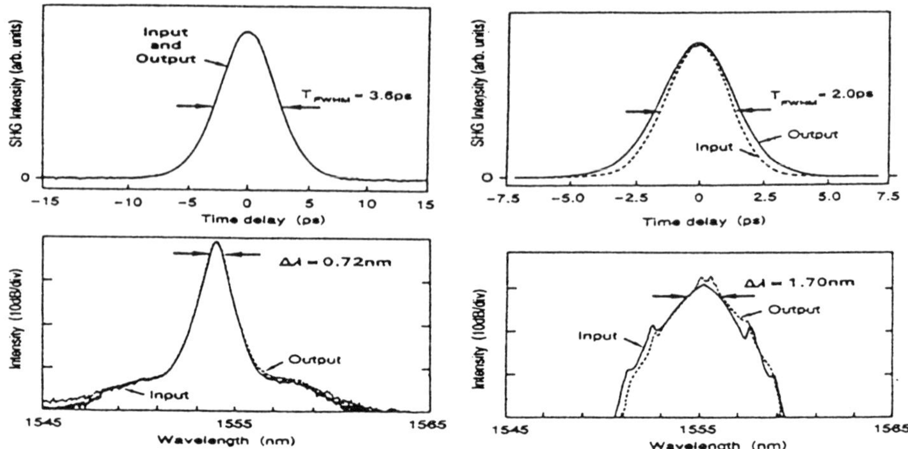

Figure 8. (a) ACF and spectra of 3.6 ps pulses at the input and output of LCDDF1 for the fundamental soliton input power, illustrating high-quality soliton loss compensation. The input and output autocorrelations are indistinguishable. The 20 km of fibre is equivalent to 18 z_0. (Marked input pulse widths are deconvolved halfwidths.) (b) Autocorrelation traces and spectra of 2.0 ps pulses at the input and output of LCDDF1 for a soliton input pulse power close to that of the fundamental. A 10% temporal broadening of the pulses is obtained and a small degree of spectral narrowing, lobing and shifting is observed. The 20 km of fibre is equivalent to 44 z_0. (Marked input pulse widths are deconvolved halfwidths.)

length but the deviations in pulse form between input and output are still small. The maximum ripple is <1 dB over a 4 nm bandwidth centred on the pulse illustrating the quality of the loss compensation.

In conclusion we have demonstrated the fabrication of a 38 km long soliton loss compensating transmission fibre. An excellent match to the dispersion variation to loss profile has been obtained. The quality of the fibre has been confirmed in a series of single pulse propagation experiments. Fundamental soliton propagation of 2.0 ps pulses over $\approx$ 44 soliton periods (20 km) and 3.5 ps pulses over 18 soliton periods (38 km) with minimal temporal and spectral deformation have been obtained. Our results clearly indicate that high-quality, soliton loss compensation can be achieved for a total distributed loss of $\approx$ 10 dB (40 km of fibre) and illustrate the pos-

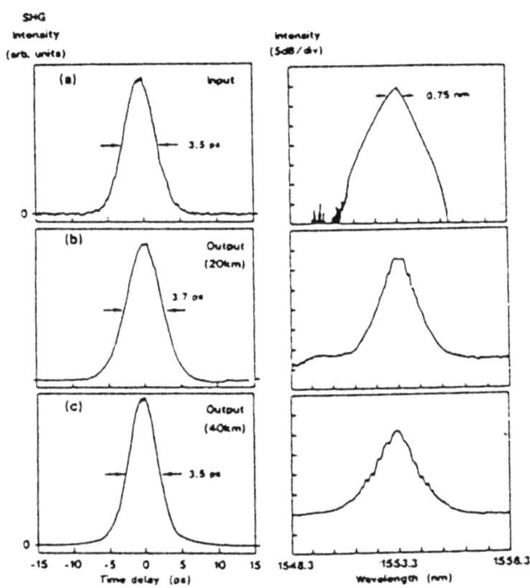

Figure 9.   Spectra and autocorrelation plots of 3.5 ps, N=1 solitons at (a) system input (0 km), (b) output of LCDDF1 (20 km) and (c) at output of LCDDF2 (38 km). (Marked pulse widths are deconvolved pulse halfwidths.)

sibility of amplifier spacings of $\geq$ 40 soliton periods for use in ultra-high speed transmission lines. Such span lengths (expressed in soliton periods) are around two orders of magnitude greater than those that can be envisaged for use in conventional soliton transmission systems limited by the constraints of average soliton dynamics. LCDDFs should therefore enable access to single channel system bit-rates well in excess of 40 Gbit/s providing the average dispersion of such fibres can be kept low enough to keep deleterious effects such as ASE induced timing jitter, soliton self frequency shift and soliton interaction effects to an acceptable level. It has also recently been pointed out that such fibres also seem offer advantages relative to conventional DSF in alleviating the problems due to soliton collisions in WDM based soliton systems [26]. Dispersion shifted fibres produced from a commercial preform typically exhibit a maximum variation in dispersion of $\approx$ 0.2 ps/(nm.km) along the entire pull length. With care improvements should be possible to enable $\approx$ 0.1 ps/(nm.km) accuracy along a 40 km fibre length and would enable a scaling of the dispersion characteristics of our fibre by $\approx$ 1/5, thereby reducing the average dispersion to $\approx$ 0.5 ps/(nm.km), a value commensurate with current conventional soliton systems.

5. Conclusions

Technology for the fabrication of dispersion decreasing fibre has developed to the extent that accurate and controllable dispersion variation can readily be obtained over fibre lengths ranging from 10 m to 40 km. A wide range of applications have been demonstrated in the laboratory and a number extended to real system tests. Whether such components will ever find real world application remains to be seen, however, it is already evident that the technology has opened a significant new dimension to what can be achieved within the fibre environment, undoubtedly new applications will open up.

6. Acknowledgements

The authors would like to acknowledge the considerable contribution of A. D. Ellis, D. M. Spirit, T. Widdowson and W. A. Pender during collaborative work on the 40 Gbit/s soliton transmitter. The contribution of Pirelli (Cavi) through supply of preforms used within this work and that of Fibre-Core UK for advising and assisting in the use of the 7 m pulling tower is gratefully acknowledged.

In addition, the assistance of York Technology and Pirelli U. K. for providing access to dispersion measuring instrumentation is greatly appreciated. This work is funded in part by European Union RACE project R2015 "ARTEMIS", European Union ACTs project "MIDAS", EPSRC and the Royal Society.

References

1. Tajima, K.: *Opt. Lett.* **12** (1987), 54-56.
2. Kuehl, H. H: *J. Opt. Soc. Am. B* **5** (1988), 709-713.
3. Mamyshev, P. V., Chernikov, S. V. and Dianov, E. M.: *IEEE J. Quantum Electron.* **27** (1991), 2347-2351.
4. Richardson, D. J., Chamberlin, R. P., Dong, L. and Payne, D. N.: *Electron. Lett.* **30** (1994), 1326-1327.
5. Chernikov, S. V. and Mamyshev, P. V.: *J. Opt. Soc. Am. B* **8** (1991), 1634-1641.
6. Bogatyrev, V. A. et al.: *IEEE J. of Lightwave Technol.* **9** (1991), 561-565.
7. Chernikov, S. V., Richardson, D. J., Laming, R. I., Dianov, E. M. and Payne, D. N.: *Electron. Lett.* **28** (1992), 1220-1221; Chernikov, S. V. and Taylor, J. R.: *Electron. Lett.* **28** (1992), 931-932.
8. Chernikov, S. V., Richardson, D. J., Laming, R. I., Dianov, E. M. and Payne, D. N.: *Electron. Lett.* **28** (1992), 931-932.
9. Richardson, D. J., Chamberlin, R. P., Dong, L., Payne, D. N., Ellis, A. D., Widdowson, T. and Spirit, D. M.: *Electron. Lett.* **31** (1995), 470-472.
10. Ellis, A. D., Pender, W. A., Widdowson, T., Richardson, D. J., Dong, L. and Chamberlain, R. P.: *Electron. Lett.* **31** (1995), 1362-1364.
11. Richardson, D. J., Chamberlin, R. P., Dong, L. and Payne, D. N.: *Electron. Lett.* **31** (1995), 1681-1682.
12. See e.g. Ainslie, B. J. and Day, C. R.: *IEEE J. of Lightwave Technol.* **8** (1986), 967; Snyder, A. W. and Love, J. D.: *Optical Waveguide Theory*, Chapman and Hall,

(1983).

13. Chernikov, S. V., Taylor, J. R. and Kayshap, R.: *Opt. Lett.* **19** (1993), 539-542.
14. Swanson, E. A., Chin, S. R., Hall, K., Rauschenbach, K. A., Bondurant, R. S. and Miller, J. W: *Technical Digest Optical Fibre Communications*, Post deadline paper, **PD15-1**, San Jose, (1994).
15. Ellis, A. D., Widdowson, T., Shan, X., Wickens, G. E. and Spirit, D. M.: *Electron. Lett.* **29** (1993), 990-991.
16. Blow, K. J. and Doran, N. J.: *Photon Tech. Lett.* **3** (1991), 369-371.
17. Hasegawa, A. and Kodama, Y.: *Opt. Lett.* **15** (1990), 1443-1445.
18. Smith, N. J., Blow, K. J. and Andonovic, I.: *J. Lightwave Technol.* **10** (1992), 1329-1333.
19. Treitinger, L. et al.: *Proceedings European Conference on Optical Communications*, *Vol.1* **Tu. A.2.2**, Brussels, (1995), 189.
20. Gordon, J. P. and Haus, H. A.: *Opt. Lett.* **11** (1986), 665-667.
21. Mollenauer, L. F., Gordon, J. P. and Evangelides, S. G.: *Opt. Lett.* **17** (1992), 1575-1577.
22. Nakazawa, M., Yamada, E., Kubota, H. and Suzuki, E.: *Electron. Lett.* **27** (1991), 1270-1271.
23. Mollenauer, L. F., Gordon, J. P. and Islam, M. N.: *IEEE J. Quantum Electron.* **QE-22**, 157-173, (1986).
24. Nakazawa, M. and Kurokawa, K.: *Electron. Lett.* **27** (1991), 1369-1371.
25. Evans, A. F., Stentz, A. J. and Boyd, R. W.: *Proceedings European Conference on Optical Communications ECOC'94*, *Vol.1*, Florence, (1994), 323-326.
26. Stentz, A. J., Boyd, R. W. and Evans, A. F.: *Opt. Lett.* **20** (1995), 1770-1772.

EFFECTS OF NONLINEAR GAIN ON SOLITON TRANSMISSION IN FIBERS

M. MATSUMOTO AND A. HASEGAWA

*Department of Communication Engineering,
Osaka University
2-1 Yamada-Oka, Suita, Osaka 565, Japan*

Abstract. In this paper we discuss the effect of nonlinear gain in suppressing the linear-wave growth in soliton transmission systems. We also show that bi- or multi-stable soliton transmission may be achieved when higher-order dependency of the gain on the intensity is considered. Some properties of nonlinear (amplifying) loop mirrors as the nonlinear-gain element are presented.

1. Introduction

In long-distance soliton transmission systems, some gain should be provided to overcome fiber loss. If the gain exactly compensates for the loss, the ideal nonlinear Schrödinger equation is satisfied and the ideal soliton propagation is achieved under the condition that higher-order effects can be neglected. In this case the power of noise emitted from the optical amplifiers is accumulated at most linearly with transmission distance. When band-pass filters are inserted periodically in the transmission line to stabilize the soliton frequency [1],[2] or finite gain bandwidth of the optical amplifiers is considered [3],[4], some excess gain should be given to compensate for the extra loss that solitons suffer at wings of their frequency spectrum. In this case the noise whose frequency lies in the filter pass band grows exponentially. The amplified noise degrades the SN ratio and the interaction of the noise with the soliton eventually destroy the soliton [5],[6]. Aside from the soliton transmission, the problem of the stability of background is also an important issue in the operation of mode-locked lasers [7],[8].

One method to reduce the noise amplification or the background instability is to give the excess gain in the form of a nonlinear gain that selectively amplifies solitons with the noise unamplified or attenuated. In this paper we summarize the effect of nonlinear gain in eliminating the

293

A. Hasegawa (ed.), Physics and Applications of Optical Solitons in Fibres '95, 293–305.
© 1996 *Kluwer Academic Publishers. Printed in the Netherlands.*

low-power radiation that coexists with solitons. We further show that multi-stable soliton transmission may be achieved when the effect of gain nonlinearity with higher orders is considered.

2. Elimination of low-power radiation by means of nonlinear gain

The pulse propagation in optical fibers with linear and nonlinear amplification and narrow-band filtering can be described by the averaged nonlinear Schrödinger equation with additional terms

$$i\frac{\partial q}{\partial Z} + \frac{1}{2}\frac{\partial^2 q}{\partial T^2} + |q|^2 q = i\delta(|q|^2)q + i\beta\frac{\partial^2 q}{\partial T^2}. \tag{1}$$

Here $\delta(|q|^2)$ is an excess gain (difference between the averaged amplifier gain and the fiber loss), that may be a function of $|q|^2$ and β is a curvature of the frequency response of the narrow-band filter at the filter center frequency. We first consider the evolution of low-power radiation. The low-power amplitude $a_\Omega(Z)$, with which $q(Z,T)$ is given by $q(Z,T) = a_\Omega(Z)\exp(-i\Omega T)$, evolves according to $|a_\Omega(Z)|^2 = |a_\Omega(0)|^2 \exp[2(\delta_0 - \beta\Omega^2)Z]$, where δ_0 is a linear-gain coefficient for small signals, that is, $\delta_0 = \lim_{|q|\to 0}\delta(|q|^2)$. Using this solution we can evaluate the accumulation and amplification of the noise generated by optical amplifiers inserted in the transmission line. Each amplifier is assumed to emit white noise with power spectrum density N_0. The noise power emitted from an amplifier grows to

$$P_N = \int_{-\infty}^{\infty}\frac{N_0}{2\pi}\exp[2(\delta_0 - \beta\Omega^2)Z]d\Omega = \frac{N_0}{2\pi}\sqrt{\frac{\pi}{2\beta Z}}\exp(2\delta_0 Z) \tag{2}$$

after propagation over a distance Z. If amplifiers emitting uncorrelated noise are inserted in the line at a rate of M amplifiers per unit distance, the noise power at the receiver end is accumulated to

$$P = \int_0^L MP_N dZ = \frac{N_0 M}{\sqrt{2\pi\beta}}\int_0^{\sqrt{L}}\exp(2\delta_0 x^2)dx \tag{3}$$

where L is the normalized total distance of the transmission system [9]. The noise power is expressed in unnormalized quantities as

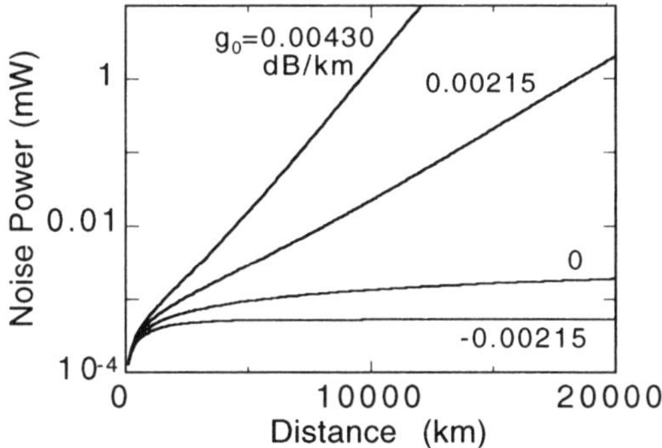

Figure 1. Growth of noise power in a transmission system with narrow-band filtering.

$$P[W] = \frac{\sqrt{\pi z_f}\, cn_{sp} \hbar \omega \alpha \Delta \lambda}{\lambda^2} \int_0^{\sqrt{l}} \exp(2g_0 x^2)\,dx \tag{4}$$

where amplification is assumed to be distributed to compensate for fiber loss α while filters with Lorentzian shape with half-power bandwidth $\Delta\lambda$ are inserted at every z_f. In (4), n_{sp} is the spontaneous emission factor of the amplifier, and l and g_0 are related to L and δ_0, respectively, through $l = L z_d$ and $g_0 = \delta_0 / z_d$, where z_d is the dispersion distance. Figure 1 shows an example of the noise power versus the transmission distance. n_{sp}, α, $\Delta\lambda$, and z_f used in the calculation are 2, 0.25dB/km, 0.58nm, and 50.5km, respectively. The values of $\Delta\lambda$ and z_f correspond to the filter strength β=0.3, if a group-velocity dispersion parameter D=0.5 ps/nm/km is assumed. The linear gain coefficient g_0=0.00430dB/km in the figure corresponds to δ_0=0.1 for a soliton with pulse width of 20ps. The noise power is shown to grow significantly even when the excess gain is very small. The noise power saturates at low levels, on the other hand, when there is a net linear loss $(g_0 < 0)$. The problem of the linear-wave growth will be more severe in actual soliton transmission systems since dispersive radiations such as those generated by the discrete lumped amplification of solitons [10]-[12], by the random variation of the group-velocity dispersion [13], by the random birefringence variation [14], and by the filtering of solitons [6] also contribute to the growth.

Now we examine transmission properties of solitons in the presence of nonlinear gain. Here we consider two types of nonlinear gain. In one case, the gain coefficient is given by a second-order polynomial of $|q|^2$ as [5],[15],[16]

$$\delta(|q|^2) = \delta_0 + \gamma_1 |q|^2 + \gamma_2 |q|^4. \tag{5}$$

In the other case, the property of fast saturable absorption is explicitly contained in the gain coefficient as [17]

$$\delta(|q|^2) = \delta_0' - \frac{\alpha_0}{1 + |q|^2 / I_s}, \tag{6}$$

where I_s is a saturation intensity. In (6), the gain coefficient for small signals is $\delta_0 = \delta_0' - \alpha_0$. One-soliton dynamics governed by (1) together with (5) and (6) can be analyzed by evaluating the dynamical evolution of the soliton parameters, the amplitude η and the frequency κ, with which the one-soliton solution is given by

$$q(T,Z) = \eta(Z)\,\mathrm{sech}\{\eta(Z)[T + \kappa(Z) - \theta]\}$$
$$\exp\left\{-i\kappa(Z)T + \frac{i}{2}\left[\eta^2(Z) - \kappa^2(Z)\right]Z - i\sigma\right\}. \tag{7}$$

From the energy and momentum conservation relations modified by the presence of the perturbation [18], we have [9]

$$\frac{d\eta}{dZ} = \begin{cases} 2\delta_0\eta - 2\beta\eta\left(\frac{1}{3}\eta^2 + \kappa^2\right) + \frac{4}{3}\gamma_1\eta^3 + \frac{16}{15}\gamma_2\eta^5 & \text{(8a)} \\[0.5em] \qquad\qquad \text{(for the nonlinear gain given by (5))} \\[0.5em] 2\delta_0'\eta - 2\beta\eta\left(\frac{1}{3}\eta^2 + \kappa^2\right) - \frac{2\alpha_0 I_s}{\sqrt{\eta^2 + I_s}}\cosh^{-1}\sqrt{1 + \frac{\eta^2}{I_s}} & \text{(8b)} \\[0.5em] \qquad\qquad \text{(for the nonlinera gain given by (6))} \end{cases}$$

and
$$\frac{d\kappa}{dZ} = -\frac{4}{3}\beta\eta^2\kappa. \tag{9}$$

From Eq.(9) it is apparent that the frequency approaches asymptotically to a stationary value $\kappa = 0$ if $\beta > 0$. Stationary values for the amplitude, on the other hand are given by minimum points of the potential function ϕ defined by

$$\frac{d\eta}{dZ} = -\frac{\partial\phi}{\partial\eta}. \tag{10}$$

Figure 2 shows examples of the potential versus η when the frequency is kept at $\kappa = 0$. Both potentials have minima at $\eta = 0$ and $\eta \cong 1$. The presence of

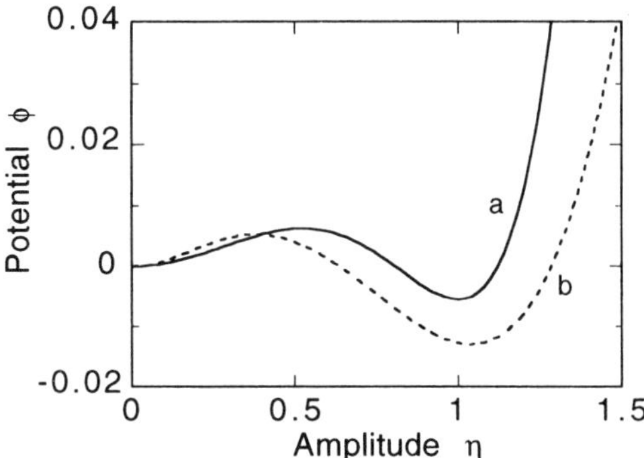

Figure 2. Potential functions for soliton transmission systems with nonlinear gain. (a : nonlinear gain is given by (5) with $\beta = 0.3$, $\delta_0 = -0.05$, $\gamma_1 = 0.5$, and $\gamma_2 = -0.34375$, b : nonlinear gain is given by (6) with $\beta = 0.3$, $\delta_0' = 0.15$, $\alpha_0 = 0.25$, and $I_s = 0.1$)

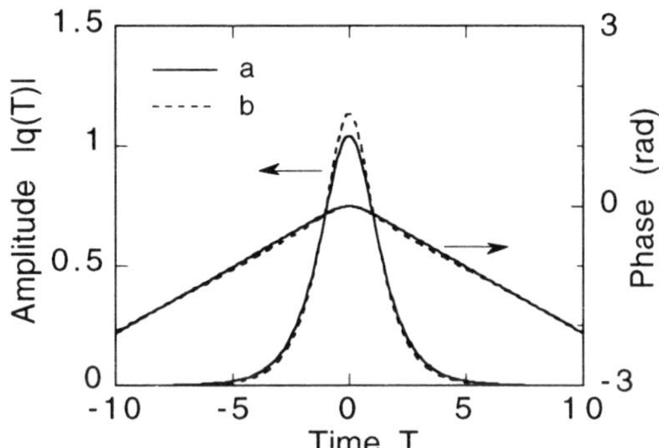

Figure 3. Asymptotic pulse shapes and phase distributions in soliton transmission systems with nonlinear gain. a and b correspond to those given in Figure.2.

minima at $\eta = 0$ implies that the small signal gain δ_0 is negative, which is needed for the suppression of the linear-wave growth. The minima at $\eta \cong 1$ attract solitons, realizing stable soliton transmission.

For the nonlinear gain (5), the condition for the existence of minima at $\eta = 0$ and $\eta = 1$ is $\delta_0 < 0$, $\gamma_2 < 0$, $\beta < 2\gamma_1$, $8\gamma_2 < 15\delta_0$, and $15\delta_0 - 5(\beta - 2\gamma_1) + 8\gamma_2 = 0$. It is noted that the inclusion of γ_2 is necessary to have the double-minimum potential. If γ_2 is absent, γ_1 must be larger than $\beta / 2$ for the small signal gain δ_0 to be negative. In this instance, however, the amplitude of soliton goes to zero or infinity because the effect of loss saturation owing to γ_1 surpasses the effect of equivalent gain saturation given by the action of the filter.

Figure 3 shows the asymptotic pulse shapes and phase distributions obtained by solving (1) numerically for both types of the nonlinear gain. The asymptotic pulses have approximately sech shapes and positive frequency chirp around the pulse center. Recently it has been reported that the phase variation across the pulse may play an important role in reducing the pulse-to-pulse interaction through the formation of out-of-phase bound states [19],[20].

The above consideration of the effect of the gain nonlinearity on the soliton transmission is rather qualitative. Here we present simulation results based on more realistic situation assuming a specific device for the nonlinear-gain element [21]. One natural candidate for such a device that can operate at relatively low power is a nonlinear amplifying loop mirror (NALM)[22]. The NALM has been successfully employed as an effective saturable absorber in mode-locked fiber lasers [23],[24].

The amplitude transmission coefficient of the NALM that incorporates a coupler with a coupling ratio of 1:1 and an amplifier with power gain G_N at one end of the loop is approximately given by

$$t = \sqrt{G_N}\,\sin\!\left(\frac{G_N-1}{4}L_N|q|^2\right)\exp\!\left(i\frac{G_N+1}{4}L_N|q|^2\right), \tag{11}$$

where L_N is the loop length.   When the argument of the sinusoidal function is small, the magnitude of the coefficient becomes $|t| \cong \sqrt{G_N}\,(G_N-1)L_N|q|^2/4$, which indicates that the NALM behaves as a saturable absorber.

Figure 4(a) is a result of our simulation of soliton transmission in a system with NALM's periodically inserted. The insertion period of the NALM in Figure 4(a) is $Z=10$ and the values of G_N and L_N are 28 (14.5dB) and 0.05, respectively. Filters and excess gain ($\beta = 3\delta = 0.3$) are distributed throughout the transmission line. In this figure it is shown that the pulse is compressed at the output of NALM's since the NALM amplifies the center portion of the pulse and remove pedestals. Although the compressed and phase-modulated pulse changes its shape with distance, it recovers its shape at the output of a series of NALM's. Figure 4(b) shows the propagation of a soliton in the absence of NALM's. The soliton waveform is significantly distorted and destroyed after propagating approximately $Z=30$ as a result of the interaction with the amplified linear waves generated by the filtering of the soliton.

In addition to the use of the NALM as a noise filter as shown above, a NALM can also be used as an adiabatic amplifier of solitons, which can recover the soliton decayed adiabatically in propagating down the fiber [25],[26]. The

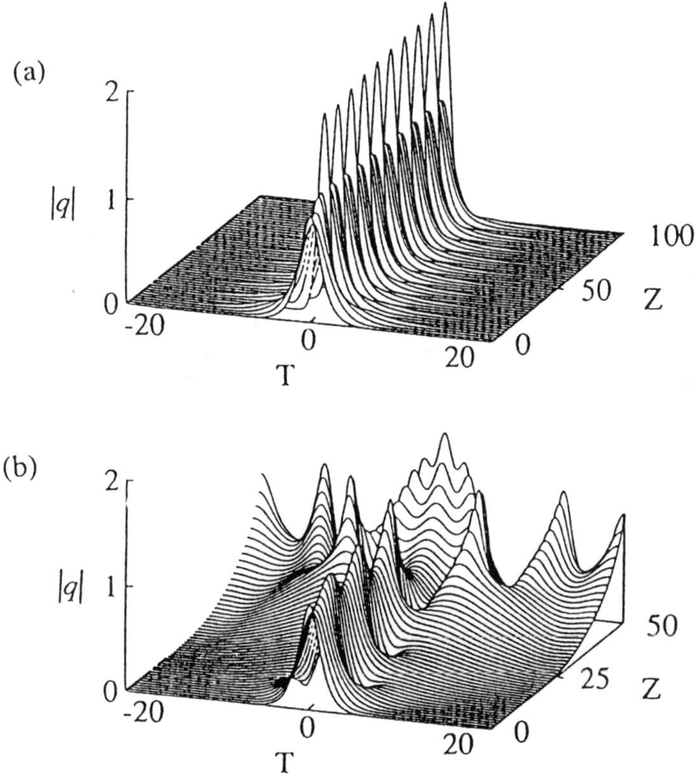

Figure 4.   Propagation of a soliton in a system with bandwidth-limited amplification.
(a) : NALM's are periodically inserted. (b) : NALM's are not inserted.

necessity of the adiabatic soliton amplifier will arise in transmitting picosecond
or subpicosecond solitons, for which the dispersion distance is much smaller
than the loss distance of the fiber.

The property of a nonlinear-gain element or a saturable absorber similar to that
of the NALM can also be realized by using the effect of nonlinear polarization
rotation with a polarizer[27], by using a nonlinear directional coupler[28],[29],
or by using quantum-well saturable absorbers[30].

3. Multi-stable soliton transmission

In the previous Section we considered a gain coefficient given by a second-order polynomial of $|q|^2$ as (5). Including higher-order terms in the expression, we have a generalized form of the gain coefficient as

$$\delta(|q|^2) = \delta_0 + \sum_{n=1}^{\infty} \gamma_n |q|^{2n} . \tag{12}$$

The evolution of the soliton amplitude is then governed by

$$\frac{d\eta}{dZ} = 2\delta_0 \eta - 2\beta\eta\left(\frac{1}{3}\eta^2 + \kappa^2\right) + \sum_{n=1}^{\infty} \frac{2^{2n+1}(n!)^2}{(2n+1)!} \gamma_n \eta^{2n+1} , \tag{13}$$

which gives a potential function

$$\phi(\eta) = -\delta_0\eta^2 + \frac{1}{6}\beta\eta^4 - \sum_{n=1}^{\infty} \frac{2^{2n+1}(n!)^2}{(2n+2)!} \gamma_n \eta^{2n+2} \equiv \sum_{n=1}^{\infty} a_{2n}\eta^{2n} , \tag{14}$$

where the soliton frequency is again assumed to be kept at $\kappa = 0$. The potential can have multiple minima at designed locations if the values of the coefficients a_{2n} are carefully determined. Multi-stable soliton transmission is then achieved since a soliton can stay stably at any one of these potential minima. The existence of bistable soliton states under the effect of higher-order dependency of the refractive index on intensity has been discussed in the literature [31],[32]. In that case the bistability means the existence of solitons with different pulse shapes or different peak amplitudes for the same pulse energy or the same pulse width. In our case of gain nonlinearity, the transmission medium itself presents bi- or multi-stability. Solitons with different amplitudes are transmitted stably even in the presence of perturbations such as amplifier noise.

Figure 5 shows an example of the potential having multiple minima. Solitons with amplitude $\eta \cong 1.2$ or 1.9 will propagate stably. Figure 6 shows a numerical example of pulse propagation (numerical solution of (1)) under the effect of nonlinear gain with higher-order terms that yields the potential shown in Figure 5. Up- and down-switching between the two stable states can be achieved merely by adding and extracting some energy to and from the soliton, respectively. In Figure 6, a linear lumped amplification with amplitude coefficient 1.2 and a linear lumped attenuation with coefficient 0.8 are applied at $Z=50$ and 100, respectively.

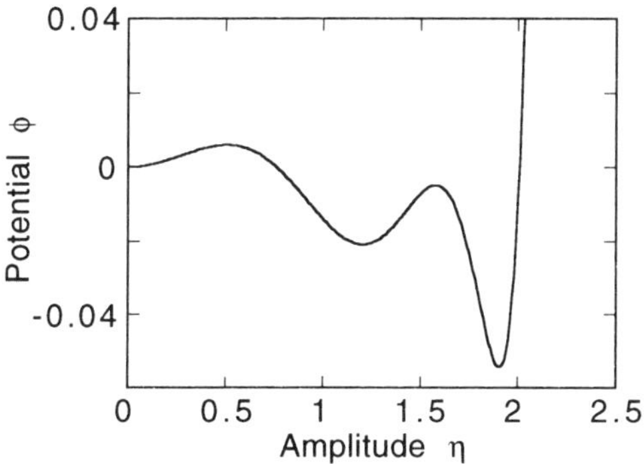

Figure 5.   Potential function having multiple minima. ($\beta = 0.3$, $a_2 = 0.05$, $a_4 = -0.12$, $a_6 = 0.06$, $a_{10} = -0.004$, and $a_{14} = 0.00012$).

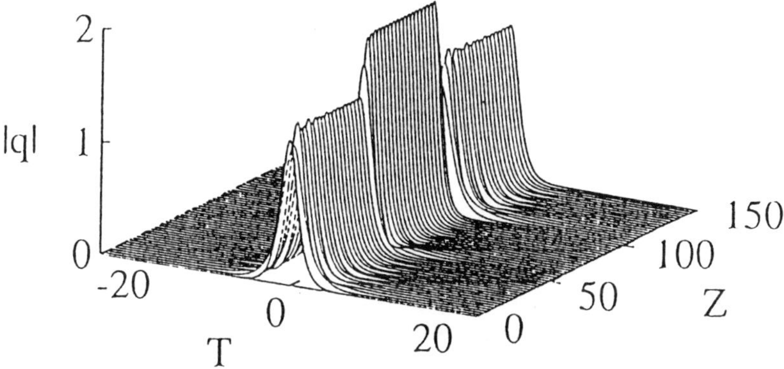

Figure 6.   Propagation of a soliton under the presence of nonlinear gain whose potential function is given in Figure5.   Switching between bistable states is performed at $Z=50$ and 100.

One possible way to obtain the nonlinear gain that gives multi-stable transmission states is to use nonlinear interferometric devices such as nonlinear loop mirrors [33], which are discussed as a saturable absorber in the previous section. Here we examine the soliton-transmission property of a nonlinear loop mirror exhibiting multiple transmission peaks. The nonlinear loop mirror we analyze here is depicted in Figure 7.

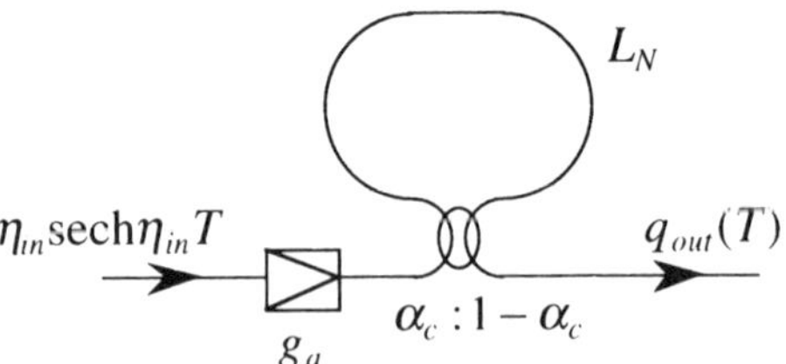

Figure7.   A nonlinear loop mirror with a preamplifier.

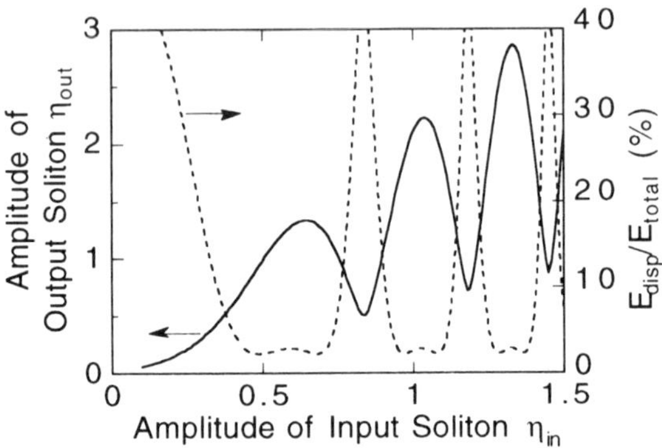

Figure 8.   Output soliton amplitude and percentage of dispersive-wave energy versus input soliton amplitude for a nonlinear loop mirror ($g_a = 1.65$, $\alpha_c = 0.2$, and $L_N = 5$).

The incoming soliton with amplitude η_{in} is first amplified with amplification coefficient g_a and then fed into the loop mirror. The loop mirror consists of a directional coupler with power coupling ratio $\alpha_c : 1 - \alpha_c$ and a fiber loop with length L_N. If $N = 1$ solitons are formed in the loop for both propagation directions, the output pulse can be approximately given by

$$
\begin{aligned}
q_{out}(T) = {}& \sqrt{\alpha_c}\,(2g_a\sqrt{\alpha_c} - 1)\eta_{in}\,\mathrm{sech}\!\left[(2g_a\sqrt{\alpha_c} - 1)\eta_{in}T\right] \\
& \exp\!\left[\frac{i}{2}(2g_a\sqrt{\alpha_c} - 1)^2\,\eta_{in}^2 L_N\right] \\
& - \sqrt{1-\alpha_c}\,(2g_a\sqrt{1-\alpha_c} - 1)\eta_{in}\,\mathrm{sech}\!\left[(2g_a\sqrt{1-\alpha_c} - 1)\eta_{in}T\right] \\
& \exp\!\left[\frac{i}{2}(2g_a\sqrt{1-\alpha_c} - 1)^2\,\eta_{in}^2 L_N\right]
\end{aligned}
\tag{15}
$$

We then numerically evaluate the eigenvalue of the soliton component contained in q_{out} and the energy of dispersive radiation. Figure 8 shows an example of the amplitude of the output soliton η_{out} (two times the imaginary part of the eigenvalue) and the percentage of the dispersive radiation contained in the output pulse. The energy of the dispersive radiation is calculated from

$$E_{disp} = E_{total} - E_{sol} = \int_{-\infty}^{\infty} \left| q_{out}(T) \right|^2 dT - 2\eta_{out}. \qquad (16)$$

When $\eta_{in} \cong 0.65$, 1.05, or 1.35, two pulses propagating in the loop in opposite directions interfere constructively at the output coupler, so that soliton-like pulses are produced at the output port. For input amplitudes between these transmission peaks, destructive interference takes place, so that the output waveform differs significantly from that of the $N=1$ soliton. In Refs. [34] and [35] the first transmission peak is proposed to be used for obtaining stable soliton transmission. If the soliton can also be stabilized near the second- or higher-order peaks with the condition $\left| d\eta_{out} / d\eta_{in} \right| < \eta_{out} / \eta_{in}$ being satisfied, multi-stable-state soliton transmission will be achieved. The bi- or multi-stability of soliton transmission in fibers can be utilized to enhance the capacity of soliton transmission systems or soliton storage rings.

4. Conclusion

In this paper we discussed the effect of nonlinear gain on soliton transmission in fibers. By using a suitable combination of linear gain, which should be negative, and nonlinear gain, we can stabilize both the background and the soliton. We also showed that bistable soliton transmission can be realized by including higher-order gain nonlinearity and switching between these stable states is easily carried out. The existence of the bistable or multi-stable states under the effect of gain nonlinearity may be utilized to enhance the capacity of soliton transmission systems or soliton storage rings. Candidates for the nonlinear devices discussed in this paper include nonlinear interferometric devices such as nonlinear loop mirrors and nonlinear directional couplers. For these devices to be actually used, their ability to operate at low powers and stability should be improved.

Although we are concerned only with bright solitons in this paper, the use of nonlinear gain has been shown to be effective to stabilize transmission of dark solitons in the normal dispersion regime as well [36],[37].

References

1.   Mecozzi, A., Moores, J.D., Haus, H.A., and Lai,Y. :Soliton transmission control, *Opt. Lett.* **16** (1991), 1841-1843.

2. Kodama, Y. and Hasegawa, A. :Generation of asymptotically stable optical solitons and suppression of the Gordon-Haus effect, *Opt. Lett.* **17** (1992), 31-33.

3. Blow, K.J., Doran, N.J., and Wood, D. :Trapping of energy into solitary waves in amplified nonlinear dispersive systems, *Opt. Lett.* **12** (1987), 1011-1013.

4. Nakazawa, M., Kurokawa, K., Kubota, H., and Yamada, E. :Observation of the trapping of an optical soliton by adiabatic gain narrowing and its escape, *Phys. Rev. Lett.* **65** (1990), 1881-1884.

5. Kodama, Y., Romagnoli, M., and Wabnitz, S. :Soliton stability and interactions in fibre lasers, *Electron. Lett.* **28** (1992), 1981-1983.

6. Matsumoto, M. and Hasegawa, A. :Numerical study of the reduction of instability in bandwidth-limited amplified soliton transmission, *Opt. Lett.* **18** (1993), 897-899.

7. Haus, H.A., Fujimoto, J.G., and Ippen, E.P. : Structures for additive pulse mode locking, *J. Opt. Soc. Am. B* **8** (1991), 2068-2076.

8. Chen, C.-J., Wai, P.K.A., and Menyuk, C.R. : Stability of passively mode-locked fiber lasers with fast saturable absorption, *Opt. Lett.* **19** (1994), 198-200.

9. Matsumoto, M., Ikeda, H., Uda, T., and Hasegawa, A. :Stable soliton transmission in the system with nonlinear gain, *J. Lightwave Technol.* **13** (1995), 658-665.

10. Hasegawa, A. and Kodama, Y. : Guiding-center soliton in optical fibers, *Opt. Lett.* **15** (1990), 1443-1445.

11. Mollenauer, L.F., Evangelides, Jr., S.G., and Haus, H.A. :Long-distance soliton propagation using lumped amplifiers and dispersion sifted fiber, *J. Lightwave Technol.* **9** (1991), 194-197.

12. Blow, K.J. and Doran, N.J. :Average soliton dynamics and the operation of soliton systems with lumped amplifiers, *IEEE Photonics Technol. Lett.* **3** (1991), 369-371.

13. Ohhira, R., Hasegawa, A., and Kodama, Y. : Methods of constructing a long-haul soliton transmission system with fibers having a distribution in dispersion, *Opt. Lett.* **20** (1995), 701-703.

14. Wai, P.K.A., Menyuk, C.R., and Chen, H.H. :Stability of solitons in randomly varying birefringent fibers, *Opt. Lett.* **16** (1991), 1231-1233.

15. Moores, J.D. : On the Ginzburg-Landau laser mode-locking model with fifth-order saturable absorber term, *Opt. Commun.* **96** (1993), 65-70.

16. Malomed, B.A. : Evolution of nonsoliton and "quasi-classical" wavetrains in nonlinear Schrödinger and Korteweg de-Vries equations with dissipative perturbations, *Physica* **29D** (1987), 155-172.

17. Afanasjev, V.V., Loh. W.H., Grudinin, A.B., Atkinson, D., and Payne, D.N. :Unlimited soliton propagation and noise suppression in a system with spectral filtering and saturable absorption, *Tech. Dig. Conf. Lasers and Electro-Optics* (1994), 366-367.

18. Kodama, Y. and Hasegawa, A. : Nonlinear pulse propagation in a monomode dielectric guide, *IEEE J. Quantum Electron.* **QE-23** (1987), 510-524.

19. Uzunov, I.M., Musehall, R., Gölles, M., Lederer, F., and Wabnitz, S. : Effect of nonlinear gain and filtering on soliton interaction, *Opt. Commun.* **118** (1995), 577-580.

20. Afanasjev, V.V. and Akhmediev, N. : Soliton interaction and bound states in amplified-damped fiber systems, *Opt. Lett.* **20** (1995), 1970-1972.

21. Matsumoto, M, Ikeda, H., and Hasegawa, A. : Suppression of noise accumulation in bandwidth-limited soliton transmission by means of nonlinear loop mirror, *Opt. Lett.* **19** (1994), 183-185.

22. Fermann, M.E., Haberl, F., Hofer, M., and Hochreiter, H. : Nonlinear amplifying loop mirror, *Opt. Lett.* **15** (1990), 752-754.

23. Duling, III, I.N. : Subpicosecond all-fibre erbium laser, *Electron. Lett.* **27** (1991), 544-545.

24. Richardson, D.J., Laming, R.I., Payne, D.N., Phillips, M.W., and Matsas, V.J. : 320fs soliton

generation with passively mode-locked erbium fibre laser, *Electron. Lett.* **27** (1991), 730-732.

25. Matsumoto, M, Kodama, Y., and Hasegawa, A. : Adiabatic amplification of solitons by means of nonlinear amplifying loop mirrors, *Opt. Lett.* **19** (1994), 1019-1021.

26. Rottwitt, K., Margulis, W., and Taylor, J.R. : Soliton recovery using a nonlinear amplifying loop mirror, *Electron. Lett.* **31** (1995), 395-397.

27. Stolen, R.H., Botineau, J., and Ashkin, A. : Intensity discrimination of optical pulses with birefringent fibers, *Opt. Lett.* **7** (1982), 512-514.

28. Winful, H.G. and Walton, D.T. : Passive mode locking through nonlinear coupling in a dual-core fiber laser, *Opt. Lett.* **17** (1992), 1688-1690.

29. Chu, P.L., Peng, G.D., Malomed, B.A., Hatami-Hanza, H., and Skinner, I.M. : Time-domain soliton filter based on a semidissipative dual-core fiber, *Opt. Lett.* **20** (1995), 1092-1094.

30. Atkinson, D., Loh, W.H., Afanasjev, V.V., Grudinin, A.B., Seeds, A.J., and Payne, D.N. : Increased amplifier spacing in a soliton system with quantum-well saturable absorbers and spectral filtering, *Opt. Lett.* **19** (1994), 1514-1516.

31. Enns, R.H., Edmundson, D.E., Rangnekar, S.S., and Kaplan, A.E. : Optical switching between bistable soliton states : a theoretical review, *Optical and Quantum Electron.* **24** (1992), s1295-s1314.

32. Herrmann, J. : Bistable bright solitons in dispersive media with a linear and quadratic intensity-dependent refractive index change, *Opt. Commun.* **87** (1992), 161-165.

33. Doran, N.J. and Wood, D. : Nonlinear-optical loop mirror, *Opt. Lett.* **13** (1988), 56-58.

34. Smith, N.J. and Doran, N.J. : Picosecond soliton propagation using nonlinear optical loop mirrors as intensity filters, *Electron. Lett.* **30** (1994), 1084-1085.

35. Smith, N.J. and Doran, N.J. : Picosecond soliton transmission using concatenated nonlinear optical loop-mirror intensity filters, *J. Opt. Soc. Am. B* **12** (1995), 1117-1125.

36. Matsumoto, M., Ikeda, H., and Hasegawa, A. : Reduction of Gordon-Haus effect on dark solitons by means of nonlinear gain, *Electron. Lett.* **31** (1995), 482-483.

37. Ikeda, H., Matsumoto, M., and Hasegawa, A. : Transmission control of dark solitons by means of nonlinear gain, *Opt. Lett.* **20** (1995), 1113-1115.

NONLINEAR POLARIZATION MODE DISPERSION IN OPTICAL FIBERS WITH RANDOMLY VARYING BIREFRINGENCE

P. K. A. WAI
Department of Electronic Engineering
The Hong Kong Polytechnic University
Hung Hom, Kowloon, Hong Kong

AND

C. R. MENYUK
Department of Computer Sciences and Electrical Engineering
University of Maryland Baltimore County
Baltimore, Maryland 21228-5398, USA

Abstract. The evolution of polarization states are studied in fibers with randomly varying birefringence. A physical model in which the birefringence orientation varies arbitrarily while the birefringence strength is fixed is used to study the evolution of polarization states in fibers with randomly varying birefringence. We show that the coupled nonlinear Schrödinger equation which describes wave evolution over a long length along a communication fiber can be reduced to the Manakov equation with additional terms that describe linear and nonlinear polarization mode dispersion. The coefficients of the linear and nonlinear polarization mode dispersion are the Stokes parameters and the squares of the Stokes parameter. The implications for numerical simulations of transoceanic communications are discussed.

1. Introduction

A single-mode fiber can support two polarization modes. Early studies on light propagation in fibers typically ignore the effect of birefringence by assuming that the two modes have the same refractive index. In real fibers, because of perturbations such as core ellipticity, stress, bending, and twist,

A. Hasegawa (ed.), Physics and Applications of Optical Solitons in Fibres '95, 307–317.
© 1996 *Kluwer Academic Publishers. Printed in the Netherlands.*

the two polarization modes have different refractive indices. As a result, the phase accumulated by the two polarization modes will vary along the fiber. If the birefringence is constant, the two polarization modes will accumulate a 2π phase difference in a beat length, $L_B = \lambda/\Delta n$, where Δn is the refractive index difference and λ is the wavelength. The two polarization modes also have different group velocities because of the birefringence. The differential group velocity delay between the two polarizations is called polarization mode dispersion (PMD) [1]- [14] which, similar to chromatic dispersion, will lead to broadening of signals in time [14].

In communication fibers, the relative index difference $\Delta n/n$ between the two polarization modes is typically in the range $\sim 10^{-7} - 10^{-5}$, and the polarization mode dispersion is on the order of 1 psec/$\sqrt{\text{km}}$ so that even in the best communication fibers $L_B \sim 10$ m and a differential group delay of ~ 100 picoseconds will be accumulated in 10,000 km. This is potentially devastating to communications because dual images can be generated from a single input signal. An important characteristic of fiber birefringence is that its orientation and magnitude change randomly on a length scale that is on the order of 100 m. Very little is known about the details of the random variation. It is generally assumed that the birefringence is locally linear. The random birefringence couples the two polarization modes. Light in one state of polarization will be scattered into the other. It is common to represent the state of polarization of light on the Poincaré sphere [15]. In this representation, each point on the sphere corresponds to a polarization state. The equator corresponds to linearly polarization,the north and south poles correspond to left and right circular polarization,and the points in between correspond to varying degrees of ellipticity. Points diametrically opposite to each other correspond to orthogonally polarized states. If the fiber birefringence is constant, the polarization state of light will describe a circle on the Poincaré sphere around the fiber eigenstate with a period of L_B. When the fiber birefringence varies randomly, the polarization state of light will be scattered all over the surface of the Poincaré sphere and eventually reach a uniform distribution even if it is initially polarized along one of the fiber eigenaxes. The evolution of polarization state of light depends on both the fiber autocorrelation length and the beat length. Since the fiber birefringence is assumed to vary on the equator only, the diffusion length of the polarization states in the azimuthal and equatorial direction will in general be different [16]. The asymmetric evolution of the polarization states would impact on both NRZ (non- return-to-zero) and soliton communications. In work by ourselves and others [17, 18], it has been shown that when averaged over the randomly varying birefringence, the coupled nonlinear Schrödinger equation is reduced to the Manakov's equation with additional terms due to linear and nonlinear polarization mode dispersion.

The polarization mode dispersion terms have random coefficients whose detailed structure depends on the the evolution of the polarization states on Poincaré sphere.

In this paper, we study the evolution of polarization states using a physical model in which we allow birefringence orientation to varying randomly while keeping the birefringence strength fixed. Previously [16, 19], we have shown that including the variation of the strength of the birefringence does not have a significant effect on the evolution of polarization states. We then derive the Manakov-PMD equation that describe the evolution. The reminder of the paper is organized as follows: In Section 2, we present the physical model that we will study. In Section 3, we derive the Manakov-PMD equation from the coupled nonlinear Schrödinger equation. Section 4 contains the conclusions.

2. Physical Model

In the plane wave approximation, the evolution of light propagating in an optical fiber is governed by the equation

$$\frac{d\mathbf{E}}{dz} = i\mathsf{K}\mathbf{E},\tag{1}$$

where $\mathbf{E} = (E_1, E_2)^t(z)$ is the complex amplitudes of the electric fields, z is distance along the optical fiber, and

$$\mathsf{K} = k_0\mathsf{I} + \kappa_1\Sigma_1 + \kappa_2\Sigma_2 + \kappa_3\Sigma_3.\tag{2}$$

The matrix I is the identity matrix and the Σ_i, $i = 1, 2, 3$ are the Pauli matrices

$$\Sigma_1 = \begin{pmatrix} 0 & 1 \\ 1 & 0 \end{pmatrix}, \quad \Sigma_2 = \begin{pmatrix} 0 & -i \\ i & 0 \end{pmatrix}, \quad \Sigma_3 = \begin{pmatrix} 1 & 0 \\ 0 & -1 \end{pmatrix}.\tag{3}$$

Assuming that there is no polarization-dependent loss or gain, then k_0 may be complex, but all the κ_i must be real. We eliminate k_0 from Equation (1) by making the transformation

$$\mathbf{A} = \mathbf{E}\exp\left[-i\int_0^z k_0(z')\,dz'\right],\tag{4}$$

and we note that communication fibers are nearly linearly birefringent so that we may set $\kappa_2 = 0$. We now rewrite Equation (1) as

$$\frac{d\mathbf{A}}{dz} = ib\Gamma\mathbf{A},\tag{5}$$

where $\Gamma = \cos\theta\Sigma_3 + \sin\theta\Sigma_1$, $2b$ is the birefringence strength, and θ is the orientation angle of the birefringence axis. In this model, we assume that the strength of birefringence is constant but the rate of change of the birefringence orientation is driven by a white noise process $g_\theta(z)$,

$$\frac{d\theta}{dz} = g_\theta(z), \tag{6}$$

$$\langle g_\theta(z)\rangle = 0, \qquad \langle g_\theta(z)g_\theta(z')\rangle = \sigma_\theta^2 \delta(z - z'). \tag{7}$$

The fiber autocorrelation length h_{fiber} equals $2/\sigma_\theta^2$. Recently [16, 19], we have shown that the variation of the strength of birefringence does not have a significant effect on the evolution of the polarization states. Furthermore, we show that if the length scale of the rate of change of birefringence orientation is much shorter than that of the orientation angle which in turn is much shorter than the dispersion length, one would obtain the same expression for the polarization mode dispersion.

We determine the effect of randomly varying birefringence on light propagation in fibers by determining the evolution of the Stokes parameters defined as

$$\begin{aligned}
S_1 &= A_1 A_1^* - A_2 A_2^*, \\
S_2 &= A_1 A_2^* + A_2 A_1^*, \\
S_3 &= -i(A_1 A_2^* - A_2 A_1^*).
\end{aligned} \tag{8}$$

Note that without loss of generality, we have assumed $A_1 A_1^* + A_2 A_2^* = 1$. The Stokes parameters obey the equation,

$$\frac{\partial}{\partial z}\begin{pmatrix} S_1 \\ S_2 \\ S_3 \end{pmatrix} = 2b \begin{pmatrix} 0 & 0 & \sin\theta \\ 0 & 0 & -\cos\theta \\ -\sin\theta & \cos\theta & 0 \end{pmatrix} \begin{pmatrix} S_1 \\ S_2 \\ S_3 \end{pmatrix}. \tag{9}$$

If we choose the local polarization eigenstates as reference axes by defining $\tilde{\mathbf{S}} = \mathsf{R}(z)\mathbf{S}$ where

$$\mathsf{R}(z) = \begin{pmatrix} \cos\theta & \sin\theta & 0 \\ -\sin\theta & \cos\theta & 0 \\ 0 & 0 & 1 \end{pmatrix}, \tag{10}$$

Equation (9) becomes

$$\frac{\partial}{\partial z}\begin{pmatrix} \tilde{S}_1 \\ \tilde{S}_2 \\ \tilde{S}_3 \end{pmatrix} = \begin{pmatrix} \tilde{S}_2 \\ -\tilde{S}_1 \\ 0 \end{pmatrix} g_\theta + \begin{pmatrix} 0 \\ -2b\tilde{S}_3 \\ 2b\tilde{S}_2 \end{pmatrix}. \tag{11}$$

3. Nonlinear Evolution

In this section, we determine the effect of the rapidly varying birefringence on nonlinear pulse propagation in optical fibers. The equation describing the evolution in an optical fiber of the electric field $\mathbf{A}$ defined in Equation (4) is [20],

$$i\frac{\partial \mathbf{A}}{\partial z} + b\Gamma\mathbf{A} + ib'\Gamma\frac{\partial \mathbf{A}}{\partial t} \pm \frac{1}{2}\frac{\partial^2 \mathbf{A}}{\partial t^2} + |\mathbf{A}|^2\mathbf{A} - \frac{1}{3}(\mathbf{A}^\dagger\Sigma_2\mathbf{A})\Sigma_2\mathbf{A} = 0, \quad (12)$$

where $\mathbf{A}^\dagger = (A_1{}^*, A_2{}^*)$, $b' = db/d\omega$, and ω is the angular frequency. The second term and third term on the left-hand side of Equation (12) are the phase-velocity and group-velocity birefringence. The fourth term describes the chromatic dispersion. The plus sign corresponds to anomalous dispersion and the minus sign corresponds to normal dispersion. The last two terms are due to the Kerr nonlinearity. We transform Equation (12) to the local axes of birefringence with the following transformation,

$$\mathbf{\Psi} = \begin{pmatrix} U \\ V \end{pmatrix} = \begin{pmatrix} \cos\theta/2 & -\sin\theta/2 \\ \sin\theta/2 & \cos\theta/2 \end{pmatrix}\begin{pmatrix} A_1 \\ A_2 \end{pmatrix}. \quad (13)$$

Equation (12) becomes

$$i\frac{\partial \mathbf{\Psi}}{\partial z} + \tilde{\Gamma}\mathbf{\Psi} + ib'\Sigma_3\frac{\partial \mathbf{\Psi}}{\partial t} \pm \frac{1}{2}\frac{\partial^2 \mathbf{\Psi}}{\partial t^2} + |\mathbf{\Psi}|^2\mathbf{\Psi} - \frac{1}{3}(\mathbf{\Psi}^\dagger\Sigma_2\mathbf{\Psi})\Sigma_2\mathbf{\Psi} = 0, \quad (14)$$

where

$$\tilde{\Gamma} = \begin{pmatrix} b & -i\theta_z/2 \\ i\theta_z/2 & -b \end{pmatrix}. \quad (15)$$

Equations (12) and (14) are written in dimensionless form. The distance is normalized to the dispersion length scale l_d which is typically hundreds of kilometers. Since the orientation of the birefringence is randomly varying on a length scale of the order of 100 m, the ratio $\epsilon = h_{\text{fiber}}/l_d$ is very small. With this normalization, the phase birefringence Γ is typically very large, of the order ϵ^{-1}. The short fiber autocorrelation length, however, means that Γ changes many times in a distance l_d. If we consider a weak cw wave so that all but the first two terms in Equation (14) can be ignored, we find that the effect of the randomly varying birefringence is the linear randomization of the polarization of the electric field. We can remove the rapid variation of the state of polarization of the electric field by the following transformation

$$\mathbf{\Psi}(z,t) = \mathsf{T}(z)\overline{\mathbf{\Psi}}(z,t), \quad (16)$$

where

$$\mathsf{T}(z) \equiv \begin{pmatrix} u_1 & u_2 \\ -u_2^* & u_1^* \end{pmatrix} \quad (17)$$

is a unitary matrix with coefficients u_1 and u_2, so that $|u_1|^2 + |u_2|^2 = 1$ and $\mathsf{T}(z)$ satisfies the following equation

$$i\frac{\partial \mathsf{T}}{\partial z} + \tilde{\Gamma}\mathsf{T} = 0. \tag{18}$$

Substituting Equations (16), (17), and (18) into Equation (14), we obtain

$$i\frac{\partial \overline{\Psi}}{\partial z} + ib'\overline{\Sigma}\frac{\partial \overline{\Psi}}{\partial t} \pm \frac{1}{2}\frac{\partial^2 \overline{\Psi}}{\partial t^2} + |\overline{\Psi}|^2\overline{\Psi} - \frac{1}{3}(\overline{\Psi}^\dagger \Pi \overline{\Psi})\Pi \overline{\Psi} = 0, \tag{19}$$

where, letting $\overline{\Psi}(z,t) = (\overline{U}, \overline{V})^t(z,t)$, we set

$$\overline{\Sigma} = \mathsf{T}^\dagger \Sigma_3 \mathsf{T} = \begin{pmatrix} a_1 & a_4{}^* \\ a_4 & -a_1 \end{pmatrix},$$

$$\Pi = \mathsf{T}^\dagger \Sigma_2 \mathsf{T} = -\begin{pmatrix} a_3 & a_6{}^* \\ a_6 & -a_3 \end{pmatrix}. \tag{20}$$

The coefficients a_i, $i = 1, \ldots, 6$ are defined in terms of u_1 and u_2 as

$$\begin{aligned}
a_1 &= |u_1|^2 - |u_2|^2, & a_4 &= 2u_1 u_2{}^*, \\
a_2 &= -(u_1 u_2 + u_1{}^* u_2{}^*), & a_5 &= u_1{}^2 - u_2{}^{*2}, \\
a_3 &= i(u_1 u_2 - u_1{}^* u_2{}^*), & a_6 &= -i(u_1{}^2 + u_2{}^{*2}).
\end{aligned} \tag{21}$$

From this definition, we find that the coefficients a_1, a_2 and a_3 are real, while a_4, a_5 and a_6 are complex. One can further show that

$$\begin{aligned}
a_1{}^2 + a_2{}^2 + a_3{}^2 &= 1, \\
a_4{}^2 + a_5{}^2 + a_6{}^2 &= 0.
\end{aligned} \tag{22}$$

The coefficients $\{a_1, a_2, a_3\}$ satisfy Equation (11), the equation of motion for the Stokes parameters in the local axes if one replaces $\tilde{S}_i$ by a_i, $i = 1, 2,$ 3. Similarly, the coefficients $\{a_4, a_5, a_6\}$ also satisfy Equation (11) with the same white noise source g_θ. Since the coefficients of Equation (11) are real, the real parts and imaginary parts of $\{a_4, a_5, a_6\}$ satisfy Equation (11) separately. From the definition of the transfer matrix T, we find that at $z = 0$, we have $u_1 = 1$ and $u_2 = 0$. Equivalently we have $(a_1, a_2, a_3) = (1, 0, 0)$ and $(a_4, a_5, a_6) = (0, 1, -i)$. Since the fiber birefringence varies randomly, the polarization state of light will be scattered all over the surface of the Poincaré sphere and eventually reach a uniform distribution, so that $\langle a_i^2 \rangle = \langle \mathrm{Re}(a_j)^2 \rangle = \langle \mathrm{Im}(a_j)^2 \rangle = 1/3$, where $i = 1, 2, 3$ and $j = 4, 5, 6$. We also have $\langle a_i a_j \rangle = 0$ for $i \neq j$. Consequently,

$$(\overline{\Psi}^\dagger \Pi \overline{\Psi})\Pi \overline{\Psi} \longrightarrow \frac{1}{3}|\overline{\Psi}|^2 \overline{\Psi}. \tag{23}$$

The spatial average $\langle f \rangle$ of a function f is defined as

$$\langle f \rangle = \lim_{z \to \infty} \frac{1}{z} \int_0^z ds f(s). \tag{24}$$

Separating the averages of the coefficients in Equation (19) from their randomly varying parts, we obtain

$$i\frac{\partial \overline{\Psi}}{\partial z} \pm \frac{1}{2}\frac{\partial^2 \overline{\Psi}}{\partial t^2} + \frac{8}{9}|\overline{\Psi}|^2 \overline{\Psi} = -ib'(\overline{\Sigma} - \langle \overline{\Sigma} \rangle)\frac{\partial \overline{\Psi}}{\partial t} - \frac{1}{3}\widehat{\mathbf{N}}, \tag{25}$$

where $\widehat{\mathbf{N}} = (\widehat{\mathbf{N}}_1, \widehat{\mathbf{N}}_2)^t$ and

$$
\begin{aligned}
\widehat{\mathbf{N}}_1 &= (a_3{}^2 - 1/3)(2|\overline{V}|^2 - |\overline{U}|^2)\overline{U} - a_3 a_6^*(2|\overline{U}|^2 - |\overline{V}|^2)\overline{V} - a_3 a_6 \overline{U}^2 \overline{V}^* \\
&\quad - a_6^{*2}\overline{V}^2 \overline{U}^*, \\
\widehat{\mathbf{N}}_2 &= (a_3{}^2 - 1/3)(2|\overline{U}|^2 - |\overline{V}|^2)\overline{V} + a_3 a_6(2|\overline{V}|^2 - |\overline{U}|^2)\overline{U} + a_3 a_6^* \overline{V}^2 \overline{U}^* \\
&\quad - a_6{}^2 \overline{U}^2 \overline{V}^*.
\end{aligned}
\tag{26}
$$

When the right hand side equals zero, Equation (25) is known as the Manakov equation. The first term on the right hand side of Equation (25) corresponds to the usual linear polarization mode dispersion. The second term will lead to a nonlinear polarization mode dispersion which describes the effect of the fluctuation of the state of polarization from the uniform distribution. The coefficients in the right hand side of Equation (25) have zero mean when averaged over the Poincaré sphere and are of order one in magnitude. Their integrated effect however is small because they change sign on a length scale given by h_{fiber} which is much shorter than the dispersion length scale l_d. We refer to the couple equation as the Manakov-PMD equation.

To study the effect of the randomly varying terms in Equation (25), we carry out a multiple length scale expansion [21] of Equation (25) in which we assume that the field $\overline{\Psi}(z, t)$ depends on both fast (η_0) and slow (η_1) variables, defined as $\eta_0 = z/\epsilon$ and $\eta_1 = z$ where $\epsilon = h_{\text{fiber}}/l_d$. We further assume that the coefficient $a_i = a_i(\eta_0)$ and write $\overline{\Psi}(\eta_0, \eta_1, t)$ as a series in the small parameter ϵ:

$$\overline{\Psi}(\eta_0, \eta_1, t) = \sum_{n=0}^{\infty} \epsilon^n \overline{\Psi}^{(n)}(\eta_0, \eta_1, t). \tag{27}$$

Substituting these assumptions into Equation (25), we find that the zeroth order solution depends on the slow variable only, i.e. $\overline{\Psi}^{(0)}(\eta_1, t)$, and it satisfies the Manakov equation. The first order correction is given by,

$$\boldsymbol{\Psi}^{(1)}(\eta_0, \eta_1, t) \;=\; \mathsf{T}(\eta_0)\left\{-b'\left[\int_0^{\eta_0} d\eta_0'\,\overline{\Sigma}\right]\frac{\partial\overline{\boldsymbol{\Psi}}^{(0)}}{\partial t}\right.$$

$$\left.+i\frac{1}{3}\int_0^{\eta_0} d\eta_0'\,\widehat{\mathbf{N}}^{(0)}\right\}. \tag{28}$$

Note that the integrals in Equation (28) are over the fast variable η_0 only. The coefficients on the right hand side of Equation (28) are the summation of the random variables a_1 and a_4 in the linear term and $a_3{}^2$, $a_6{}^2$ and $a_3 a_6$ in the nonlinear term rather than the variables themselves. Physically, the pulse envelope does not respond to the rapidly varying a_i but rather to their cumulative effects. The polarization mode dispersion terms are a generalization of those we found earlier [17]. As we showed, the linear and nonlinear polarization mode dispersion can lead to radiation and depolarization of solitons.

The relative strength of the linear and nonlinear polarization mode dispersion depends on the strength of the the birefringence b' and the variances of the random coefficients appear in Equation (28). The variance of the random coefficients is typically proportional to hZ/l_d^2, where h is the decorrelation length of a_i or $a_i a_j$ and Z is the fiber length. The decorrelation lengths of a_1 and a_4 both approximately equal d_1 the equatorial diffusion length [16], while the decorrelation length of $a_3{}^2$, $a_3 a_6$ and $a_6{}^2$ all approximately equal d_3 the azimuthal diffusion length [16]. The relative contribution from the linear and the nonlinear polarization mode dispersion in Equation (28) to the pulse deformation will depend on the relative strength of d_1 and d_3. In Reference [16], we showed that the azimuthal diffusion length and the equatorial diffusion length are different in general. When the fiber autocorrelation length is larger than the beat length $h_{\text{fiber}} \gtrsim L_B$, then one finds $d_1 \gtrsim d_3$. When the fiber autocorrelation length is much shorter than the beat length, $h_{\text{fiber}} \ll L_B$, then one finds $d_1 \ll d_3$. Therefore, in the limit where $h_{\text{fiber}} \ll L_B$, the contribution from the nonlinear polarization mode dispersion will dominate.

As an example, let us assume that the electric field is initially aligned with one of the local polarization eigenaxes such that $\overline{\boldsymbol{\Psi}}(0, t) = [q(t), 0]$, where $q(t)$ is the input waveform, so that the nonlinear term in Equation (26) can be simplified to $\widehat{\mathbf{N}}^{(0)} = [-(a_3{}^2 - 1/3)|q|^2 q, -a_3 a_6 |q|^2 q]^t$. Using Equation (11) and the theory of stochastic differential equation [22], we derive the equations of motions for the random coefficients of the linear and nonlinear polarization mode dispersion terms and then solve the equations numerically. We find that $\text{var}\left[\int^z dz'(a_1 - \langle a_1\rangle)\right] = \text{var}\left[\int^z dz'\text{Re}(a_4)\right] = \text{var}\left[\int^z dz'\text{Im}(a_4)\right]$ and $\text{var}\{\int^z dz'[\text{Re}(a_3 a_6)]^2\} = \text{var}\{\int^z dz'[\text{Im}(a_3 a_6)]^2\}$. In

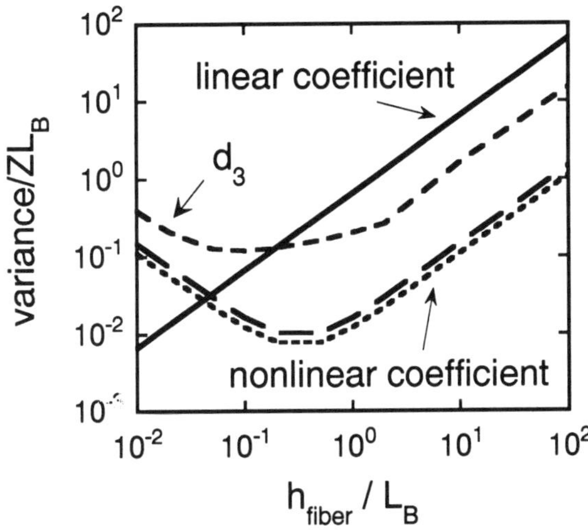

Figure 1.   The variances of the random coefficients of linear and nonlinear polarization mode dispersion versus h_{fiber}/L_B at large distance. The solid curve gives var$\left[\int^z dz'(a_1 - \langle a_1\rangle)\right]$, the long-dashed curve gives var$(\int^z dz' a_3{}^2)$, and the dotted curve gives var$\{\int^z dz'[\text{Re}(a_3 a_6)]^2\}$. The short-dashed curve is the azimuthal diffusion length d_3.

Figure 1, we plot the variance of the random coefficients versus h_{fiber}/L_B at large distance. The solid curve gives var$\left[\int^z dz'(a_1 - \langle a_1\rangle)\right]$, the long-dashed curve gives var$(\int^z dz' a_3{}^2)$, and the dotted curve gives var$\{\int^z dz'[\text{Re}(a_3 a_6)]^2\}$. The short-dashed curve is the azimuthal diffusion length. The variances are normalized by the distance and the beat length.

We note that [16] the equatorial diffusion length $d_1 \simeq h_{\text{fiber}}$. From Figure 1, var$\left[\int^z dz'(a_1 - \langle a_1\rangle)\right] \simeq d_1 Z$, while var$(\int^z dz' a_3{}^2)$ and var$\{\int^z dz'$ $[\text{Re}(a_3 a_6)]^2\} \propto d_3 Z$. When $h_{\text{fiber}} \ll L_B$ the variance of the random coefficients of the nonlinear term dominate, hence the nonlinear polarization mode dispersion determines the pulse deformation. When $h_{\text{fiber}} \gtrsim L_B$, the variance of the random coefficients of the linear term dominate. The relative contribution of the linear and nonlinear polarization mode dispersion to pulse deformation will be determined by the strength of the birefringence b'.

In simulations to date, it has been the practice to use a coarse-step approach in which one simply randomizes the electric fields at fixed intervals separated by a length z_{step} that is long compared to h_{fiber} and the d_j, while treating the field evolution inside the intervals deterministically. As long as the dispersive and nonlinear scale lengths are very long compared to

z_{step}, this procedure will accurately reproduce the averaging that yields the Manakov equation. However, it will exaggerate the variance of the linear polarization mode dispersion by a factor that approximately equals $z_{\text{step}}/h_{\text{fiber}}$ and will exaggerate the variance of the nonlinear polarization mode dispersion by a factor that approximately equals z_{step}/d_3. In the former case, one can compensate for the exaggeration by appropriately lowering b', but that is not possible in the latter case, and the coarse-step approach will significantly overestimate the contribution of nonlinear polarization mode dispersion when $d_3 \ll z_{\text{step}}$.

4. Conclusions

Using a physical model in which we allowed the birefringence orientation to varying randomly while keeping the birefringence strength fixed, we have determined the effect of the rapidly varying birefringence on nonlinear pulse propagation in optical fibers. The length over which the linear term averages is given by the equatorial diffusion length measured with respect to the local axes of birefringence which is also the fiber autocorrelation length. The length over which the nonlinear term averages is given by the diffusion length in the azimuthal direction on the Poincaré sphere. At distances much larger than the longer of the two lengths, the effect of the random birefringence is that the electric field polarization varies rapidly on the Poincaré sphere, and the pulse envelope evolves according to Manakov equation on the dispersive and nonlinear length scales. When one takes into account the finite fiber decorrelation length, one finds that additional randomly varying terms are added to the Manakov equation, leading to the Manakov-PMD equation. These additional terms are both linear and nonlinear in the field strength. The linear terms lead to the usual linear polarization mode dispersion, while the nonlinear terms will lead to a nonlinear polarization mode dispersion.

The usual practice of simply scrambling the electric fields at fixed intervals while follow their evolution inside the intervals deterministically using the coupled nonlinear Schrödinger equation, will not properly weight the contributions from the linear and nonlinear polarization mode dispersion. When $h_{\text{fiber}} \ll L_B$ it will seriously overestimate the nonlinear polarization mode dispersion.

References

1. Poole, C. D. and Wagner, R. E.: Phenomenological approach to polarization dispersion in long single-mode fibers, *Electron. Lett.* **22** (1986), 1029-1030.
2. Andresciani, D., Curti, F., Matera, F. and Daino, B.: Measurement of the group-delay difference between the principal states of polarization on a low-birefringence terrestrial fiber cable, *Opt. Lett.* **12** (1987), 844-846.

3. Bergano, N. S., Poole, C. D. and Wagner, R. E.: Investigation of polarization dispersion in long lengths of single-mode fiber using multilongitudinal mode lasers, *J. Lightwave Tech.* **5** (1987), 1618-1622.

4. Poole, C. d.: Statistical treatment of polarization dispersion in single-mode fiber, *Opt. Lett.* **13** (1988), 687-689.

5. Curti, F., Daino, B., Mao, Q., Matera, F. and Someda, C. G.: Concatenation of polarization-dispersion in single-mode fibers, *Electron. Lett.* **25** (1989), 290-291.

6. C. D. Poole, C. D.: Measurement of polarization-mode dispersion in single-mode fibers with random mode coupling, *Opt. Lett.* **14** (1989), 523-525.

7. Curti, F., Daino, B., De Marchis, G. and Matera, F.: Statistical treatment of the evolution of the principal states of polarization in single-mode fibers, *J. Lightwave Tech.* **8** (1990), 1162-1165.

8. Poole, C. D., Winters, J. H. and Nagel, J. A.: Dynamical equation for polarization dispersion, *Opt. Lett.* **16** (1991), 372-374.

9. Betti, S., Curti, F., De Marchis, G., Iannone, E. and Matera, F.: Evolution of the bandwidth of the principal states of polarization in single-mode fibers, *Opt. Lett.* **16** (1991), 467-469.

10. Foschini, G. J. and Poole, C. D.: Statistical theory of polarization dispersion in single mode fibers, *J. Lightwave Tech.* **9** (1991), 1439-1456.

11. Namihara, Y., Kawazawa, T. and Wakabayashi, H.: Polarization mode dispersion measurements in 1520 km EDFA system, *Electron. Lett.* **28** (1992), 881-882.

12. Galtarossa, A. and Schiano, M.: Complete characterization of polarization mode dispersion in erbium doped optical amplifiers, *Electron. Lett.* **28** (1992), 2143-2144; Galtarossa, A., Comment: Polarization mode dispersion in 1520 EDFA system, *Electron. Lett.* **29** (1993), 564-565.

13. Menyuk, C. R. and Wai, P. K. A.: Polarization evolution and dispersion in fibers with spatially varying birefringence, *J. Opt. Soc. Am. B* **11** (1994), 1288-1296.

14. Poole, C. D. and Giles, C. R.: Polarization-dependent pulse compression and broadening due to polarization dispersion in dispersion-shifted fiber, *Opt. Lett.* **13** (1988), 155-157.

15. Born, M. and Wolf, E.: *Principles of Optics.* Pergamon, Oxford, (1984), Chap. 1.

16. Wai, P. K. A. and Menyuk, C. R.: Anisotropic evolution of the state of polarization in optical fibers with random varying birefringence, to appear in *Optics Letters.*

17. Wai, P. K. A., Menyuk, C. R. and Chen, H. H.: Stability of solitons in randomly varying birefringent fibers, *Opt. Lett.* **16** (1991), 1231-1233.

18. Evangelides, Jr., S. G., Mollenauer, L. F., Gordon, J. P. and Bergano, N. S.: Polarization multiplexing with solitons, *J. Lightwave Tech.* **10** (1992), 28-35.

19. Wai, P. K. A. and Menyuk, C. R.: Polarization decorrelation in fibers with randomly varying birefringence, *Opt. Lett.* **19** (1994), 1517-1519.

20. Menyuk, C.R.: Pulse propagation in an elliptically birefringent Kerr medium, *IEEE J. Quantum Electron.* **25** (1989), 2674-2682.

21. Nayfeh, A.: *Perturbation Methods.* Wiley, New York, (1973).

22. See, e.g. Arnold, L.: *Stochastic Differential Equations, Theory and Applications,* Wiley, New York, (1974).

THE APPLICATION OF OPTICAL PHASE CONJUGATION IN SOLITON COMMUNICATIONS

SIEN CHI

Professor, Institute of Electro-Optical Engineering
National Chiao Tung University
1001 Ta Hsuen Rd., Hsinchu, Taiwan, Republic of China

SENFAR WEN

Associate Professor, Department of Electrical Engineering
Chung-Hua Polytechnic Institute
30 Tung Shiang, Hsinchu, Taiwan, Republic of China

AND

JENG-CHERNG DUNG

Ph.D. Candidate, Institute of Electro-Optical Engineering
National Chiao Tung University
1001 Ta Hsuen Rd., Hsinchu, Taiwan, Republic of China

Abstract. Recent theoretical works on the application of optical phase conjugation in the soliton communications are reviewed and the effect of the in-line filter on the optical phase conjugation is numerically studied. It is shown that, by properly applying the conjugators and choosing the filter bandwidth, the soliton transmission can be significantly improved and its performance is better than applying the sliding-frequency filter for the considered system of 14.3 Gbits/s bit rate.

1. Introduction

The optical phase conjugation (OPC) has been proposed to compensate for the chromatic dispersion to improve the bit rate of the optical transmission system [1]. Recently, the OPC is demonstrated to improve the bit rate of the transmission of the 1.55 μm signal in the standard fiber with the loss compensated by the erbium-doped fiber amplifiers [2, 3]. The conjugator

A. Hasegawa (ed.), Physics and Applications of Optical Solitons in Fibres '95, 319–331.

can be made by using either the dispersion-shifted fiber (DSF) [2] or the semiconductor laser amplifier (SLA) [3].

This technique in fact can also be applied to improve the performance of the soliton transmission system [4-8]. For the soliton system, large separation between neighboring solitons is required to avoid the nonlinear interaction between them and the bit rate is reduced [9]. On the other hand, when the soliton is amplified by the optical amplifier to compensate for the fiber loss, the introduced amplified spontaneous emission noise (ASEN) will randomly modulate the carrier frequency of the soliton and cause the timing jitter of the soliton [10]. Such an effect is known as the Gordon-Haus effect. Usually the sliding-frequency filter which is placed after every amplifier is used to reduce the soliton interaction and Gordon-Haus effect [11, 12].

With the OPC, it is shown that in the lossless fiber the effects of the second-order dispersion, Kerr effect, and Raman effect on the pulse propagation can be completely undone [13, 14]. Since the soliton interaction is a combined effect of the second-order dispersion and Kerr effect, it can be completely undone by the OPC in the lossless fiber. In the presence of the power perturbation, the soliton interaction can be well undone by applying the conjugator when the pulse shapes of the solitons have not yet changed significantly [6]. As to the Gordon-Haus effect, it is shown that, in the absence of the soliton interaction, the standard deviation of the timing jitter can be reduced to a half when a conjugator is applied at the middle of the system [4, 5, 7]. Since the soliton interaction depends on the separation of the solitons, the Gordon-Haus effect affects the soliton interaction and complicates the problem [15].

In this paper, we will review the previous works on the soliton communications using the OPC and consider the effect of the in-line optical filter on such a system. On the contrary to the technique by using the optical filter to suppress the soliton interaction and Gordon-Haus effect, the filter is not favorable by using the OPC because it may distort the soliton and phase-conjugate soliton. As the ASEN may saturate the optical amplifier, it is necessary to use the optical filter to suppress the ASEN.

2. Theoretical Model

The wave equation which describes the soliton propagation in the single mode fiber can be written as

$$i\frac{\partial \phi}{\partial z} - \frac{1}{2}\beta_2\frac{\partial^2 \phi}{\partial \tau^2} - i\frac{1}{6}\beta_3\frac{\partial^3 \phi}{\partial \tau^3} + n_2\beta_0|\phi|^2\phi - c_r\frac{\partial |\phi|^2}{\partial \tau}\phi = -\frac{1}{2}i\alpha\phi, \quad (1)$$

where β_2 and β_3 represent the second-order and third-order dispersion, respectively; n_2 is the Kerr coefficient; c_r is the coefficient of the self-frequency shift (SFS); α is the fiber loss. From the Equation (1), when $\beta_3 = \alpha = 0$, it can be proved that the effects of the second-order dispersion, Kerr effect, and SFS can be completely recovered by the OPC [13, 14].

Therefore, in the lossless fiber without the third-order dispersion, the soliton interaction can be completely undone by the OPC. We will numerically solve Equation (1) with the fiber loss periodically compensated by the amplifier with the amplifier spacing $L_a = 30$ km. The ASEN power per unit frequency generated by an amplifier is $p_a = n_{sp}(G - 1)h\nu$, where n_{sp} is the spontaneous emission factor, $G = \exp(\alpha L_a)$ is the gain of the amplifier, and $h\nu$ is the photon energy. The soliton wavelength λ is assumed to be 1.55 μm and the pulsewidth (FWHM) $t_s = 20$ ps. Because the effects of the third-order dispersion and SFS are negligible for the considered pulsewidth, these two effects are neglected in this paper for simplicity. The other coefficients are taken as $\beta_2 = -0.64\text{ps}^2/\text{km}$ (0.5ps/km/nm), $n_2 = 3.2 \times 10^{-20}\text{m}^2/\text{W}$, $\alpha = 0.2$ dB/km, and $n_{sp} = 1$. The effective fiber cross section $A_{eff} = 35\ \mu\text{m}^2$. To enhance the soliton interaction, the separation of the neighboring solitons is taken to be 3.5 pulse widths and the corresponding bit rate is 14.3 Gbits/s.

The OPC can be generated by the nearly degenerate four-wave mixing in the DSF or the SLA [2, 3]. It is shown that ideal phase conjugation can be achieved by using the conjugator made by either the DSF or the SLA [16, 17]. Because the wavelength separation between the signal soliton and the conjugate soliton is small, the dispersions for the two solitons can be taken to be the same. Therefore, we assume the conjugator is ideal and the operation of the conjugator is to take the complex conjugate of the field of the signal soliton.In Ref. [2] and [3], the conversion efficiencies including the coupling loss for the conjugators made by using the DSF and SLA are about -25.1 dB and -12.5 dB, respectively, which are equivalent to 125.5 km and 62.5 km fiber loss (0.2 dB/km), respectively.

Using the DSF, higher conversion efficiency can be obtained by using higher pump power [18]. Since the considered separation between the conjugators in this paper is much larger than the equivalent length, the ASEN introduced by the amplifiers which are used to compensate for the losses due to the conjugators are neglected.

3. Reduction of the Soliton Interaction

It has been found that, to undo the soliton interaction well by the OPC, for the case with short pulsewidth, where the fiber loss is periodically com-

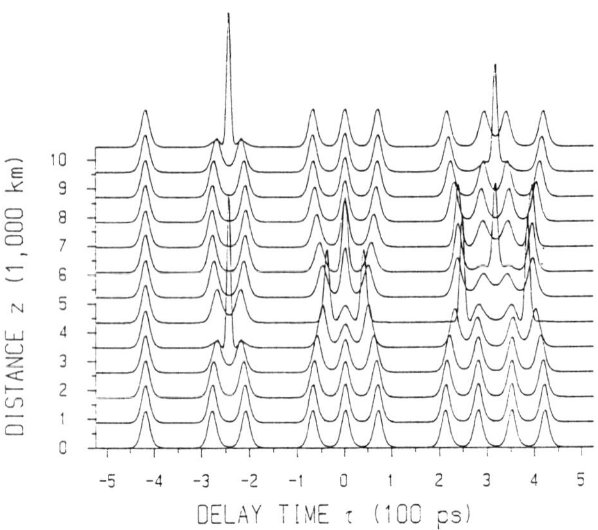

Figure 1. Power envelope of the soliton bit stream along the fiber without the ASEN.

pensated by the distributed amplifier and the soliton period is shorter than the amplifier spacing, the conjugator should be applied before there are significant changes in the pulse shapes [6]. For the case with long pulsewidth, where the fiber loss can be periodically compensated by the lumped amplifier and the soliton period is much longer than the amplifier spacing, the undoing is better and it is not well when the conjugator is applied near the coalescence distance [7, 8]. In this paper, the considered 20 ps soliton is the long pulsewidth case.

In the following of this section, the cases considered in Ref. [8] are shown. In the lossless case, if there are only two solitons, they periodically coalesce and the first coalescence distance is at $L_i = 3,470$ km. Figure 1 shows the evolution of a soliton bit stream (010110111011110) along the periodically amplified fiber. One can see that, after propagating about L_i distance, the soliton interaction becomes apparent and the interaction depends on the bit pattern. Figure 2 shows the standard deviation σ of the timing jitter of the solitons caused by the interaction. The simulated soliton bit stream consists of 640 bits which are pseudo-random and include 320 zeros and 320 soliton pulses. Because of the initial overlap of the solitons, the initial standard deviation is not zero. In the Figure 2, as the distance increases, the standard deviation increases until at about $z = L_i$ where the standard deviation is maximum. After this distance, the solitons separate and the standard deviation decreases. When the conjugators are applied, if the

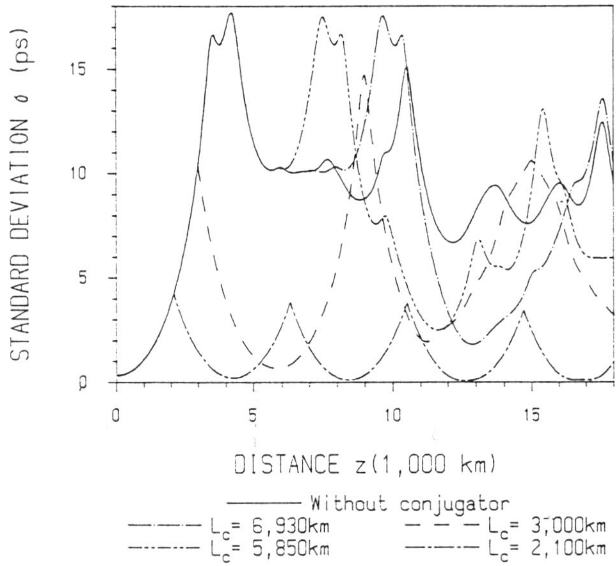

Figure 2.   Standard deviation σ of the timing jitter of the soliton along the fiber. The j-th conjugator is applied at $(2j-1)L_c$.

standard deviation decreases to the initial value, the undoing of the soliton interaction is complete. The conjugator is applied at $z = (2n-1)L_c$ and the soliton interaction may be undone at $z = 2nL_c$, where n is a positive integer.

One can see that for the cases with $L_c < L_i$, the undoing of the soliton interaction is good and it is better for shorter L_c. For the cases with $L_c = L_i$, the undoing of the soliton interaction is not so good as the cases with $L_c < L_i$. The case with $L_c = 6,930$ km is better than the case with $L_c = 5,850$ km because the separation of the solitons is better at $L_c = 6,930$ km. However, it is found that the timing jitter is more complicated in the presence of the ASEN and it is improper to apply the conjugator after the soliton coalescence [8].

4. Reduction of the Gordon-Haus Effect

When the soliton is periodically amplified by the optical amplifiers, every amplifier introduces the ASEN to the soliton. It has been shown that, due to the randomly modulated carrier frequency of the soliton by the ASEN, the variance of the carrier frequency of the soliton caused by an amplifier is [10, 19]

$$\langle \bar{\Omega}^2 \rangle_0 = \frac{2\pi n_2 n_{sp} hc (G-1)^2}{3\tau \lambda^2 |\beta_2| A_{eff} G L_a \alpha},\tag{2}$$

where $\tau = t_s/1.763$ is the soliton characteristic time. After the solitons propagate L distance with $M = L/L_a$ amplifiers, a conjugator is applied and both the frequencies of the solitons and the ASEN are inverted. For further propagating ΔL distance, where $\Delta L = \Delta M L_a$, the total variance of the timing jitter of the soliton at $z = L + \Delta L$ becomes [4,5,7]

$$\langle \delta t^2 \rangle = \beta_2^2 \langle \bar{\Omega}^2 \rangle_0 \left\{ \sum_{j=1}^{M} [L - jL_a - \Delta L]^2 + \sum_{j=1}^{\Delta M} [\Delta L - jL_a]^2 \right\},\tag{3}$$

where the terms with first and second summations represent the variances caused by the amplifiers which are used before and after the conjugator, respectively. If the conjugator is applied at the middle of the system, $\Delta L = L$ and $\Delta M = M$. In Ref. [4], it is first found that, for this case, the variance can be reduced to $1/4$ of the value obtained without using the conjugator or the standard deviation is reduced to $1/2$ of the value obtained without using the conjugator.

In the following of this section, the cases considered in Ref. [7] are shown. Figure 3 shows the standard deviation $\sigma = (\langle \delta t^2 \rangle)^{1/2}$ calculated from the Equation (3) along the fiber for 9,000 km transmission distance and the conjugator is applied at 4,500 km. In the Figure 3, the case without the conjugator and the numerically simulated standard deviation of the timing jitter for the transmission of 256 solitons in the presence of the ASEN are shown. One can see that the standard deviation at 9,000 km for the case with the conjugator is reduced to a half to compare with the case without the conjugator. It is noticed that the minimum standard deviation is $\sigma = 2.1$ ps at 6,360 km from the Equation (3). To show the reduction of the timing jitter, by assuming $L_a \ll L, \Delta L$, Equation (3) can be approximated by

$$\langle \delta t^2 \rangle = \beta_2^2 \langle \bar{\Omega}^2 \rangle_0 \left\{ \frac{1}{3L_a} [L^3 - 3\Delta L L^2 + 3\Delta L^2 L] + \frac{1}{3L_a} \Delta L^3 \right\}.\tag{4}$$

From the Equation (4), the minimum total variance is at $\Delta L = (\sqrt{2} - 1)L$ or at the distance $z = \sqrt{2}L$, where the total variance is reduced to 0.12 of the value obtained without using the conjugator or the standard deviation is reduced to 0.35 of the value obtained without using the conjugator. This result agrees with the case shown in the Figure 3.

In Reference [5], it is found that, to reduce the Gordon-Haus effect, the optimal position to apply the conjugator is at $2/3$ from the transmitter

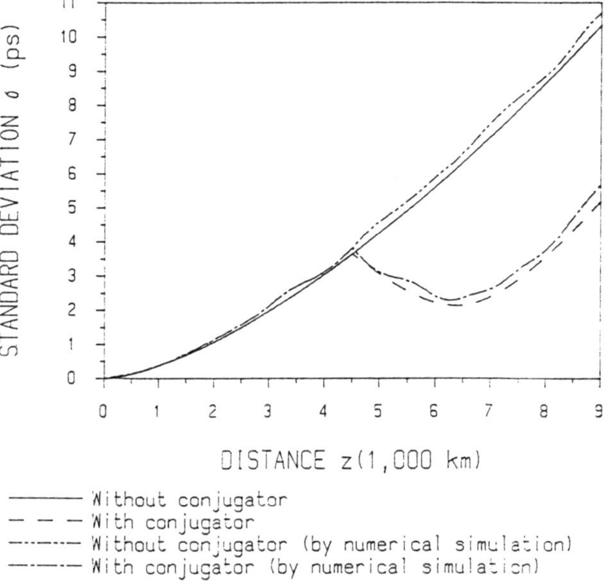

Figure 3. Standard deviation σ of the timing jitter of the soliton along the fiber. A conjugator is applied at 4,500 km.

end. A more accurate result Ref. [7] shows, the optimum position is at $1/\sqrt{2}$ from the transmission end, which can be obtained from the Equation (4). For a total transmission distance of L_t, the optimal position is at $(L/\sqrt{2}L)L_t = 0.707L_t$. Figure 4 shows the same case as the Figure 3 except the conjugator is applied at 6,360 km, which is approximately $0.707L_t$. In the Figure 4, the meanings of the data lines are the same as the Figure 3. At $L_t = 9,000$ km, the standard deviations are 3.6 ps and 3.7 ps for the results calculated from the Equation (3) and obtained by the numerical simulations, respectively. The standard deviation is reduced to about 0.35 of the value obtained without using the conjugator. However, when the soliton separation is as small as the cases considered in the section 3, the effect of the soliton interaction on the timing jitter must be considered.

5. Reduction of the Soliton Interaction and Gordon-Haus Effect

It has been found that, to reduce both the soliton interaction and the Gordon- Haus effect, the conjugator must be applied before the solitons coalesce [8]. In this section, the cases considered in Ref. [8] are shown.

Including the ASEN, Figure 5 shows an example of the evolution of the solitons with the same bit stream as the Figure 1. To clearly show the

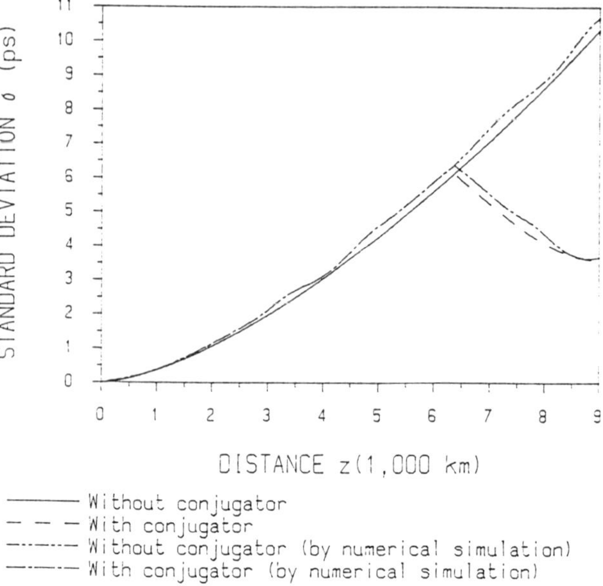

Figure 4. Standard deviation σ of the timing jitter of the soliton along the fiber. A conjugator is applied at 6,360 km.

pulse shapes of the solitons shown in the Figure 5, the ASEN has been filtered out by a Lorentzian transfer function with 50 GHz bandwidth. The ASEN changes the separation between the neighboring solitons through the Gordon-Haus effect and the soliton interaction in the bit stream shows a random way. Thus, the conjugator should be applied before the combined effect of the soliton interaction and timing jitter is out of control.

Figure 6 shows the standard deviation σ of the timing jitter of the solitons caused by the combination of the soliton interaction and the Gordon-Haus effect for the same 640 pseudo-random bits considered in the Figure 2. For comparison, the theoretical curve for the timing jitter without the soliton interaction and OPC is shown. The cases with the conjugators are also shown in the Figure 6 where the conjugators are applied every L_c to more effectively reduce the Gordon-Haus effect. It is shown that, when the conjugator is applied after L_i, the reduction of the timing jitter is poor. When the conjugator is applied before L_i, the reduction of the timing jitter is more effective. For the case with shorter L_c, the reduction is better.

From the section 4, one can deduce that the distance for the best reduction of the timing jitter due to the Gordon-Haus effect lies between the conjugator which is clearly shown in Figure 6. Therefore, the transmission distance should be chosen so that the corresponding standard deviation is minimum.

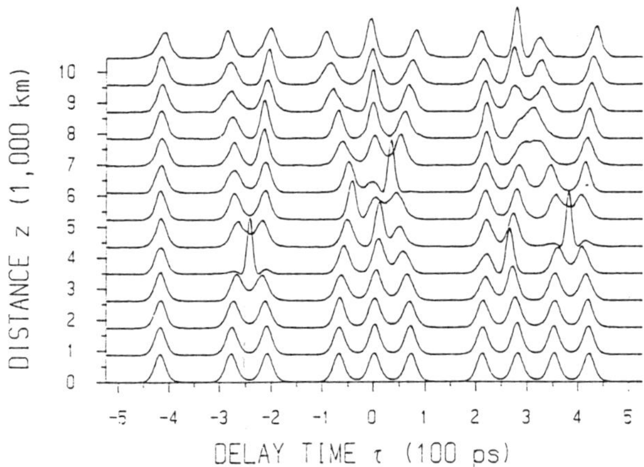

Figure 5. Power envelope of the soliton bit stream along the fiber without the ASEN. The ASEN has been filtered out by a 50 GHz filter to show the soliton pulse shapes.

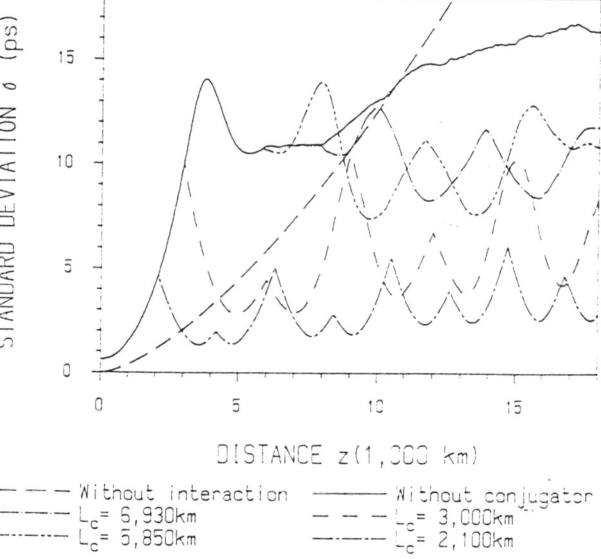

Figure 6. Standard deviation σ of the timing jitter of the solitons along the fiber with the ASEN. The j-th conjugator is appled at jL_c.

The sliding-frequency filter has been used to reduce the soliton inter-action and Gordon-Haus effect [11, 12]. It is found that, if we place the sliding Fabry-Perot filter after every amplification period without using the OPC, the soliton separation can not be well maintained even without the ASEN because the considered soliton separation is small [8]. The solitons may coalesce or repel each other depending on the bit pattern and filter. For example, for the bit stream shown in the Figure 1, there are solitons that repel each other only after propagating about 6,900 km with 190 GHz bandwidth and +3 GHz/Mm sliding rate where the filter is optimized to maintain the soliton separation without considering the ASEN. Therefore, for the considered system using the OPC, the allowed transmission length is longer than the system using the sliding-frequency filter.

6. Effect of Optical Filter

From the theory of OPC, it recovers the effects of the second-order dispersion and Kerr effect on the soliton before the conjugator. These two effects modulate the phase of the soliton. As the in-line optical filter changes the magnitudes and phases of the spectral components of the electric field envelopes of the soliton and phase-conjugate soliton, it may deteriorate the reduction of the soliton interaction and the Gordon-Haus effect. Therefore, it is not favorable to use the filter. However, the filter is required to avoid the saturation of the optical amplifier by the ASEN. In this paper, we take the Fabry-Perot filter as an example and its transfer function can be written as

$$H(\nu) = \frac{1}{1 + i\frac{2\nu}{\Delta\nu_f}},\tag{5}$$

where $\Delta\nu_f$ is the bandwidth.

To avoid the distortion caused by the filter, it requires $\Delta\nu_f \gg \Delta\nu_s$, where $\Delta\nu_s$ is the spectral width of the electric field envelope of the soliton and $\Delta\nu_s = 23.6$ GHz for the 20 ps soliton. Another requirement for the filter bandwidth is set by the signal-to-noise ratio and the saturation of optical amplifier by the ASEN. It requires the soliton power $P_s \gg Np_a\Delta\nu_f$, where $P_s = 1.763^2 A_{eff}|\beta_2|/\beta_0 n_2 t_s^2$ and N is number of amplifiers used in the transmission system. This requirement leads to $\Delta\nu_f \ll (1.48 \times 10^5/N)\Delta\nu_s$ with the considered numerical parameters. For the examples shown in the following, this requirement can be satisfied.

Figure 7(a) shows an example of the evolution of the soliton bit stream with the ASEN, where the filter of bandwidth $\Delta\nu_f = 10\Delta\nu_s$ is applied after every amplifier and the conjugator is applied at every 2,100 km. The moving of the solitons in the figure is due to the filter. One can see that

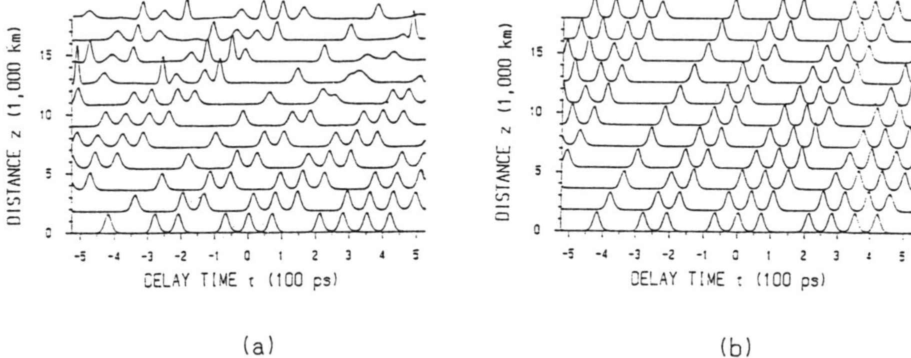

Figure 7.   Power envelope of the soliton bit stream along the fiber with the ASEN, where a conjugator is applied every 2,100 km and a filter is applied every 30 km. The filter bandwidth $\Delta\nu_f =$ (a) $10\Delta\nu_s$, (b) $20\Delta\nu_s$. The ASEN has been filtered out by a 50 GHz filter to show the soliton pulse shapes.

the soliton separation can not maintain in this case. To reduce the soliton interaction and the Gordon-Haus effect with the filter, larger filter bandwidth is required.

Figure 7(b) shows the same case as the Figure 7(a) except $\Delta\nu_f = 20\Delta\nu_s$. From the figure, the soliton separation maintains well. It is found that the effect of the filter is negligible when the filter bandwidth larger than about $26\Delta\nu_s$. However, it is not necessary to use such a filter bandwidth. For the considered soliton transmission system with 10^{-9} bit error rate (BER), the corresponding standard deviation of the timing jitter is 3.8 ps. Narrower filter bandwidth can be used if this requirement is satisfied. Figure 8 shows the standard deviation σ of the timing jitter of the solitons with filter and conjugator. To compare with Figure 6 for the case with $L_c = 2,100$ km, the standard deviation shown in Figure 8 is larger. By properly applying the conjugators and choosing the filter bandwidth, the allowed transmission distance is still longer than the case using the sliding-frequency filter. From Figure 8, for 10^{-9} BER and $\Delta\nu_f = 20\Delta\nu_s$, the allowed transmission lengths are over 18,000 km with either $L_c = 2,100$ km or 1,500 km. Even with narrower filter bandwidth $\Delta\nu_f = 15\Delta\nu_s$, the allowed transmission lengths are 9,600 km and 17,190 km with $L_c = 2,100$ km and 1,500 km,

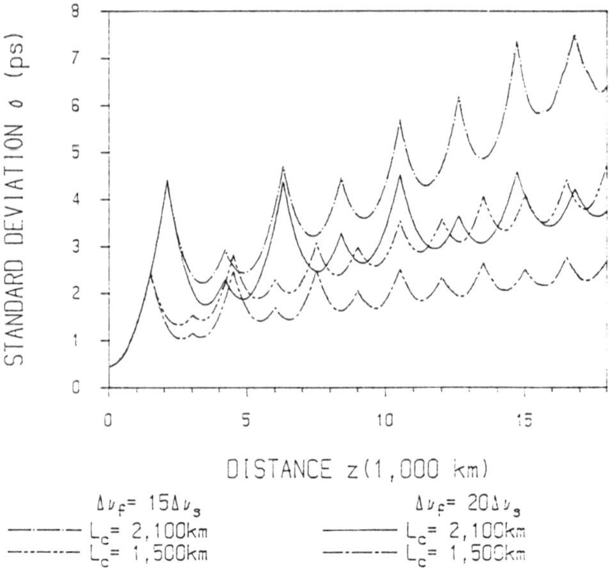

Figure 8. Standard deviation σ of the timing jitter of the solitons along the fiber with the ASEN, where the filter of bandwidth $\Delta\nu_f$ is applied every 30 km and the j-th conjugator is applied at jL_c.

respectively.

7. Conclusions

Recent theoretical works on the application of OPC in the soliton communications are reviewed and the effect of the in-line filter on the OPC is numerically studied. To reduce the soliton interaction and the Gordon-Haus effect by the OPC, the conjugator should be applied before the soliton coalescence otherwise the combined effect of the soliton interaction and the Gordon-Haus effect will render the pulse shape and pulse arrival time uncontrollable. The filter distorts the soliton and phase-conjugate soliton. This leads to deteriorate the system. It is shown that, by properly applying the conjugators and choosing the filter bandwidth, the soliton transmission can be significantly improved and its performance is better than applying the sliding-frequency filter for the considered example.

References

1. Yariv, A., Fekete, D. and Pepper, D. M.: Compensation for channel dispersion by nonlinear optical phase conjugation, *Opt. Lett.* **4** (1979), 52-54.

2. Watanabe, S., Naito, T. and Chikama, T.: Compensation of chromatic dispersion in a single-mode fiber by optical phase conjugation, *IEEE Photon. Technol. Lett.* **5** (1993), 92-95.

3. Tatham, M. C., Sherlock, G. and Westbrook, L. D.: Compensation fibre chromatic dispersion by optical phase conjugation in a semiconductor laser amplifier, *Electron. Lett.* **29** (1993), 1851-1852.

4. Forysiak, W. and Doran, N. J.: Conjugate solitons in amplified optical fibre transmission systems, *Electron. Lett.* **30** (1994), 154-155.

5. Forysiak, W. and Doran, N. J.: Phase conjugation for jitter and soliton-soliton compensation in soliton communications, *Conference on Lasers and Electro-Optics*, **CThN2**.

6. Wen, S. and Chi, S.: Undoing of soliton interaction by optical phase conjugation, *Electron. Lett.* **30** (1994), 663-664.

7. Wen, S. and Chi, S.: Improving soliton transmission by optical phase conjugation, *Proc. Soc. Photo-Opt. Instrum. Eng.* **2321** (1994), 6-9.

8. Wen, S. and Chi, S.: Reduction of the soliton interaction and the Gordon-Haus effect by optical phase conjugation, *Opt. Lett.* **20** (1995), 976-978.

9. Chu, P. L. and Desem, C.: Optical fiber communication using solitons, *Technical Digest of IOOC'83* (1983), 27-30.

10. Gordon, J. P. and Haus, H. A.: Random walk of coherently amplified solitons in optical fiber transmission, *Opt. Lett.* **11** (1986), 665-667.

11. Kodama, Y. and Wabnitz, S.: Reduction and suppression of soliton interactions by bandpass filters, *Opt. Lett.* **18** (1993), 1311-1313.

12. Mollenauer, L. F., Gordon, J. P. and Evangelides, S. G.: The sliding-frequency guiding filter: an improved form of soliton jitter control, *Opt. Lett.* **17** (1992), 1575-1577.

13. Fisher, R. A., Suydam, B. R. and Yevick, D.: Optical phase conjugation for time-domain undoing of dispersive self-phase-modulation effects, *Opt. Lett.* **8** (1983), 611-613.

14. Chi, S. and Wen, S.: Recovery of soliton self-frequency shift by optical phase conjugation, *Opt. Lett.* **19** (1994), 1705-1707.

15. Georges, T. and Favre, F.: Influence of soliton interaction on amplifier noise-induced jitter: a first-order analytical solution, *Opt. Lett.* **16** (1991), 1656-1658.

16. Wen, S., Chi, S. and Chang, T.-C.: Effect of cross-phase modulation on optical phase conjugation in dispersion-shifted fiber, *Opt. Lett.* **19** (1994), 939-941.

17. Wen, S. and Chi, S.: Effect of carrier depletion on optical phase conjugation in semiconductor laser amplifier, *Opt. Lett.* **20** (1995), 590-592.

18. Watanabe, S. and Chikama, T.: Highly efficient conversion and parametric gain of nondegenerate forward four-wave mixing in a singlemode fiber, *Electron. Lett.* **30** (1994), 163-164.

19. Marcuse, D.: An alternative derivative of the Gordon-Haus effect, *IEEE J. Lightwave Technol.* **10** (1992), 273-278.

EFFECT OF MUTUAL INTERACTIONS ON THE BIT-RATE IN SOLITON COMMUNICATION SYSTEMS

A. N. PILIPETSKII AND C. R. MENYUK

Department of Electrical Engineering and Computer Sciences
University of Maryland Baltimore County
Baltimore, Maryland 21228-5398, USA

Abstract. We discuss the influence of soliton interactions on information transmission. It is shown that the short-range mutual interaction of neighboring time slots can lead to non-Gaussian tails on the distribution function and a substantial increase in the bit error rate. The long-range interactions via the acoustic effect significantly increase the timing jitter of solitons in communication lines. We calculate the correlation in near-neighbor soliton time shifts due to the acoustic interaction and show that the acoustic effect can cause correlated errors that cannot be corrected using standard, simple error correction codes like the Hamming code. We argue that similar effects will appear in NRZ systems.

1. Introduction

In modern-day optical fiber communication systems, one aspires to achieve bit error rates that are lower than 10^{-12}. The limits to error-free soliton transmission, both in bit rate and distance, are set by the timing jitter in pulse arrival times. This timing jitter is caused by several effects: the Gordon-Haus effect, the polarization effect, and the acoustic effect. In the past, nearly all theoretical calculations of the bit error rate have relied that the timing jitter of solitons is Gaussian-distributed. With this assumption, one need merely know the standard deviation of the timing jitter and the maximum allowable jitter to determine the bit error rate.

In this paper, we calculate the probability distribution function for the soliton temporal location in a filtered system, and we will show that the distribution function is substantially affected by the soliton-soliton inter-

333

A. Hasegawa (ed.), Physics and Applications of Optical Solitons in Fibres '95, 333–344.

actions, leading to non-Gaussian tails in the distribution function, significantly changing the bit error rates without substantially changing the standard deviation [1]. It should be mentioned that the shape of the guiding filter itself can also lead to non-Gaussian distribution of the timing jitter [2].

At bit rates in excess of 10 Gbit/sec, the acoustic effect becomes the dominant cause of the timing jitter over trans-oceanic distances [3, 4]. The acoustic effect is created by the large transverse gradient of the electric field in the optical fiber due to the soliton pulses. These large electric field gradients electrostrively excite acoustic waves that influence later solitons. The acoustic wave perturbs the effective refractive index of the fiber, leading to changes in the frequencies and temporal locations of the solitons. The excited acoustic wave propagates transversely to the fiber axis; so, the perturbation of the effective refractive index of the fiber changes on a time scale of 1 nsec — the time that it takes the acoustic wave to cross the fiber core area. We thus expect that neighboring solitons in a high-bit rate transmission system, operating at more than 5 Gbit/sec, will experience correlated time shifts. Obviously the correlated time shifts of solitons can cause correlated errors in information transmission. The second goal of this paper is to calculate the correlation between time shifts of solitons and to discuss the consequences for information transmission. Thus in this paper we discuss the influence of relatively short-range soliton-soliton interaction on the distribution function and the correlations in timing jitter due to the relatively long-range acoustic interaction.

2. Non-Gaussian Corrections to the Gordon-Haus Distribution Due to Soliton Interactions.

The system that we will study is of current experimental interest [4]; however, the basic purpose of this paper is not to explain the details of the experiments but to illustrate with a simple, yet realistic example the importance of non-Gaussian tails and to show how to calculate the actual distribution function. For this reason, we will take into account only the contribution of spontaneous emission that leads to Gordon-Haus jitter [5], leaving aside for the present the contributions due to acoustic and polarization effects, although these are quite important in the experiments [4].

The basic physical mechanism leading to non-Gaussian tails that we will consider is the mutual interaction between solitons in neighboring time slots [6]. This interaction is typically small, but when by chance two solitons in neighboring time slots are displaced toward each other by roughly half the temporal distance to the time slot boundary, then the mutual interaction becomes significant and the solitons are pulled closer to the time slot boundary.

The stochastic differential equations that describe two filtered solitons with the same phase in neighboring time slots may be written [7]

$$\ddot{q}_1 = -\gamma\dot{q}_1 + \alpha\mathcal{S}_1 + 4\exp\left[-(2\Delta + q_2 - q_1)\right], \tag{1a}$$

$$\ddot{q}_2 = -\gamma\dot{q}_2 + \alpha\mathcal{S}_2 - 4\exp\left[-(2\Delta + q_2 - q_1)\right], \tag{1b}$$

where we are using normalized soliton units and assuming that the solitons have unity amplitude. The $q_{1,2}$ are the time deviations from the soliton initial positions, and Δ is the half-duration of the time slot. The super-dots indicate derivatives with respect to z, the normalized distance along the fiber. The quantity γ gives the effect of the guiding filter, $\mathcal{S}_{1,2}$ are Gaussian-distributed, independent white noise sources with zero mean and unity variance, and α is the noise figure. The expression that we use for the mutual interaction was first derived by Gordon [6]. In this simple example we are focusing on non-sliding filters [8].

It is convenient to transform Equations (1) using sum and difference coordinates, $q_+ = (q_1 + q_2)/\sqrt{2}$ and $q_- = -(q_2 - q_1)/\sqrt{2}$ so Equations (1) becomes

$$\ddot{q}_- = -\gamma\dot{q}_- + \alpha\mathcal{S}_- + 4\sqrt{2}\exp\left[-(2\Delta - \sqrt{2}q_-)\right], \tag{2a}$$

$$\ddot{q}_2 = -\gamma\dot{q}_+ + \alpha\mathcal{S}_+. \tag{2b}$$

The approach that we will use to determine the probability distribution function is a variant of the method of characteristics or the path integral method [9]. The random process generated by Equations (2) is a Markov process in the variables $\mathbf{Q} = (q_+, \dot{q}_+, q_-, \dot{q}_-)$. Indeed, for distances $\gamma z \gg 1$, the reduced process $\mathbf{q}(z) = [q_+(z), q_-(z)]$ is a Markov process which is a significant simplification. This reduction is not possible in the unfiltered system so that the filtered system is actually easier to analyze. One can now proceed as follows: We first note that the equation for q_+ is a linear stochastic differential equation whose solution can be immediately expressed in terms of Gaussian distribution functions, so that we need purely determine the distribution function for q_-. In the absence of the mutual interaction, the probability distribution function $f_{GH}(q_-)$ is known. There are different paths over which $q_- \to q_{GH-}$ as $z \to z_f$, the final z-value, corresponding to different realizations of $\mathcal{S}_-$. When the mutual interaction is present, the modified q_--value $q_{M-}(q_{GH-}, \mathcal{P})$ does not depend uniquely on q_{GH-} but also on the path $\mathcal{P}$ by which q_{GH-} is reached. The probability distribution function $f_{M-}(q_{M-})$ is given by

$$f_M(q_{M-}) = \int dq_{GH-} \int d\mathcal{P}\, f_{GH-}(q_{GH-})\delta\left[q_{M_\mp} - q_{M-}(q_{GH-}, \mathcal{P})\right], \tag{3}$$

where $\int d\mathcal{P}\ldots$ implies an appropriately weighted integral over all possible paths [9]. A discretization of Equation (3) leads to a computationally intensive, yet tractible and accurate calculation of the distribution function and ultimately the bit error rate. This approach will be described in detail elsewhere.

Here instead we use a maximum-likelihood approach that is less accurate but reduces the problem to one that can be solved in a few minutes on a workstation. There is one path $\mathcal{P}_{\max}$ leading to $\mathbf{q}_{GH-}$that is most likely to occur. Since any path that is reasonably likely to occur will be fairly close to this path, it is reasonable to eliminate the integral $\int d\mathcal{P}\ldots$ in Equation (3) by using only $\mathcal{P}_{\max}$ with each $\mathbf{q}_{GH-}$, defining a unique relationship $\mathbf{q}_M(\mathbf{q}_{GH-}) = \mathbf{q}_{M-}(\mathbf{q}_{GH-}, \mathcal{P}_{\max})$. For the most likely way to achieve a separation q_- the Equations (2) becomes

$$\ddot{q}_- = -\gamma\dot{q}_- + \alpha\mathcal{S}_{\max} + 4\sqrt{2}\exp\left[-(2\Delta - \sqrt{2}q_-)\right], \tag{4}$$

where $\mathcal{S}_{\max}$ is the contribution of the noise source along $\mathcal{P}_{\max}$.

To determine $\mathcal{P}_{\max}$, we first note that the probability distribution function in the absence of the mutual interactions is given by

$$f_{GH}(q_{GH-}) = \frac{\gamma}{\alpha\sqrt{2\pi z}}\exp(-\gamma^2 q_{GH-}^2/2\alpha^2 z), \tag{5}$$

when $\gamma z \gg 1$. The distribution function for the q_+ has the same form. Since the random process that generates f_{GH} is a Markov process, it follows that

$$f_{GH-}(q_-, z; q_f-, z_f) = \frac{\gamma^2}{2\pi\alpha^2 z^{1/2}(z_f - z)^{1/2}}$$
$$\exp\left[-\frac{\gamma^2}{2\alpha^2}\left(\frac{q_-^2}{z} + \frac{(q_{f-} - q_-)^2}{z_f - z}\right)\right], \tag{6}$$

where $f_{GH}(q_-, z; q_{f-}, z_f)$ is the joint probability distribution function for the soliton having the coordinate q_- after a distance z and then having q_{f-} after a distance z_f. Completing the square for q_-, one finds

$$f_{GH-}(q_-, z; q_f-, z_f) = \frac{\gamma^2}{2\pi\alpha^2 z^{1/2}(z_f - z)^{1/2}}$$
$$\exp\left[-\frac{\gamma^2}{2\alpha^2}\frac{z_f}{z(z_f - z)}\left(q_- - \frac{z}{z_f}q_{f-}\right)^2 - \frac{\gamma^2}{2\alpha^2 z_f}q_{f-}^2\right]. \tag{7}$$

Hence, one finds $q_- = zq_{f-}/z_f$ along $\mathcal{P}_{\max}$, which implies that $\dot{q}_- = q_{f-}/z_f$ and $\ddot{q}_- = 0$ along $\mathcal{P}_{\max}$. Using Equation (4) and noting that in the absence of the mutual interaction $\ddot{q}_- = -\gamma\dot{q}_- + \alpha\mathcal{S}_{\max}$, we conclude that

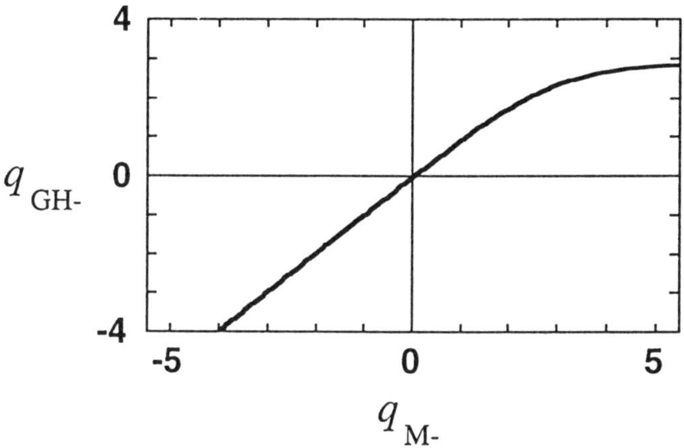

Figure 1.   Solution for $q_{GH-}(z_f)$ as a function of $q_{M-}(z_f)$. This function is calculated by first finding $q_{M-}(q_f)$ at z_f and then inverting after recalling that $q_{GH-}(z_f) = q_f$.

$S_{\max} = \gamma q_{f-}/\alpha z_f$. Using this value of $S_{\max}$ in Equation (4) allows one to compute $q_{M-}(q_{GH-}, \mathcal{P}_{\max})$.

We now consider a specific set of parameters, $\gamma = 0.4$, $\Delta = 5.5$ and $z_f = 250$, that correspond closely to current experimental values at 10 Gbit/sec over 35 Mm [4]. We also set $\alpha = 10^{-2}$ which is a bit higher than the theoretically predicted value, but we neglect the acoustic and polarization effects that enhance the jitter. In Figure 1, we show $q_{GH-}(q_{M-})$ which is obtained by inverting $q_{M-}(q_{GH-}, \mathcal{P}_{\max})$ at $z = z_f$. In the range $q_{f-} = 0\text{--}2$, $q_{M-}(z_f)$ hardly differs from q_{f-}, but beyond $q_{f-} = 2$ the deviations become significant. Now we can calculate the modified difference distribution function $f_{M-}(q_{M-} = f_{GH-}[q_{GH-}(q_{M-})]dq_{GH-}/dq_{M-}$ and the full distribution function (noting that $q_{M+} = q_{GH+}$) $f_M(q_{M1}, q_{M2}) = f_M(q_{M-})f_M(q_{M+})$. Finally, we calculate the one dimensional distribution function $f_M(q_{M1}) = \int_{-\infty}^{\infty} f_M(q_{M1}, q_{M2})dq_{M2}$ and the escape distribution function $E_M(q_{M1}) = 2\int_{-q_{M1}}^{\infty} f_M(q'_{M1})dq'_{M1}$. It can be seen from Figure 2 that a significant tail is visible beyond $q_M \simeq 2.5$, as shown in Figure 1. Assuming that there is an error when $|q| > q_{\mathrm{error}}$, $E_M(q_{\mathrm{error}})$ equals the bit error rate. For the parameters of Reference 4, $q_{\mathrm{error}} = 4.5$. In the example that we are considering, $E_M(4.5)$ is in excess of 10^{-13} which is measurable while the assumption that the probability distribution function is Gaussian yields a number that is too small to measure.

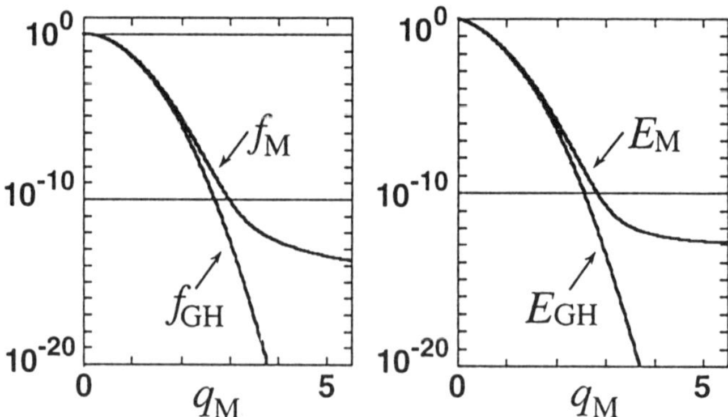

Figure 2. The probability distribution function $f_M(q_M)$ and the escape distribution function.

3. Acoustic Effect and Correlated Errors in Soliton Transmission

The physical source of the Gordon-Haus and polarization effects is spontaneous emission in the erbium-doped fiber amplifiers. From the standpoint of communication theory, both these effects are sources of additive white noise [10]. While chromatic dispersion, polarization-mode dispersion, and in-line filtering, in combination with the channel nonlinearity seriously complicate the calculation of the bit error rate due to these noise sources [1], these noise sources lead to only a very weak inter-symbol interference in a soliton system and nearly uncorrelated errors from bit to bit. By contrast, the acoustic effect is not a noise source at all, but it is rather a source of inter-symbol interference [10]. The impact of the acoustic effect on any particular bit depends in a completely deterministic way on the bits that preceded it, and errors are highly correlated from bit to bit.

To date, there has been little or no thought given to the possible impact of methods such as error-control coding or feedback (equalization) in eliminating errors due to the optical fiber transmission line in soliton communications. Given the power of these techniques, it is apparent that more thought must be given to their potentialities in the future. We open here the discussion by pointing out the inadequacy of a simple Hamming code in dealing with the highly correlated errors that are due to the acoustic effect.

Each soliton that propagates in a communication line excites an acoustic wave and perturbs the fiber refractive index, and this perturbed refractive index $\delta n(t)$ affects the subsequent solitons [3]. If one pulse follows another at an interval T, then the first pulse changes the mean frequency of the other by

$$\frac{d\omega}{dz} = -\frac{\omega}{c}\frac{d\left(\delta n\right)}{dt}\bigg|_{t=T}. \tag{8}$$

The index perturbation $\delta n(t)$ is proportional to the energy of the first pulse and its functional form may be found in Reference 3. This frequency shift leads to a temporal shift of the pulses relative to each other. A data stream that consists of an arbitrary sequence of ones and zeros is physically represented in a soliton communication line by a sequence of solitons that are separated from each other by an interval T and appear with probability equal to $1/2$. The i-th soliton in the pulse sequence is then described by the temporal position t_i and a deviation from the central signal frequency $\delta\Omega_i$,

$$\frac{dt_i}{dz} = -\frac{\left(\delta\Omega_i\right)\lambda^2}{2\pi c}D, \tag{9a}$$

$$\frac{d\left(\delta\Omega_i\right)}{dz} = -\frac{\omega}{c}\sum_l \frac{d\left(\delta n\right)}{dt}\bigg|_{t_i-t_l}, \tag{9b}$$

where λ is the soliton wavelength, D is the average dispersion in the communication line, and z is the propagation distance. The sum in Equation (9b) is taken over the acoustic responses of all the preceding pulses.

We wish to calculatethe correlation function between time shifts of the pulses:

$$f\left(N\right) = \frac{\langle\tau_i\tau_{i+N}\rangle - \langle\tau_i\rangle\langle\tau_{i+N}\rangle}{\sigma^2}, \tag{10}$$

where $\sigma \equiv \langle\tau_i^2\rangle - \langle\tau_i\rangle^2$, $\tau_i = t_i(z) - t_i(0)$ is the time shift of the pulse in the time slot with number i from its initial position, and σ is the variance of the acoustically induced timing jitter. We assume that the process distribution for the 1-bits and 0-bits is stationary so that f does not depend on i. Taking into account that the probability of having a soliton in each bit is equal to $1/2$ and that soliton time shifts are much smaller than 1 nsec, one finds

$$f(N) = \frac{\sum_{l=1}^{\infty} \frac{d\delta n}{dt}\bigg|_{lT}\frac{d\delta n}{dt}\bigg|_{(l+N)T}}{\sum_{l=1}^{\infty}\left(\frac{d\delta n}{dt}\bigg|_{lT}\right)^2}, \tag{11}$$

where T is the bit period. The correlation function of the soliton time shifts at a bit rate of 20 Gbit/sec is presented in Figure 3. The long term correlations shown in Figure 3 on a time scale of about 20 nsec are due to reflections from the fiber cladding boundary, while short term correlations

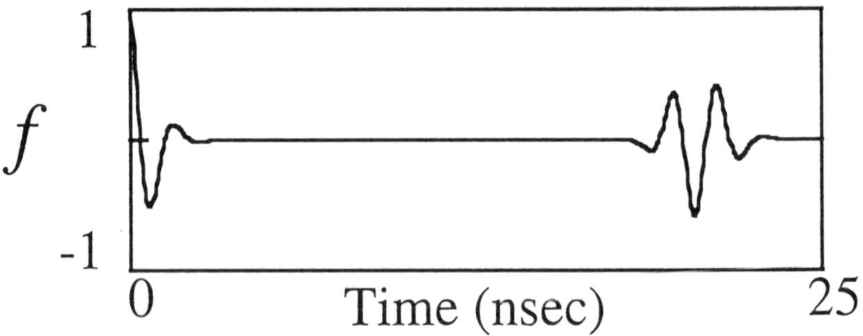

Figure 3. The correlation function for the acoustically-induced timing jitter at a bit rate of 20 Gbit/sec.

on a time scale of 2 nsec are due to the finite transit time of the acoustic waves through the core. In an unfiltered soliton system, the acoustic effect leads to timing jitter of the pulses with a variance of

$$\sigma = 4.8 \frac{D^2 F^{1/2}}{\tau} z^2,$$

(12)

where D is fiber dispersion in psec/nm/km, z is the propagation distance in Mm, and F is the bit rate [3, 4]. With guiding filters, the timing jitter is reduced by a factor $2/\beta z$ where β is the frequency damping coefficient [3, 4] but, in any case, the time shifts of neighboring solitons are strongly correlated. The correlation between time shifts of two neighboring solitons for a bit rate more than 10 Gbit/sec can be approximately expressed as

$$f \approx 1 - \frac{1.4}{F},$$

(13)

where F is the bit rate in Gbit/sec. The timing jitter that accumulates along the transmission line can cause errors when a soliton leaves its time slot.

In the case of acoustically–induced timing jitter, one finds that when one soliton moves out of its time slot, the soliton in neighboring time slot will also have a large enough time shift to move out of its time slot with high probability, leading to correlated errors. This correlation can have devastating consequences for simple error correction codes that assume that errors are independent. We illustrate this possibility with an example using a simple Hamming code [10]. In this code, k parity bits are added to an n–bit

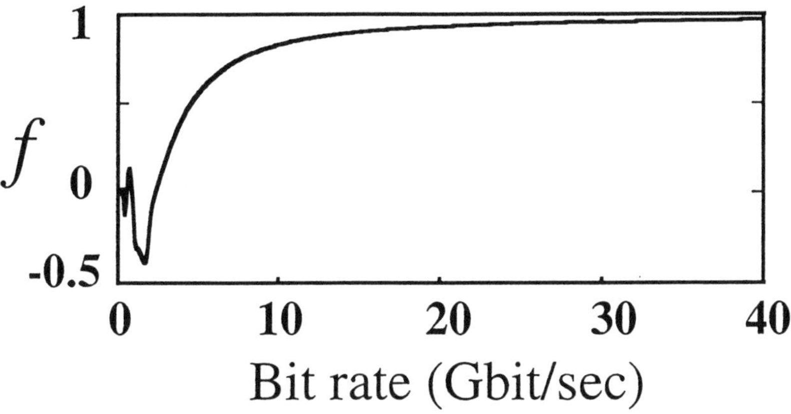

Figure 4.   The correlation between time shifts of neighboring solitons depending on the bit rate.

word, resulting in a new word with $n+k$ bits. The positions in the new word that correspond to powers of 2 are assigned the parity bits. The remaining positions are assigned the data bits. We simulated the propagation of a data stream with 10,000 words through an optical fiber, calculating the changes in the pulse frequencies and pulse positions using Equations (9). The initial data stream consisted of 8–bit words following each other with arbitrarily chosen zeros and ones. To each 8–bit word, we added 4 parity bits, so that the data stream that we simulated was represented by a sequence of 12 bit words. Our simulation parameters were $D = 0.2$ psec/nm/km, $\tau_{sol} = 10$ psec, $F = 20$ Gbit/sec, and $z = 10$ Mm. For these parameters, the probability of a single soliton moving out of its time slot approximately equaled 10^{-2}, which would be unacceptably high in a real communication system but allowed us to easily study the correlated time shifts. For the actual string that we studied, we found that the actual probability that a soliton leaves its time slot is $0.96 \cdot 10^{-2}$, while the actual probability of soliton leaving its time slot given that the preceding soliton leaves its time slot is 0.44. These results are consistent with the theoretically expected values.

A typical example of a word in which an error occurred is shown in Figure 5. Before transmission, this word was an 8–bit word 00011011. To use the Hamming code the four parity bits P_1 =XOR of bits $(3,5,7,9,11) = 1$, P_2 =XOR of bits $(3,6,7,10,11) = 0$, P_4 =XOR of bits$(5,6,7,12) = 0$, and P_8 =XOR of bits $(9,10,11,12) = 1$ were assigned to the 1st, 2nd, 4th and 8th places. The XOR operation equals 1 when there is an odd number of 1's in the variables and to 0 when there is an even number of 1's. We thus

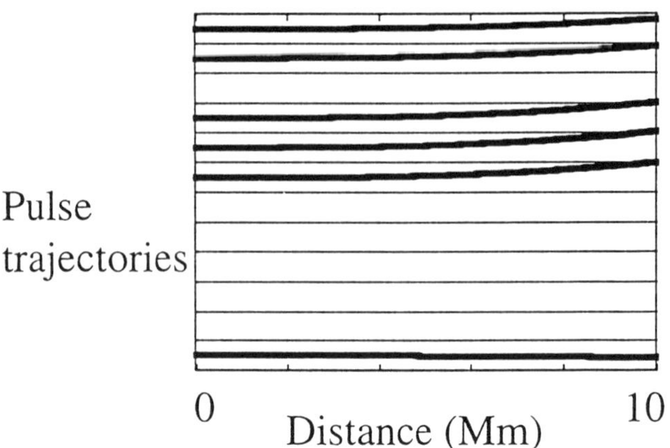

Figure 5. Evolution of the soliton maxima in one 12 bit word in our simulated bit stream. Our parameters are: bit rate $F = 20$ Gbit/sec, dispersion $D = 0.25$ psec/nm/km and soliton width $\tau = 10$ psec. The trajectories of the pulse maxima are shown by thick solid lines, and the grid indicates the boundaries of the time slots.

obtained a twelve bit word 100000111011. When the 12 bits are received after passing through the transmission line they are checked for the errors as follows: C_1 =XOR of bits (1,3,5,7,9,11), C_2 =XOR of bits (2,3,6,7,10,11), C_4 =XOR of bits (4,5,6,7,12), and C_8 =XOR of bits (8,9,10,11,12). For a single error the binary result $C = C_8 C_4 C_2 C_1 \neq 0$ will indicate the error and the bit position where the error occurs. In soliton transmission, a 1 is represented by a soliton and a 0 is represented by the absence of a soliton. Figure 5 shows that after the propagation distance $z = 10,000$ km there are already three solitons that have moved out of their time slots, and the 12-bit word is represented by the sequence 10000011111. When this word is received after the transmission line, the bits are checked for errors as follows: C_1 =XOR of bits (1,3,5,7,9,11), C_2 =XOR of bits (2,3,6,7,10,11), C_4 =XOR of bits (4,5,6,7,12), C_8 =XOR of bits (8,9,10,11,12). With a single error, the binary result $C = C_8 C_4 C_2 C_1 \neq 0$ indicates the error and the bit position where the error is located. Here $C = 1101$ equals 13 which incorrectly indicates the wrong bit. We conclude from this example that simple error correction codes that work well when bits are subject to independent errors can fail in the case of the acoustically-induced timing jitter.

4. Conclusions

In the first part, we have shown that the mutual interaction of solitons in neighboring time slots can lead to non-Gaussian tails in the distribution of the soliton jitter and a substantial increase in the bit error rate. While the physical mechanism being considered here is specific to solitons, the basic physical idea is not. Optical fibers are nonlinear channels, and there is no reason that Gaussian noise should lead to a Gaussian distribution of the quantities of interest. Indeed, one intuitively expects the deviations from a Gaussian distribution to become important for those rare events in which the noise-induced changes in the physical quantities of interest are large—just the sort of events that are likely to cause errors!

Also in this paper, we showed that errors due to the acoustic effect are highly correlated and that this correlation will have a devastating effect on simple error correction codes such as a simple Hamming code when the acoustic effect dominates, which occurs at data rates above 10 Gbit/sec. By contrast, errors due to spontaneous emission noise are nearly independent, and we therefore expect a simple Hamming code to work well at data rates below 10 Gbit/sec.

A variety of techniques exist that could be of potential use in reducing errors due to the acoustic effect. Since the acoustic effect is a source of inter-symbol interference, feedback (equalization) techniques are promising [10]. Some error-control coding techniques that deal with bursts are promising as well [10]. We believe that we have only scratched the surface of this important area of investigation.

Near-neighbor pulse interactions, filtering and the acoustic effect arepresent in NRZ systems as well as soliton systems. Thus, it seems reasonable to anticipate that these effects will be of importance in NRZ systems as well—particularly the high-power systems that are being used for long-distance repeaterless links.

5. Acknowledgment

Work at the University of Maryland was supported by NSF and ARPA through AFOSR.

References

1. Menyuk, C. R.: *Opt. Lett.* **20** (1995), 285; Georges, T.: *Electron. Lett.* **31** (1995), 1174.
2. Georges, T.: private communication.
3. Dianov, E. M., Luchnikov, A. V., Pilipetskii, A. N. and Prokhorov, A. M.: *Appl. Phys. B* **54** (1992), 175; Golovchenko, E. A. and Pilipetskii, A. N.: *J. Lightwave*

Tech. **12** (1994), 1052.

4. Mollenauer, L. F., Mamyshev, P. V. and Neubelt, M. J.: *Opt. Lett.* **19** (1994), 704.
5. Gordon, J. P. and Haus, H. A.: *Opt. Lett.* **11** (1986), 665.
6. Gordon, J. P.: *Opt. Lett.* **8** (1983), 596.
7. Mecozzi, A., Moores, J. D., Haus, H. A. and Lai, Y.: *Opt. Lett.* **16** (1991), 1841; Kodama, Y. and Hasegawa, A.: *Opt. Lett.* **17** (1992), 31.
8. Mollenauer, L. F., Lichtman, E., Harvey, G. T., Neubelt, M. J. and Nyman, B. M.: *Electron. Lett.* **28** (1992), 792.
9. Feynman, R. P. and Hibbs, A. R.: *Quantum Mechanics and Path Integrals*, McGraw-Hill, New York, (1965); Arnold, L.: *Stochastic Differential Equations, Theory and Applications*, Wiley, New York, (1974).
10. See, e.g. Proakis, J. G.: *Digital Communications*, McGraw-Hill, New York, (1989), Chaps. 5 and 6 for a basic discussion of noise, intersymbol interference, error-control coding and equalization.

FUTURE UNDERSEA SYSTEMS -
WILL SOLITONS BE NEEDED?

PETER K. RUNGE

AT&T Bell Laboratories
AT&T Submarine Systems, Inc.
101 Crawfords Corner Road
Holmdel, NJ, USA

Abstract. Transoceanic systems incorporating the newest technology for optical undersea systems, erbium doped fiber amplifiers, have now been installed and taken into service. These amplifiers compensate for the entire transmission loss of a transoceanic system (some 1,800 dB) and are capable in principle of amplifying any optical signal, not only the presently employed Non Return to Zero (NRZ) transmission signal format, but Optical Solitons as well. The question, therefore, arises, under what conditions may future optical undersea systems benefit from using the Soliton transmission format. The paper examines projected needs for future undersea systems and elaborates their requirements.

1. Introduction

Optical undersea systems have progressed very rapidly in their use of new transmission technologies. The first generation of regenerated systems operated at 280 Mb/s and at a transmission wavelength of 1.3 μm and was installed for first transoceanic application in 1988. Very rapidly thereafter, the technology moved to 560 Mb/s at 1.5 μm with first installations in 1991. Optical amplifier technology based on erbium doped fibers then burst onto the scene with first applications in 1994 for 2,000 km long systems operating at 2.5 Gb/s, as portions of the America's North and Columbus 2 systems from Florida to St. Thomas in the US Virgin Islands. These systems were followed rapidly in 1995 by transoceanic applications in TAT-12, US to UK, and the trans-pacific TPC-5 southern route from the US mainland via

345

A. Hasegawa (ed.), Physics and Applications of Optical Solitons in Fibres '95, 345–349.
© 1996 *Kluwer Academic Publishers. Printed in the Netherlands.*

Hawaii and Guam to Japan, using basically the same amplifier technology, but now operating at 5 Gb/s per fiber pair.

With optical amplifiers now established in undersea systems, it is natural to ask whether optical Solitons might find a place in future undersea networks, since the fundamental technology to overcome transmission loss already exists. We, therefore, need to examine future system needs in order to establish the criteria under which a transition from the present Non Return to Zero (NRZ) format to Optical Solitons might take place.

To be sure, we are not discussing an upgrade of existing systems from NRZ to Solitons, since that is technically not feasible mainly because of the radically different dispersion maps required for the transmission fibers to optimize either transmission format. We will discuss the conditions under which newly developed undersea systems may incorporate the Soliton transmission format.

2. Traffic Needs and Connectivity Trends

The fundamental driver for undersea systems has been the international calling volume for telephony. This traffic has been growing at a globally averaged yearly rate of about 11% and is expected to continue to grow at that rate. This, however, neglects to take into account the potential impact of new services that might require higher transmission bandwidth than simple voice or Fax service. Some of these services are Fast Fax, multi-media communications, video phone, and high speed data and video transport services. Some traffic forecasts predict a substantial acceleration in traffic growth for the next ten years. That being the case, there is reason to believe in a continuing need for high speed transoceanic transmission pipes with capabilities beyond the present 5 Gb/s per fiber pair.

However, a second major trend is apparent. It is based on the desire to provide optical connectivity in the form of networks rather than point-to-point communication pipes for a variety of advantages. These networks provide interconnection between points not presently connected into the existing undersea communications systems. They provide diversity of paths for restoration purposes. They provide multiple access points into a network and opportunities for growth for unforeseen demand and opportunities for reconfiguration and grooming of traffic. Optical networks can provide economic connectivity and preserve the sovereignty of express through traffic. Because of these benefits, optical networks have begun to appear in various markets ranging from inter-connecting multiple points within a country in a domestic undersea network to networks connecting a number of countries within a region in a regional undersea network. We are also witnessing

Single-Channel Laboratory Experiments

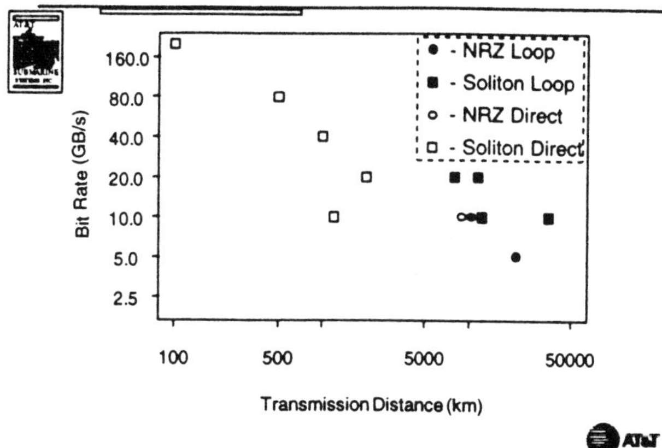

Figure 1. Single channel laboratory experiments for NRZ and Soliton transmission using straight line or loop configurations; transmission bit rate vs. transmission distance.

the emergence of inter-regional networks that tie together multiple regional economic zones and their networks and in turn tie those into the global network.

Recent examples of these new networks are the Asian Pacific Cable Network (APCN), the Fiber Lightguide Around the Globe network (FLAG), and African Optical Network (Africa One), the Central and South American network (Pan-Am). These examples illustrate the trend away from point-to-point communication pipes.

3. Technology Status

Extensive investigations of transmission properties of optical signals using NRZ as well as Soliton formats have been performed in numerous laboratories around the world. Figure 1 attempts to summarize the latest available results and plots the transmission bit rate achieved for single channel per fiber pair transmission experiments as a function of transmission distance. The figure clearly indicates superior performance of the Soliton transmission format for all transmission distances. Even for distances in excess of 5000 km, Solitons have demonstrated suitable transmission at speeds of 20 Gb/s, whereas the transmission speed using the NRZ format seems to be capped at about 10 Gb/s because of the combined effects of fiber nonlinearity and dispersion.

The situation is quite different when we examine wavelength division multiplexed (WDM) laboratory experimental transmission results. Figure

WDM Laboratory Experiments

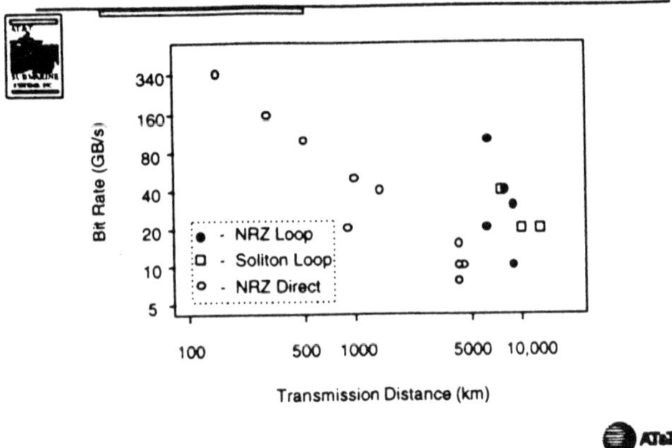

Figure 2. Wavelength division multiplexed laboratory experiments for NRZ and Soliton transmission using direct line and loop configurations; bit rate times number of wavelength vs. transmission distance.

2 shows the most recent transmission experiment results plotted as total capacity (Number of wavelength channels times bit rate per channel) as a function of transmission distance. Here the NRZ format has demonstrated an apparent superiority over the Soliton format for all transmission distances below 10,000 km. This present advantage is due to the ability of NRZ WDM to more effectively utilize the optical bandwidth available in the erbium amplified system. The most outstanding result, in terms of total bit rate distance product, is a 20 channel WDM experiment operating at 5 Gb/s per channel for a total capacity of 100 Gb/s over 6,300 km [1]. This experiment has incorporated optical gain equalization to broaden the useful optical transmission window threefold to about 11 nm.

If we compare the NRZ and Soliton transmission formats on demonstrated experimental transmission performance, we have to conclude: In terms of total through-put capacity, NRZ presently has the edge, 100 Gb/s over 6300 km for point-to-point transmission pipes. In terms of flexibility for future network architectures, NRZ has the edge as well, because of the number of available wavelength carriers (20). However, progress in technology advancement has been very rapid and is expected to continue, since neither transmission technology seems to have run into fundamental limits yet. We therefore should examine the conditions under which Solitons transmission may be incorporated in future undersea networks.

4. Future Network Needs and Requirements for Soliton Transmission

The apparent trend for undersea systems is away from large point-to-point systems with ever increasing capacity towards undersea networks with certain superior characteristics over the point-to-point system. This results in certain high level requirements for optical networks that take into account the need to provide multi-point connectivities of network nodes with diversity of paths for restoration, as well as reconfiguration capability to respond to unforeseen changes in traffic demand. Needless to say, these requirements all have to be met at the lowest possible cost. Based on these requirements, it is clear that optical Soliton transmission will find its way into undersea networks, if it can demonstrate the attributes below:

- Significant cost advantage over the present systems using the NRZ format; here we mean life cycle cost including development cost.
- Significant performance advantage for transmission and more importantly, for optical networking at reduced life cycle cost.
- Significant feature advantages for optical networking at competitive costs.

5. Conclusion

Optical amplifier undersea systems have entered service in 1994 and are now spanning the oceans in the form of TAT-12 across the Atlantic, and TPC-5 across the Pacific Ocean operating at 5 Gb/s per fiber pair We have shown that presently the NRZ modulation format used in today's optical amplifier undersea systems has demonstrated performance advantage over the Optical Soliton transmission far in excess of today's 5 Gb/s transmission rate, and have shown in laboratory experiments that 100 Gb/s may be possible. We have outlined the requirements Optical Soliton transmission needs to meet in order to find its way into future optical undersea networks.

Reference

1. Bergano, N. S. et. al.: 100 Gb/s WDM Transmission of Twenty 5 Gb/s Data Channels over Transoceanic Distances Using a Gain Flattened Amplifier Chain, *Post deadline paper, at ECOC '95* September 21, (1995), Brussels, Belgium

IMPLEMENTABLE SOLITON SYSTEM DESIGN

JUAN FARRÉ AND TAKASHI ONO

Opto-Electronics Research Laboratories

NEC Corporation

4-1-1 Miyazaki, Miyamae-ku

Kawasaki-shi, Kanagawa 216

JAPAN

Abstract. We have proposed a low dispersion soliton to expand the tolerances for dispersion and power. The characteristics of the 40 Gb/s soliton transmission at low dispersion $(D \leq 0.1\,\mathrm{ps/km/nm})$ was investigated through theoretical estimation and numerical simulation. It is found that the low dispersion soliton has advantages in point of dynamic range and robustness to dispersion fluctuation.

1. Introduction

The performance of ultra high speed long-haul communication systems strongly depends on the dispersion values and arrangement of the optical transmission line. For a soliton-based system, dispersion defines the maximum transmitted distance, the optical amplifier spacing, the quality of transmission, as well as the design tolerances and engineerability of the whole system.

Optical soliton transmission in fibers is possible due to the balance between self-phase modulation (SPM) and group velocity dispersion (GVD). Ideally, this equilibrium is only possible in loss-less media, for a specific pulse-shape (secant-hyperbolic) at the anomalous (D>0) dispersion region and at a precise peak power level (soliton power):

351

A. Hasegawa (ed.), Physics and Applications of Optical Solitons in Fibres '95, 351–363.
© 1996 *Kluwer Academic Publishers. Printed in the Netherlands.*

$$P_o = 0.777 \frac{D}{\tau_{FWHM}^2} \frac{\lambda^3 A_{eff}}{\pi^2 n_2 c},\qquad(1)$$

where all symbols can be found in Table 1.

At this power level, SPM-induced pulse compression exactly compensates for the GVD-induced pulse broadening. However, in real-world implementations, loss, non-ideal power levels and deviations from the design dispersion values are unavoidable. In this contribution, we propose a system design with improved tolerance and engineerability.

TABLE 1. Symbol table

Symbol	Description
α	power attenuation per unit length (dB/km)
λ	wavelength
τ_{FWHM}	full-width half-maximum pulse width
A_{eff}	effective fiber core area cross-section
BR	Bit-rate
c	speed of light in vacuum
D	group velocity dispersion parameter
D_i	dispersion of section i
D_{av}	average dispersion
D'	slope of dispersion parameter
L_{amp}	amplifier spacing
n_2	non-linear coefficient
P_0	soliton peak power
P_l	soliton launch power
Q	path-average factor
z_0	soliton period
z_i	length of section i

2. Theoretical Estimation

In this section, a simple estimation for a soliton system is discussed by using theoretical equations. We found the effectiveness of low dispersion soliton in high bit rate region to achieve maximum transmission distance with large tolerances.

2.1. SOLITON PERIOD AND DISPERSION

Distributed fiber loss is periodically compensated by lumped optical amplifiers regularly spaced in the link. In these systems, the power level of the transmitted pulses decays between each amplifier. Here, the excess SPM in the first part of each amplifier span due to a power level above the soliton level is compensated by the excess GVD in the second part of the span. This is the path-averaged soliton regime. The launched peak soliton power is given by

$$P_l = P_o \frac{0.23 \alpha L_{amp}}{1 - 10^{(-\alpha L_{amp}/10)}} .$$

(2)

So, a large amplifier spacing requires a larger launch power level.

Mollenauer [1] and Hasegawa [2] have found that if amplifier spacing L_{amp} is much smaller than the soliton full period $8z_o$, unperturbed soliton transmission will take place:

$$L_{amp} \leq \frac{8z_o}{10} << z_o$$

$$z_o = \frac{\tau_{FWHM}^2 \times 2\pi \, c}{1.763^2 \, D^2 \, \lambda^2},$$

(3)

where 10 is an arbitrary safety factor.

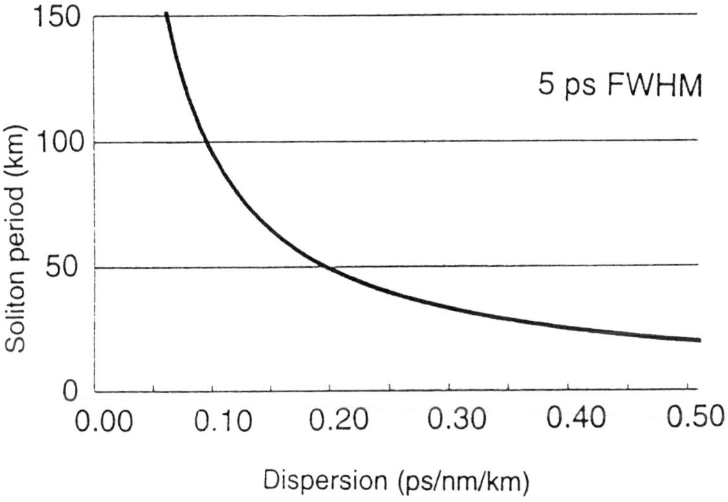

Figure 1. Soliton period for a 1,552 nm system using 5 ps pulses

We have plotted Equation (3) in Figure 1, for a system at 1,552 nm wavelength using 5 ps pulses. As can be seen, dispersion values smaller than 0.10 ps/km/nm result in soliton periods in excess of 50 km. That is, any perturbation to the link will be averaged out in such a length. Therefore, low dispersion operation increases the system tolerance with respect to power level variations, fiber geometry, local dispersion, and amplifier location[3].

This result would lead us to conclude that the lower the dispersion, the better the system design. However, the transmission power level and therefore the maximum transmission length is linked to the dispersion by Equation (1). We need therefore to look at the interplay between dispersion and the four intrinsic mechanisms which limit transmission distance in an optically repeated single-wavelength soliton-based system:

- •soliton period and stability of average or dynamic soliton approximation (Figure 1),

- •interactions between successively launched pulses,

- •amplitude noise due to amplified spontaneous emission and

- •Gordon-Haus time jitter.

2.2. SOLITON INTERACTION

Soliton pulses launched into a fiber closely in time after each other overlap partially. This causes a force between them, and after propagation, pulses will gradually deviate from their time slots and even collapse. A bit error will occur if the corresponding pulse jumps out of its bit slot.

In time-division multiplex systems, bits are extracted by a gating circuit. The time-width of this gate is -due to imperfections- smaller than the bit slot. Following the analysis in [3]-[6], we choose a window value of 2/3 of the bit slot. The propagation distance at which pulses have deviated 1/3 from the center of their time slot due to interaction is given by [7,8]

$$L_{col} = \frac{\tau^2_{FWHM} 2\pi c}{1.763^2 D \lambda^2} \frac{\exp(q/2)}{2} \arccos(\exp(-2/6q)), \qquad (4)$$

where $q = 1.763 / BR / \tau_{FWHM}$ is inversely proportional to the system duty cycle. So, higher duty cycle means shorter collision distance. The formula above represents the worst case: two isolated pulses (marks) with the same phase and amplitude. In general, in a pseudo-random bit sequence, more pulses will be present simultaneously, so the collision length will be longer [7]. If pulses are launched with alternate amplitudes or different phases, collision length will also be longer [7].

2.3. AMPLITUDE NOISE

Due to the presence of optical amplifiers, amplified spontaneous emission (ASE) adds up throughout the cascade. Then, the signal to noise ratio (SNR) is degraded at the receiver. We define the electrical SNR at the receiver as the ratio of the signal power and the beat noise product between signal and ASE within the receiver bandwidth:

$$SNR = \frac{1}{2} \frac{\left(\frac{\eta e}{h\nu} P_l 2\tau_{FWHM} \times BR \right)^2}{2 \left(\frac{\eta e}{h\nu} \right)^2 P_l\, 2\tau_{FWHM} \times BR\, (G-1) n_{sp}\, h\nu\, 0.8\, BR\, N_{amp}}, \quad (5)$$

where the soliton power P_l is given by Equations. (1) and (2), η is photo-diode quantum efficiency which is assumed to be 100%, G is the amplifier gain, n_{sp} is the amplifier noise figure.

The first factor in Equation (5) is due to the mark ratio. We use an effective receiver bandwidth which is 0.8 times the bit-rate. In that formula, we have supposed that the dominant source of noise is the beat between signal and ASE, and that shot noise, receiver noise and ASE-ASE beat are negligible. Furthermore, we have used Gaussian statistics, and we evaluate the system after the last amplifier. So, we assume that optical bandwidth is much larger than the electrical one, and that we have zero average white Gaussian noise. Furthermore, we interpret the last amplifier as a pre-amplifier.

By substituting the number of amplifiers N_{amp} with L_{snr} / L_{amp} and rearranging:

$$L_{snr} = \frac{P_o\, \tau_{FWHM}\, L_{amp}}{SNR\, (G-1) n_{sp}\, h\nu\, 0.8}. \quad (6)$$

Our criterion is to guarantee an SNR value of 20.6 dB, which is 5 dB over the 10^{-9} bit error rate (BER) requirement [4].

2.4. GORDON-HAUS TIME JITTER

Part of the ASE noise is incorporated into the soliton spectrum due to the non-linear Kerr effect. As a result, the center frequency of the soliton is partially randomized. Then, due to dispersion, the velocity of the pulses and therefore the arrival time after propagation will also be randomized [6,9,10]. This can lead to errors, when pulses arrive before or after the receiver acceptance window. The

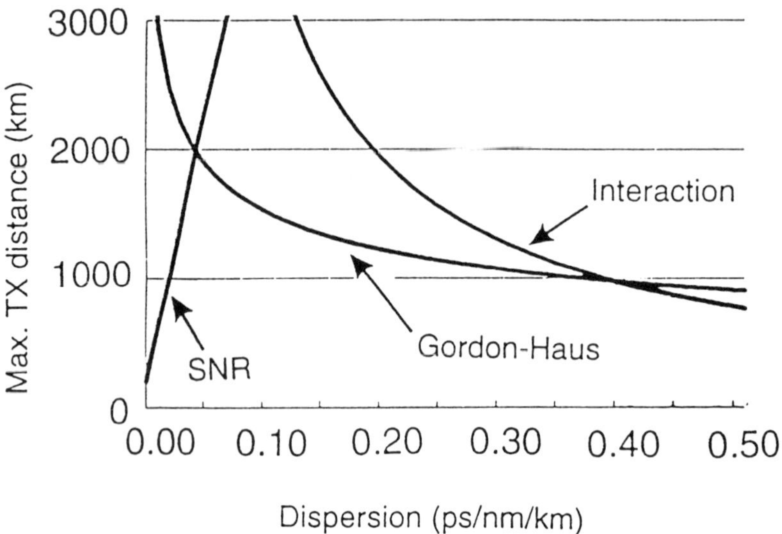

Figure 2. Optimum dispersion range for a 40 Gb/s system with 5 ps pulses

Gordon-Haus limit is the maximum transmission distance for a given fiber and optical amplifier configuration for which timing errors cause a given bit error rate (10^{-14} in our case). The closed form expression for the maximum transmission distance is [6]:

$$L_{GH} = \sqrt{\frac{1}{BR^2}\frac{\tau_{FWHM}}{7.74}\frac{A_{eff}}{1.763D\,n_2\,h}\frac{P_l\,/\,P_o\,L_{amp}}{n_{sp}\,(G-1)}}. \tag{7}$$

2.5. OPTIMUM DISPERSION

We have plotted in Figure 2 the maximum transmission distance as a function of dispersion for a 40 Gb/s system using 5 ps pulses. For low dispersion values, the soliton power is low, and therefore the SNR limit is dominant. At around 0.05 ps/km/nm, the Gordon-Haus limitation takes over. Pulse to pulse interaction is not a problem until dispersion values around 0.40 ps/km/nm. As a conclusion, the optimum dispersion range is 0.05 to 0.10 ps/km/nm. From Figure 1, in this range, the tolerance for the system design will be very large.

3. Dynamic Range: Design Tolerance

We analyze here the power tolerance and transmission distance as a function of amplifier spacing and dispersion for single-dispersion systems. The received eye-closure penalty is a measure of transmission quality: the normalized maximum opening of the worst received trace, in dB. It is a function of the launched power P_l. Dynamic range expresses the tolerance to P_l: the interval of power levels for which the eye-closure penalty is below 1dB. A large dynamic range means tolerance to transmitted power levels, increased design flexibility, better ruggedness, longer life-time of the system and immunity to inhomogeneities in the fiber core, dispersion, optical amplifier characteristics and loss deviations.

TABLE 2. Parameters for the simulations

Parameter	Value
BR	40 Gb/s
t_{FWHM}	5.0 ps
L_{amp}	25, 33, 40, 50 km
D	0.05, 0.1, 0.3, 0.5 ps/nm/km
D'	0.07 ps/nm^2/km
α	0.25 dB/km
A_{eff}	50 μm^2
n_2	$3.2{\cdot}10^{-20}$ W/m^2

The parameters for the investigated system can be found in Table 2. ASE noise was ignored in this calculation. The dynamic range can be seen in Figure 3. For D=0.05 ps/km/nm, dynamic range is over 4 dB after 5,000 km propagation distance. The influence of amplifier spacing is minimum (50 km amplifier spacing can be tolerated). The case D=0.10 ps/km/nm performs as well, with little influence of amplifier spacing. However, for D=0.30 ps/km/nm, dynamic range at 5,000 km is only 2 dB. For D=0.50 ps/km/nm, transmission with 50 km amplifier spacing is limited to 1,500 km. Dynamic range is only 4 dB for the 25 km spacing.

As a conclusion, a dynamic range of 3 dB at 5,000 km transmission is possible with L_{amp} = 50 km for a D below or equal to 0.1 ps/km/nm. We have confirmed that very low dispersion operation increases tolerance to launched power levels.

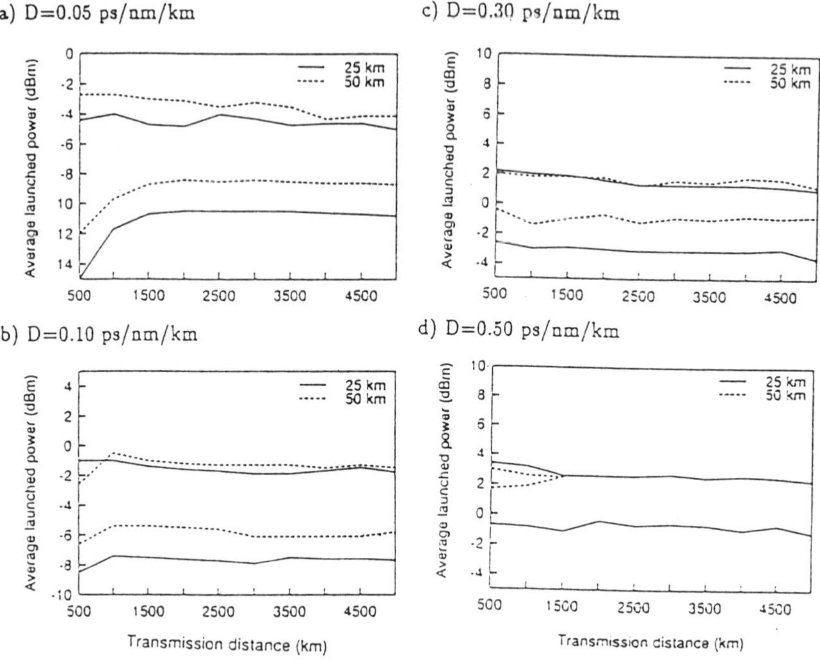

Figure 3. Dynamic range for 1 dB penalty as a function of propagation length

4. Dispersion Inhomogeneities

In a practical implementation of any long-haul system, the dispersion parameter may change significantly at each fiber draw. Here we compare the robustness of low average dispersion (D=0.05 ps/km/nm) and high average dispersion (D=0.30 ps/km/nm) against dispersion variations.

We model these variations as shown in Figure 4. We combine different dispersion sections of 50 km length in a pseudo-random order at steps of $D_{av} \pm 0.15$ ps/nm/km with D_{av} ranging between 0 and 0.30 ps/km/nm. The standard deviation of the sequence is 0.12 ps/km/nm. Let us emphasize that for the low dispersion systems, some of the long sections lay in the normal (D<0) dispersion region. This is in principle a region prohibited to soliton propagation. However, as long as the average dispersion is positive, average soliton regime will still take place[11].

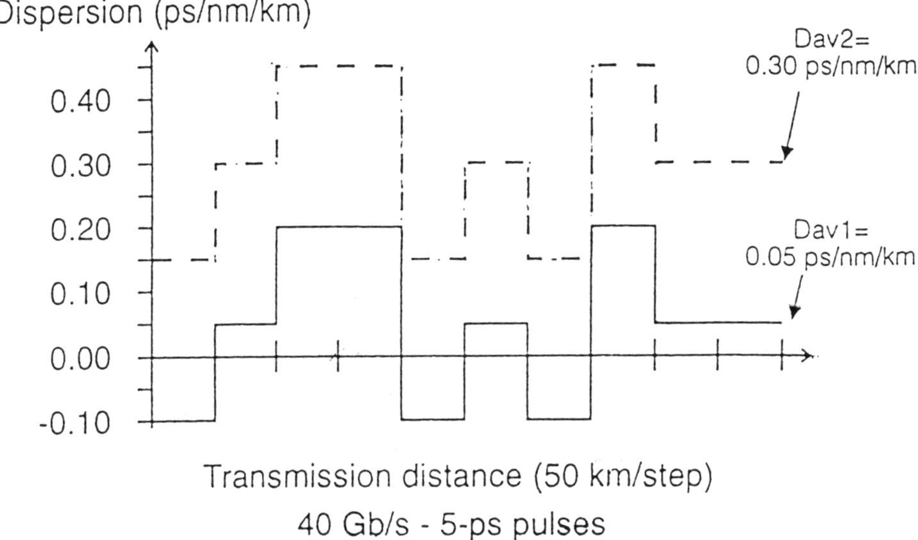

Figure 4. Random dispersion sequence arrangement. Low and high average dispersion shown

We show in Figure 5 the dynamic range for systems with inhomogeneous dispersion for low and high average dispersion. For 0.05 ps/km/nm, dynamic range is in the excess of 2 dB. However, at 0.30 ps/km/nm, the soliton regime is unsustainable.

As a conclusion, low dispersion systems (0.05 ps/km/nm) can successfully transmit even in the presence of very long (50 km) negative dispersion sections (-0.10 ps/km/nm). The same variations cannot be tolerated at higher averages. Therefore, low average dispersion systems relax the fiber selection process, and therefore make them cheaper to implement.

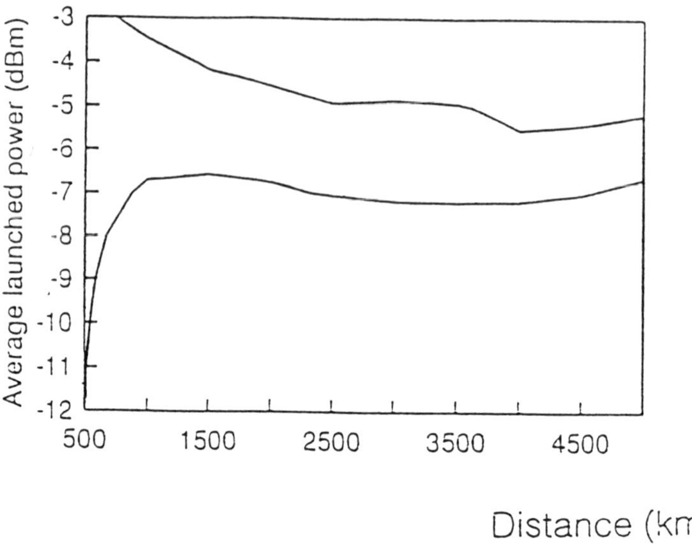

Figure 5. Dynamic range for random dispersion systems

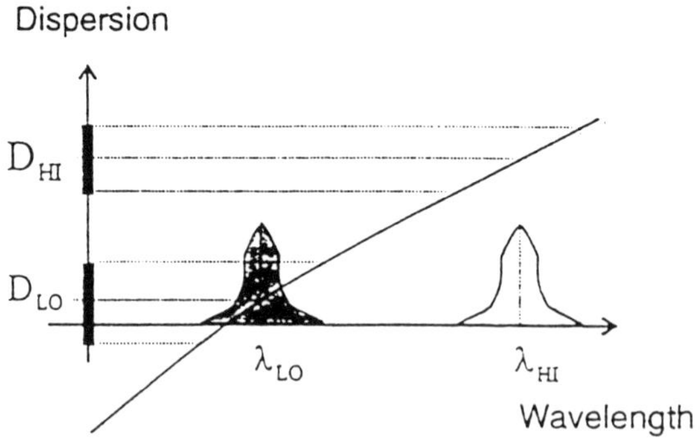

Figure 6. At low average dispersions, part of the signal spectrum invades the negative dispersion zone.

5. Optimum Dispersion for Increased Power and Dispersion Tolerances

As seen in the previous sections, low dispersion increases the system design tolerance. The question then arises of which is the lower limit to the average dispersion for an inhomogeneous dispersion system. As shown in Figure 6, the broad spectrum of the transmitted soliton can invade the negative dispersion region for low average dispersions thus degrading transmission quality. This is due to the slope of dispersion.

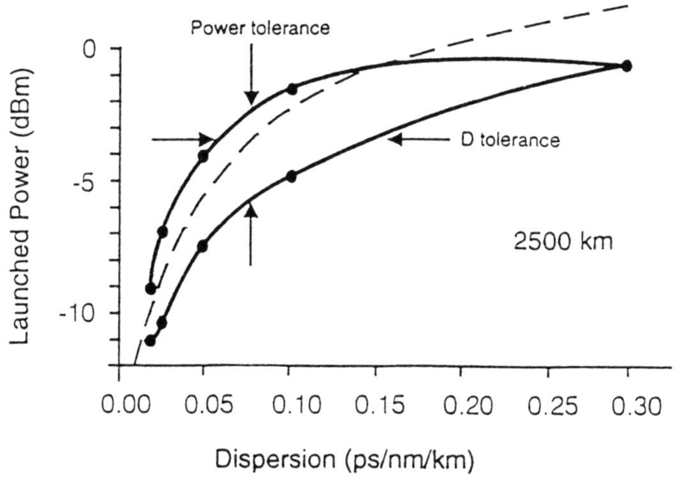

Figure 7. Optimum design. Large power and dispersion tolerance.

Figure 7 maps both the dynamic range and dispersion tolerance for a transmission distance of 2,500 km. The area between the two solid lines is the allowed operation area. The dots are results from computer simulations. The dashed line shows the theoretical soliton power level of Equation (1). As seen from the figure, both the power tolerance and the tolerance to dispersion fluctuations vary strongly with the average dispersion. The optimum design is again 0.10 ps/km/nm. This guarantees a power tolerance of over 3 dB and a tolerance to average dispersion of ±0.05 ps/km/nm.

6. Conclusion

We have investigated with numerical simulations transmission of 40 Gb/s soliton pulses over 5,000 km, in order to find the optimal dispersion operation point and configuration, and tolerances. We find that operation at low dispersion ($D \leq 0.10$ ps/km/nm) is advantageous because it provides

1. Better dynamic range (over 3 dB at 2,500 km)
2. Better immunity to soliton-soliton interactions (collision length over 5,000 km for a 50 km amplifier spacing and a $D = 0.05$ ps/km/nm)
3. Robustness to dispersion fluctuations (only the system with average $D = 0.05$ ps/km/nm tolerates standard deviations of dispersion of 0.12 ps/km/nm with 50 km correlation length)
4. Robustness to fibers with negative dispersion
5. Large tolerance to average dispersion deviation (± 0.05 ps/km/nm)
6. Larger tolerance to transmitted wavelength variation (± 1 nm).
7. Simplicity, thus lower cost (it does not require selection of different special fibers, or lossy splices between fibers of different dispersions which would require extra amplifier gain and thus increase noise)

Optimum performance in all these areas is reached for $0.05 \leq D \leq 0.10$ ps/km/nm, for which the compromise between SNR and Gordon-Haus limitation limits the transmission distance to 2,500 km at 40 Gb/s with 50 km amplifier spacing.

All in all, low dispersion operation of soliton systems relaxes the requirements to their implementation.

References

1. Mollenauer, L. F., Gordon, J. P. and Islam, M. N.: Soliton propagation in long fibers with periodically compensated loss, *IEEE J. Quantum Electron.*, **Vol. 22, No. 1** (1986), 157-173.
2. Hasegawa, A.: Numerical study of optical soliton transmission amplified periodically by the stimulated Raman proces, *Appl. Opt.*, **Vol. 23, No.** 1 (1984), 3302-3309.
3. Farré, J. and Nakaya, S.: Soliton dynamic range in dispersion-managed and low-dispersion communication systems, *Topical meeting of the IEICE,* **Vol. LQE94-40** (1994), 43-48.
4. Wright, J. V. and Carter, S.F.: Constraints on the design of long-haul soliton systems, *Nonlinear wave phenomena* (1991), MA2-1-MA2-3, Brighton, UK.
5. Mitchell, A. F., Wright, J. V., Carter, S. F., Ellis, A. D., Lord, A., Lyle, J. and Scott, J. M.: The future of optically amplified submarine systems, *Suboptic,* 49-54, Versailles, France, 1993.
6. Gordon, J. P. and Haus, H. A.: Random walk of coherently amplified solitons in optical fiber transmission, *Optics Letters,* **Vol. 11, No. 10** (1986), 665-667.
7. Taylor, J. R., Desem, C. and Chu, P. I.: *Optical Solitons - Theory and experiment, Chapter 5,* Cambridge Univ. Press, 1992.

8. Karpmann, V. I. and Solovev, V. V.: A perturbational approach to the two-soliton systems, *Physica D,* **Vol. 3D** (1981), 487-502.

9. Nakazawa, M.: Ultra-high speed soliton communication using Erbium-doped optical amplifiers, Technical report of the IEICE, **Vol. 11, No. GC-3**, (1994).

10. Blow, K. J. and Doran, N. J.: Average soliton dynamics and the operation of soliton systems with lumped amplifiers, *IEEE Photon. Technol. Lett.,* **Vol. 3, No. 4** (1991), 369-371.

11. Nakazawa, M. and Kubota, H.: Optical soliton communication in a positively and negatively dispersion-allocated optical fiber transmission line, *Electron. Lett.,* **Vol. 31, No. 3**, (1994)

BREATHING SOLITON
IN CASCADED TRANSMISSION SYSTEM
WITH PASSIVE DISPERSION COMPENSATION

I. R. GABITOV

L. D. Landau Institute for Theoretical Physics
Kosygina 2, 117940 Moscow, Russia

AND

S. K. TURITSYN

Institüt für Theoretische Physik I
Heinrich-Heine-Universität Düsseldorf
40225 Düsseldorf 1, Germany

Abstract. A theory of optical pulse propagation in cascaded transmission systems which are based on dispersion-compensating fiber technique is developed. The existence of two scales associated with fiber dispersion and system residual dispersion leads to a simple model for the averaged pulse dynamics. In the particular case of practical importance, the averaged pulse dynamics is governed by the nonlinear Schrödinger equation. The pulse transmission stability is examined.

1. Introduction

The application of dispersion-compensating fibers (DCF) to overcome fiber chromatic dispersion in optical transmission systems has been the subject of intensive investigations during the last few years [1]-[5]. This approach is very promising inasmuch as it is simple. It is compatible with the present concept of the all-optical transparency of the system and cascadable. Moreover, all system components are commercially available. All investigations to date focused primarily on point-to-point transmission lines operating in a linear regime. They showed the great potential of this method.

Recent technological achievements, such as design of a chirped fiber

A. Hasegawa (ed.), Physics and Applications of Optical Solitons in Fibres '95, 365–373.
© 1996 *Kluwer Academic Publishers. Printed in the Netherlands.*

grating [6], allow dispersion of 500 ps/nm or even more to be compensated by a grating fiber of a few decimeters in length. Moreover, the length of a semiconductor device designed to compensate the total dispersion of 50 km of standard monomode fiber would be only 20 cm [7]. Therefore, cascaded transmission systems based on the direct dispersion compensating technique are very promising especially taking into account their high-capacity, low-error bit rate and large amplifier spacing. However, the theory of optical-pulse propagation in such systems is not developed.

Here we fill this gap and present a model describing optical pulse dynamics in the cascaded transmission system based on dispersion compensating fibers. As a result of our research, a new concept of "breathing solitons" has been introduced. In many ways this concept is analogous to the center guiding solitons [9].

2. Basic Equations

A cascaded transmission system containing optical amplifiers was investigated in recent experiments [4, 5]. In this work 4×10 Gbit/s NRZ bit stream has been sent over 600 kilometers and an 8×10 Gbit/s bit stream was transmitted over 400 kilometers with an error bit-rate below 10^{-11}. In this paper we will use the system design described in [4, 5] as a typical example of cascaded system based on dispersion compensating technique.

Let us examine a transmission line composed of periodic alternating fiber sections and lumped optical amplifiers. Each section includes a dispersion-compensating fiber with normal dispersion D_{DCF} and pieces of fiber with anomalous dispersion D_{TF} as shown in Figure 1. For our future purposes we introduce the ratio $A = |D_{DCF}/D_{TF}|$. Ideally, the total dispersions should be fully compensated; however, in practice there is always some residual dispersion. For example, in the experiment presented in [4, 5], the residual dispersion was approximately equal to $\langle | D | \rangle \sim 0.5 \text{ps}/(\text{nm/km})$ while dispersion of transmission fiber was $D_{TF} \simeq -(16 \div 18) \text{ ps}/(\text{nm/km})$. Thus for the system we can indicate the three characteristic dispersion scales: the dispersion length Z_{DCF}, corresponding to the chromatic dispersion $D_{DCF} \simeq -(65 \div 80) \text{ ps}/(\text{nm/km})$ of the DCF; the dispersion length of the transmission fiber Z_{dis}; and the dispersion length which corresponds to the residual dispersion of each section Z_{DR}. Therefore, in the first approximation, pulse propagation in the transmission line can be described as follows (see Figure 1): during the first stage of the propagation through the DCF, pulses broaden dispersively and acquire a positive dispersion-induced frequency chirp $C > 0$; on the second stage - evolution through the transmission fiber - pulses compress because the sign of the dispersion has been reversed and the condition for compression $\beta_2 C < 0$ is satisfied

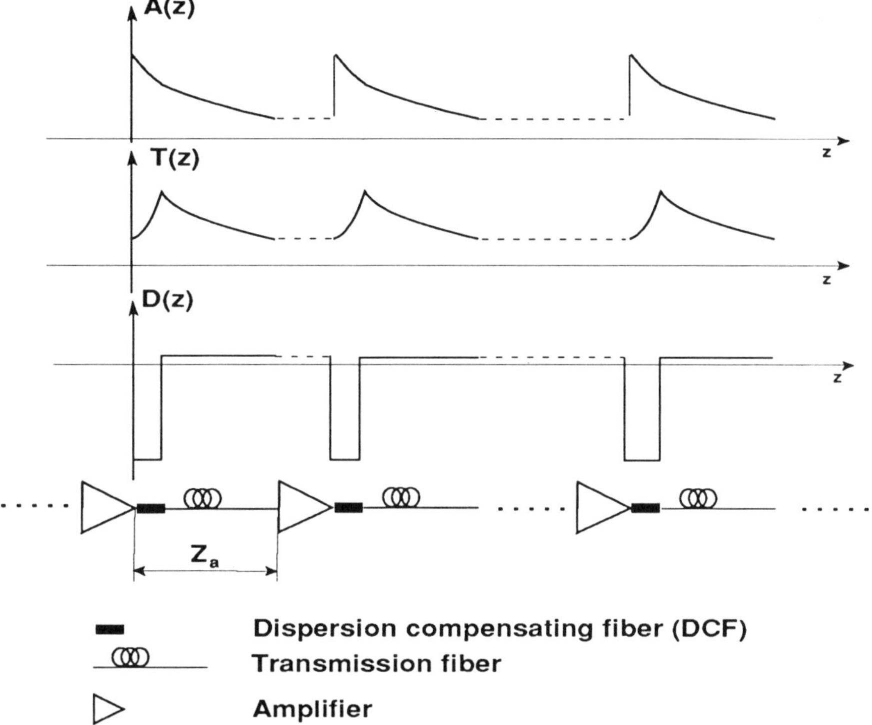

Figure 1. A schematic of a transmission system with passive dispersion compensation. Typical behavior of the pulse amplitude, pulse width and dispersion along the transmission line. Here $A(z)$, $T(z)$ and $D(z)$ are the amplitude, pulse width and dispersion along the fiber.

(see e.g. [8]). Thus, the pulse experiences breather-like oscillations. During both stages, the amplitude of the propagating pulse is reduced due to fiber losses. Therefore, after pulses propagate through both pieces of fiber, they must be amplified by an optical amplifier. This entire process, including amplification, is then repeated many times. The Kerr nonlinearity over one cycle of this process is negligibly small; therefore, dispersion and fiber losses are the main factors. The influence of the residual dispersion appears at a distance of z_{DR} and alters the shapes of pulses. In order for the transmission system to have a sufficiently small bit-error rate, the amplitude of the incident pulse must not be too small; therefore, at distances on the order of z_{DR} the influence of Kerr nonlinearity may begin to come into play. Thus, in the description of a "slow" evolution of a pulse, it is necessary to take into account not only the residual dispersion but the nonlinearity as well.

Let us introduce the following characteristic scales in this problem: Z_a - amplification period, $Z_{dis} = t_0^2/|\beta_2|$ dispersion length corresponding to the SMF (as was mentioned above, in the system under consideration there exist, in fact, several dispersion lengths) and $Z_{NL} = 1/(\alpha P_0)$ nonlinear length. Here t_0 and P_0 are an incident pulse width and peak power, β_2 is the group velocity dispersion, α is the coefficient of the Kerr nonlinearity. The evolution of optical pulses in a fiber is described by the nonlinear Schrödinger equation (NLSE):

$$iE_z + \frac{Z_{NL}}{Z_{dis}} D(z)E_{tt} + |E|^2 E = iG(z)E,$$

$$G(z)E = Z_{NL}\left(-\gamma + r \sum_{k=1}^{N} \delta(z - z_k)\right)E. \tag{1}$$

Here $E = E(t, z)$ is an envelope of electric field normalized by $\sqrt{P_0}$; t and z are time and coordinate along the fiber, normalized by t_0 and Z_{NL} respectively; γ describes fiber losses, $z_k = (1 - k)Z_a/Z_{NL}$ $(i = k, .., N)$ are the amplifiers locations; r is the coefficient of amplification that is compensating the fiber losses over the distance Z_a. Chromatic dispersion $D(z)$ is normalized to the SMF dispersion coefficient.

We would like to point out a difference between the problem that we consider and a soliton propagation in the ultra-long communication systems with varying gain and dispersion studied in [9]. Note that in the case of a guiding center soliton, the evolution of the pulse between the amplifiers also can be described by linear theory. The difference is that in the case that we analyze here, both the dispersion and the losses affect the pulse propagation. In the case of guiding center soliton only the fiber losses are significant (the dispersion and the nonlinearity can be treated as perturbations) causing the amplitude oscillations, while the form of the pulse remains unchanged. In the guiding soliton concept, the "slow" changes, analogous to the slow evolution studied in this paper, occurred under the influence of nonlinearity and dispersion at distances on the order of the soliton period (dispersion length is equal to the nonlinear length).

The concept of the guided-center solitons considered in [?] corresponds to the limit $Z_a \ll Z_{NL} \sim Z_{dis}$. In this paper, we study regime with $Z_{NL} \gg Z_a, Z_{dis}$. Note that two possibilities can be realized: $Z_a \approx Z_{dis}$ and $Z_a \ll Z_{dis} \ll Z_{NL}$. The ratio $\varepsilon_1 = Z_{DR}/Z_{dis}$ and $\varepsilon_2 = Z_a/Z_{NL}$ are two independent small parameters of the problem. The existence of these small parameters allows us to decompose the solution of Equation (1) into the sum of slowly and fast varying functions.

Below we demonstrate that in the leading order averaged dynamics of the "breathing solitons" can be described by the NLS equation and that stable pulse propagation can be achieved when $Z_{NL} \gg Z_a, Z_{dis}$. First, we analyze average pulse dynamics and then we discuss a procedure that gives a constructive description of the higher-order corrections.

Following [9] transform E to a new function q by taking out rapid oscillations of the amplitude due to periodic amplification

$$E = q(t, z) \exp(\int_0^z G(z')dz').$$ (2)

The equation for q reads

$$iq_z + \frac{Z_{NL}}{Z_{dis}} D(z)q_{tt} + c(z)|q|^2 q = 0.$$ (3)

Here

$$c(z) \equiv \exp(2 \int_0^z G(z')dz')$$

can be presented as a sum of rapidly varying and constant parts

$$c(z) = <c(z)> + \tilde{c}(z),$$

where $< \tilde{c}(z) > = 0$ and

$$<c(z)> = <\exp(2 \int_0^z G(z')dz')> = (1 - \exp(-2\gamma z_a))/(2\gamma z_a).$$

We write also $D(z)$ in a similar form

$$D(z) = <D(z)> + \tilde{D}(z),$$

where $< \tilde{D}(z) > = 0$ and $<D>$ is a small perturbation due to an average residual dispersion ($<D> \approx Z_{dis}/Z_{DR} \ll 1$). The average value is defined here as

$$<f> = \frac{1}{z_a} \int_0^{z_a} f(z)dz.$$

In the limit $Z_a, Z_{dis} \ll Z_{NL}$, one may treat the nonlinearity as a perturbation. In the lowest order, the solution of Equation (3) is

$$q(z, t) = \int_{-\infty}^{+\infty} d\omega q_\omega \exp\left(i\omega t - i\omega^2 \frac{Z_{NL}}{Z_{dis}} \int_0^z \tilde{D}(\xi)d\xi\right).$$ (4)

Here q_ω does not depend on z. Nonlinear effects come into play on a larger scale compared to Z_a, namely at the distances proportional to Z_{NL}. Therefore, to describe the influence of the nonlinearity on the pulse propagation we assume that q_ω varies slowly with z and present a solution of Equation (3) in the form

$$q(t, z) = \int_{-\infty}^{+\infty} d\omega q_\omega(z) e^{(i\omega t - i\omega^2 \int_0^z \tilde{D}(\xi)d\xi)}.$$

The evolution of $q_\omega(z)$ is governed by the equation

$$i\frac{\partial q_\omega(z)}{\partial z} - \omega^2 \frac{Z_{NL}}{Z_{dis}} < D(z) > q_\omega(z) + c(z)Q(z) = 0, \tag{5}$$

$$Q = \frac{1}{2\pi} \int_{-\infty}^{+\infty} d\omega_1 d\omega_2 d\omega_3 q_{\omega_1} q_{\omega_2} q_{\omega_3}^* \delta(\omega_1 + \omega_2 - \omega - \omega_3)$$

$$\times \exp\left[-i(\omega_1^2 + \omega_2^2 - \omega_3^2)\frac{Z_{NL}}{Z_{dis}} \int_0^z \tilde{D}(\xi)d\xi\right]. \tag{6}$$

To obtain an equation for the slow evolution of $q(t, z)$, we average Equation (5) over the interval Z_a. Since the function $q(\omega, z)$ is assumed to vary slowly on the amplification distance, it can be placed outside the averaging integral. After straightforward calculations we obtain the equation for the slowly varying envelope

$$i\frac{\partial q_\omega}{\partial z} - \omega^2 \frac{Z_{NL}}{Z_{dis}} < D(z) > q_\omega(z) + \int_{-\infty}^{+\infty} d\omega_1 d\omega_2 d\omega_3 \delta(\omega_1 + \omega_2 - \omega - \omega_3)$$

$$\times F(\omega_1, \omega_2, \omega_3) q_{\omega_1} q_{\omega_2} q_{\omega_3}^* = 0, \tag{7}$$

here the function F is given by

$$F = \frac{1 + ig - e^{-2\gamma z_a}(1 - igA) - i2g(1 + A)e^{-2\gamma z_a(1-igA)/(1+A)}}{4\pi\gamma z_a(1 - igA)(1 + ig)},$$

$$g = \frac{\omega_1^2 + \omega_2^2 - \omega_3^2}{2\gamma Z_{dis}}. \tag{8}$$

The principal results of our paper are based on Equation (7). This equation describes averaged dynamics of "breathing" pulses. The averaging of the Equation (7) leads to further simplifications in particular cases. Let us consider the evolution of a localized pulse in the framework of Equation (7). By normalizing the frequency in the Equation (7) to the characteristic spectral pulse width ω_{cr} (here $\omega_{cr}^2 = Z_{dis}/(Z_{NL} < D >) \sim Z_{DR}/Z_{NL}$), we obtain that $g \sim \omega_{cr}^2/(\gamma Z_{dis})$. In the limit

$$\omega_{cr}^2/(\gamma Z_{dis}) \ll 1 \tag{9}$$

one can approximate the function F by the first terms of the Taylor expansion ($g \ll 1$). In the leading order, F does not depend on ω and can be placed outside the integral. In the time domain, the leading order equation describing the averaged dynamics of breathing solitons reads

$$iU_z + \frac{Z_{NL}}{Z_{dis}}U_{tt} < D(z) > + < c(z) > |U|^2U = R_2. \tag{10}$$

Here $U = U(t,z) = \frac{1}{\sqrt{2\pi}} \int_{-\infty}^{+\infty} e^{i\omega t} q_\omega(z) d\omega$ is the slow varying part of the field E. In this equation the term R_2 describes higher-order corrections in $1/(\gamma Z_{dis})$ and Z_a/Z_{NL}. In the leading order, Equation (10) has the usual soliton solution $U(t,z) = U_0 \cosh^{-1}[t/\tau] \exp(iz)$, with $U_0 = 1/\sqrt{<c>}$ and $\tau = 2 <D> Z_{NL}/Z_{dis}$, if the residual dispersion $<D(z)>$ has a positive sign. It should be pointed out that, the same result can be obtained by using an asymptotic expansion in z_a/z_{NL}.

Thus, a solution of Equation (1) describing optical pulse propagation through considered cascaded transmission system is given by

$$E = \exp\left(\int_0^z G(z')dz'\right) \int_{-\infty}^{+\infty} d\omega q_\omega \exp\left(i\omega t - i\omega^2 \int_0^z \tilde{D}(\xi)d\xi\right), \qquad (11)$$

here $q_\omega = \frac{1}{\sqrt{2\pi}} \int_{-\infty}^{+\infty} d\omega U(t,z) e^{-i\omega t}$ and U is a solution of Equation (10).

To carefully derive higher-order corrections to the averaged NLS equation (left-hand-side of Equation (10)) one can use eihter the powerful Lie-transformation technique[9], or the procedure developed in [10]. The latter does not use any averaging and is based on a standard asymptotic expansion. Again we can use transformation from E to q accounting for a fast amplitude oscillations due to the periodic gain. Transformed function q can be presented as a sum of slow and fast varying parts. Rapidly varying part of q can be written as $\xi = \sum_{k=-\infty}^{+\infty} A_k \exp(i2k\pi Z z_{NL}/z_a)$. Coefficients A_k can be expanded in z_a/z_{NL}. This allows to find corrections of any order by straightforward algebraic operations. Details of the calculations will be presented elsewhere.

The solitons of the NLS equation are stable and exist if the dispersion coefficient is positive. Therefore, the sign of the residual dispersion $<D(z)>$ in the equation (10) represents a criterion for the stability of the "breathing" soliton. The amount of continuous radiation generated during the propagation of a "breathing" soliton and its dependence on system parameters can be determined by the higher-order corrections to Equation (10). We will relegate the discussion of these corrections to a future publication, in which we will address their minimization, and thus the minimization of the amount of dispersive radiation in the system. We will there show that these higher-order corrections also lead to the following algorithm of optimal dispersion management: $\sum_{j=1}^{N}(<D>_j - <D>)^2 \to$ min with respect to all possible permutations of DCF and SMF components. Here $<D>_j$ is the residual dispersion of the $j-th$ transmission element and $<D>$ is the mean value of the whole system dispersion. This algorithm is in many ways is similar to the algorithm proposed in [11], and provides the minimal variation of dispersion and, therefore, optimal system design for breathing soliton transmission.

The case which we considered here is of practical importance. The values of parameters corresponding to the pulse width $t_0 \simeq 20$ ps and the peak power $P_0 \simeq 2$ mW are $Z_a \simeq 40$ km, $Z_{dis} = t_0^2/|\beta_2| \simeq 20$ km, $Z_{NL} = (\alpha P_0)^{-1} \simeq 300$ km, here $\beta_2 \simeq -20 \text{ps}^2/\text{km}$ and $\alpha \simeq 1.66 \text{km}^{-1}\text{W}^{-1}$ are typical for conventional silica fibers at $\lambda_0 = 1550$ nm [8]. We take the value of the dispersion length corresponding to the residual dispersion as $Z_{RD} \simeq 140$ km. The value of fiber losess, if splices, connectors and presence of DCF also being taken into account, is $\gamma \simeq 0.4$ dB/km.

3. Conclusion

In conclusion, we have shown that "breathing" solitons can propagate in cascaded transmission lines with periodic lumped amplification and periodic dispersion compensation if the period of amplification and dispersion compensation is much shorter than a nonlinear length z_{NL}. In the span between two amplifiers, a pulse experiences strong attenuation and large oscillations of its width. We have derived the equation describing propagation of averaged pulses in such systems. In the limit

$$\gamma Z_{dis} Z_{NL}/Z_{DR} = \gamma D_R/(D_{TF}\alpha P_0) \gg 1$$

this propagation can be described by the NLS equation. Note that the pulse width does not enter the final form of our criterion and therefore the theory is applicable to both RZ and NRZ signal formats.

Our results complement the concept of guiding-center solitons and demonstrate that the concept of averaged dynamics can be applied in this modification of the original system. This is a further confirmation of the robust nature of optical solitons..

References

1. Onaka, H., Miyata, H., K. Otsuka, K. and Chikama, T.: 10 Gb/s-4-wave 200 km and 16-wave 150 km repeaterless transmission experiments over standard single-mode fiber, *Proceedings of ECOC '94* **4** (1994), 49.
2. Chen, C. D., Delavaux, J.-M., Hakki,P. B. W., Mizuhara, O., Nguyen,T. V., Nuyts, R. J., Ogawa, K., Park, Y. K., Tench, R. E., Tzeng, L. D. and Yeates, P. D.: Field experiment of 10 Gbit/s, 360 km transmission through embedded standard (non-DSF) fibre cables, *Electronics Letters* **30** (1994), 1159.
3. Ellis, A. D. and Spirit, D. M.: Unrepeatered transmission over 80 km standard fibre at 40Gbit/s, *Electronics Letters* **30** (1994), 72.
4. Das, C., Gaubatz, U., Gottwald, E., Kotten, K., Küppers, F., Mattheus, A. and Weiske, C.J.: Straight line 20 Gbit/s - transmission over 617 km dispersion compensated standard single mode fiber, *Electronics Letters* **31** (1995), 305.
5. Das, C., Hein, B., Gaubatz, U., Gottwald, E., Kotten, K., Küppers, F., Mattheus, A., Rapp, L. and Weiske, C. J.: 4x10 Gbit/s transmission over 617 km and 8x10 Gbit/s transmission over 412 km dispersion compensated standard - single - mode fiber, *Proceedings of OFC'95* (1995).
6. Kashyap, R.: Photosensetive Optical Fibers: Devices and Applications, *Optical Fiber Technology* **1** (1994), 17.

7. Peschel, U., Peschel, T. and Lederer, F.: Proper dispersion tailoring in semiconductors as prerequisite for soliton formation, *1995 OSA Technical Digest* **6** (1995), 139.
8. Agrawal, G.P.: *Nonlinear Fiber Optics*, Academic Press, San Diego, (1989).
9. Hasegawa, A. and Kodama, Y.: Guiding-center soliton in optical fibers, *Optics Letters* **15** (1990), 1443-1445; Hasegawa, A. and Kodama, Y.: Guiding-center soliton, *Physics Review Letters* **66** (1990), 161.
10. Kivshar, Y.S. and Turitsyn,S.K.: Semitemporal pulse collapse on periodic potentials, *Physics Review E* **49** (1994), 2536-2539.
11. Ohhira, R., Hasegawa, A. and Kodama, Y.: Methods of constructing a long-haul soliton transmission system with fibers having a distribution in dispersion, *Optics Letters* **20** (1995), 701-704.

MULTI-TEN GBIT/S SOLITON TRANSMISSION OVER TRANSOCEANIC DISTANCES

M. SUZUKI, N. EDAGAWA, I. MORITA
S. YAMAMOTO, H. TAGA AND S. AKIBA
KDD R&D Laboratories
2-1-15 Ohara, Kamifukuoka, Saitama 356, Japan

Abstract. The feasibility of 20 Gbit/s soliton-based transoceanic systems has been discussed through the 1,000 km-loop transmission experiments and the 8,100km straight-line transmission experiments. It is shown that the accumulation of the Gordon-Haus jitter can be effectively suppressed by using periodic dispersion compensation and inline optical filters. For 20 Gbit/s, 9,000 km transmission system, the Q^2 of 19 dB, the sufficient power window of 2.5 dB, and robustness to the repeater output power reduction have been confirmed, in addition to the compatibility of the conventional supervisory scheme. Future prospect toward the 100 Gbit/s system is also discussed.

1. Introdction

Evolution of optical amplifier technology based on Er-doped fiber amplifier (EDFA) has brought about a big impact to optical fiber transmission systems, and resulted in rapid increase in system capacity and non-regenerative transmission distance[1]. 5 Gbit/s transoceanic optical submarine cable systems using EDFA repeaters and NRZ signals will be in service in 1995-1996 in the Pacific and the Atlantic Ocean[2]. Research interest is now being directed toward the development of optical amplifier systems with a transmission capacity of multi-ten Gbit/s or more. To increase an aggregate system capacity, both the wavelength-division-multiplexing (WDM) transmission scheme using NRZ signals[3,4] and soliton transmission scheme proposed by Hasegawa *et al.*[5] have been intensively

A. Hasegawa (ed.), Physics and Applications of Optical Solitons in Fibres '95, 375–391.
© *1996 Kluwer Academic Publishers. Printed in the Netherlands.*

studied. Recently, the feasibility of 100 Gbit/s transmission over 6,300 km by using 20-channel NRZ 5 Gbit/s WDM signals has been demonstrated[4]. Soliton transmission systems are expected to carry higher bit rate (per channel) than NRZ systems, and thus soliton-WDM transmission with a channel bit rate more than 10 Gbit/s is also quite attractive for transmission systems with a transmission capacity of multi-ten Gbit/s or more.

Soliton transmission using EDFA-repeaters was first experimentally demonstrated by Nakazawa *et al.*[6] and followed by theoretical analysis using guiding-center theory[7] or average soliton theory[8,9]. Major constraints in soliton transmission systems are ASE-induced timing jitter (Gordon-Haus timing jitter[10]) and interaction between adjacent solitons[11,12]. To overcome these constraints, several inline soliton control techniques using frequency guiding filters[13,14], sliding-frequency guiding filters[15], and inline synchronous modulation[16] have been proposed. Post-transmission dispersion compensation[17] and the use of alternating-amplitude solitons[18] can also mitigate these constraints by controlling the optical signals at the transmitter and the receiver ends. From view point of dispersion management of the total system, we have proposed a novel transmission scheme to suppress the accumulation of Gordon-Haus jitter by using periodic dispersion compensation together with inline optical filters[19]. So far, many reports have shown the potential of soliton transmission system at over 20 Gbit/s using these soliton control techniques [16]-[29], and now the focus is moving onto their practicality[25],[30]-[33].

In this paper, we discussed the feasibility of multi-ten Gbit/s soliton transmission system with periodic dispersion compensation. 20 Gbit/s, single-channel transoceanic system was evaluated by using 1,000 km-long circulating loop and 8,100 km of straight-line transmission facility. The superimposed-tone transmission capability for possible supervisory applications and robustness to repeater output power reduction were also investigated. Finally, future prospect toward the 100 Gbit/s system was also discussed.

2. Timing Jitter Reduction by Periodic Dispersion Compensation

2.1. PRINCIPLE

A random variation of the soliton center frequency, caused by ASE noise of optical amplifiers and fiber nonlinearity, is translated into Gordon-Haus timing jitter through the non-zero fiber dispersion[10]. Zero-dispersion transmission maintaining soliton characteristics could prevent the translation of frequency variation into timing jitter. For this purpose, we have proposed novel transmission scheme using the periodic dispersion compensation[19]. Figure 1 shows the dispersion maps for the conventional soliton transmission and the proposed transmission with periodic dispersion compensation. In the conventional soliton transmission system, where the cumulative chromatic dispersion increases along the system, the variance of accumulated Gordon-Haus jitter increases by the factor of DL^3 without inline optical filters[10] or DL with inline optical filters[13,14] (D: chromatic dispersion (ps/km/nm), L: total system length(km)). On the other hand, in the proposed scheme, where the cumulative chromatic dispersion is offset periodically along the system by the dispersion compensation fiber (DCF) with negative (normal) dispersion, the accumulated timing jitter does not increase significantly because the total system dispersion is close to zero.

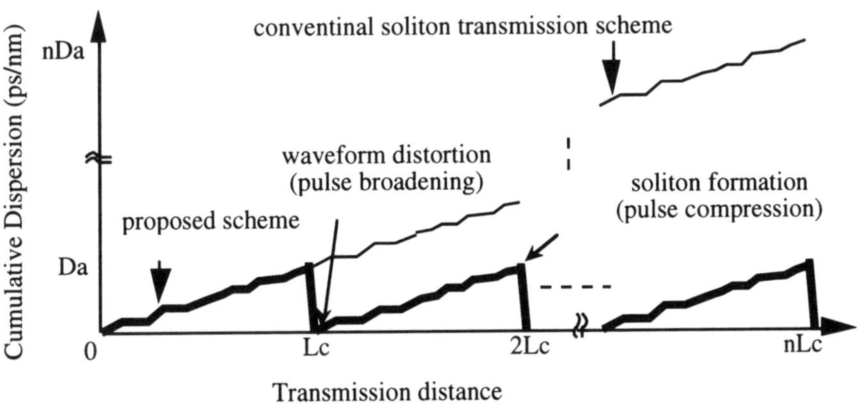

Figure 1. Cumulative dispersion versus transmission distance for conventional soliton transmission scheme and proposed transmission scheme.

The timing jitter caused by other frequency variations, such as source spectral broadening or acoustic effect, can also be suppressed in this transmission scheme. The pulse waveform is distorted by the dispersion compensation, however, it can be reshaped by soliton formation, since local dispersion of the transmission fiber is positive (anomalous) within each compensation interval. Non-soliton components can be reduced by inline optical filters.

2.2. COMPUTER SIMULATIONS

This mechanism was confirmed by the computer simulation with the following three dispersion maps as shown in Figure 2. Chromatic dispersion for soliton transmission fiber is D=0.2 ps/km/nm and amplifier spacing is 30 km. An optical bandpass filter of 6 nm FWHM is used at every EDFA repeater. The cumulative dispersion was completely compensated for by DCFs with -36 ps/nm at the end of the every 180 km-long dispersion compensation interval for Map A, and by DCFs with -72 ps/nm at the middle of the 360 km-long compensation interval for Map B, respectively. The output power of each EDFA is 3 dBm and the pulse width is 7 ps. The calculated Q^2 after electrical equalization was 19.8 dB and 19.3 dB for Map A and B, respectively. Although the normal cumulative dispersion region is included in Map B, good transmission performance similar to Map A can be obtained. These results indicate that the location of DCF in the compensation interval is not so critical. Figure 3(a) and (b) show examples of the simulated eye diagrams of 20 Gbit/s optical signals and equalized electrical signals after 9,000 km transmission for Map A, respectively. The pulse shape after transmission is not deformed although the system average dispersion is zero, and timing jitter is quite small. The effect of the length of the dispersion compensation

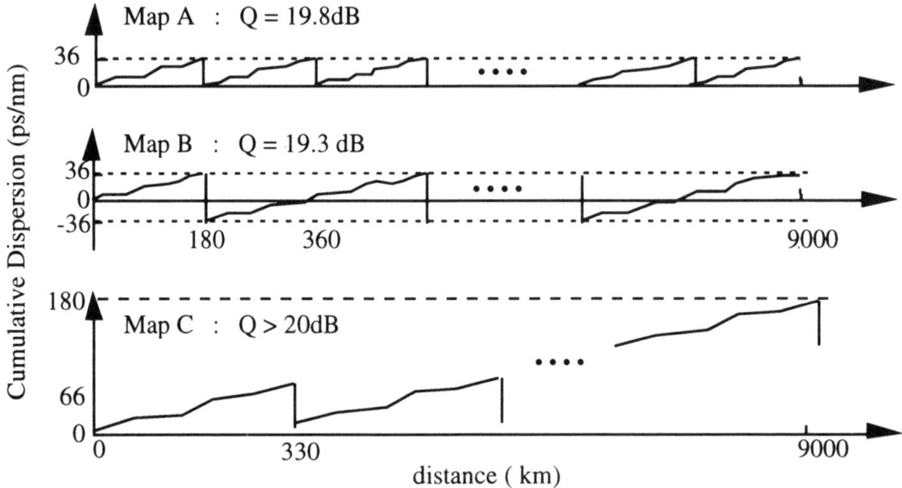

Figure 2. Possible dispersion maps for periodic dispersion compensation

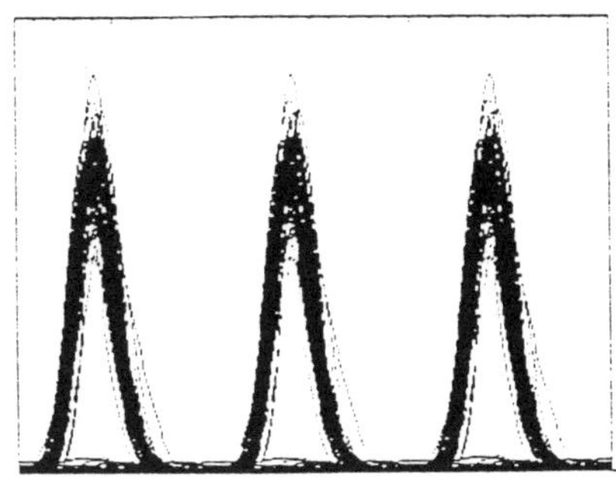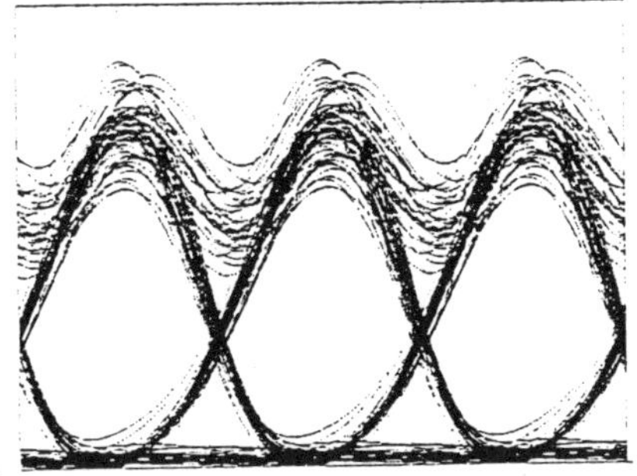

(a) received optical signals   (b) equalized electric signals (Q^2=19.8 dB)

Figure 3. Simulated eye diagrams for 20 Gbit/s signals after 9,000km transmission

has been also evaluated by changing the compensation length from 90 km to 450 km for Map A. The proposed scheme did not work well for too short dispersion compensation interval of 90 km. For 450 km compensation interval, Q^2 degradation by 2 dB was calculated. Optimum compensation interval seemed to be around 200 km to 300 km in this condition.

In Map C, almost all the cumulative dispersion of the interval is compensated for by the DCF, but, there is a small amount of residual dispersion in the interval. In such a case, transmission performance was improved and a Q^2 of over 20 dB was obtained with many different parameters, in terms of the optical pulse width, the strength of optical bandpass filter, and repeater output power. Optimum compensation rate ranged from 80 to 90% in our calculations. This tendency was confirmed by the experiments described in the next section.

3. 20 Gbit/s Transmission Experiments

3.1 LOOP EXPERIMENT

We experimentally investigate optimum dispersion compensation rate and strength of the optical filter on 20 Gbit/s soliton-based transmission system by using a 1,000 km-long recirculating loop[24],[26]. Figure 4 shows a schematic diagram of the experimental setup.

In the transmitter, a 20 Gbit/s optical soliton data stream was produced by time-division-multiplexing 10 Gbit/s RZ data pulses, which were generated with a DFB-LD, a sinusoidally-driven electroabsorption (EA) modulator[34] and two LiNbO$_3$ intensity modulators operated at 10 Gbit/s by a 2^{15}-1 pseudorandom binary sequence. The amplitude of two 10 Gbit/s subchannels was slightly alternated between large and small to reduce the soliton-soliton interaction[12],[18]. Then, the 20 Gbit/s pulses were compressed by using a high nonlinearity fiber and a high dispersion fiber. The pulse width obtained was about 10 ps and the time bandwidth product was about 0.3. A low-frequency polarization scrambler was used to suppress the polarization hole burning (PHB) effect of the EDFA repeaters. The signal wavelength (λ_s) was 1,558.7 nm.

The 1,000 km-long recirculating loop consisted of 30 spans of dispersion shifted fiber (DSF), 3 spools of dispersion compensation fiber (DCF), and 31 EDFA-repeaters, and optical bandpass filters(OBPFs) of about 5 ~ 6 nm FWHM. Average chromatic dispersion of the DSF was 0.22 ps/km/nm at λ_s and the repeater spacing was about 33 km. Each repeater comprises an erbium doped fiber, a 1,480 nm pump laser diode (LD), a WDM coupler, an optical isolator and a 10 dB monitor fiber coupler.

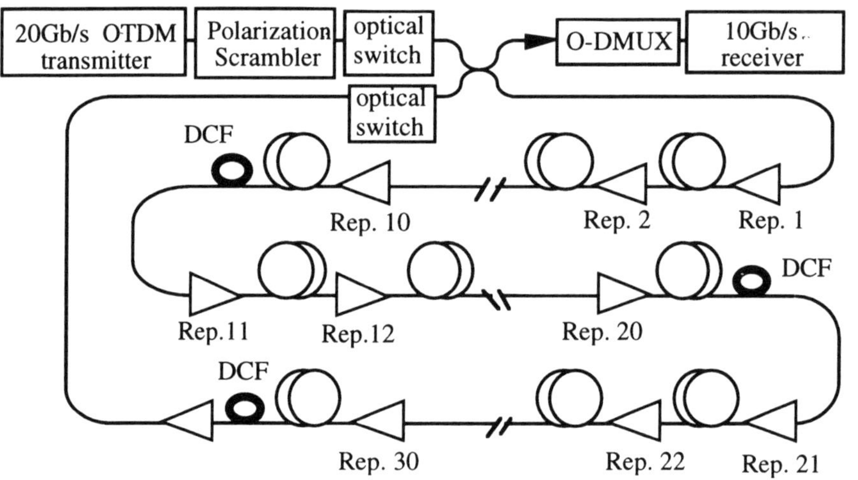

Figure 4. Schematic of experimental setup for 1,000 km-loop transmission experiment

DCF spools with dispersion of -Dc (ps/nm) were placed at every 10 repeaters to compensate for the accumulated dispersion (Da (ps/nm)) in the preceding 10 repeater spans.

In the receiver, the transmitted 20 Gbit/s signals were optically time-division-demultiplexed to 10 Gbit/s data stream with an optical gate using a sinusoidally-driven polarization insensitive EA modulator[35,36]. The gate width was about 40-50 ps. The clock recovery was done with a 20 GHz phase-locked-loop. Bit error rate (BER) or Q-factor [38] of the received 10

Gbit/s signals was measured.

3.1.1. *Effect of Dispersion Compensation Rate*

Figure 5 show the transmission distance for BER of 10^{-9} versus dispersion compensation rate defined by Dc/Da. The average launched optical power to the transmission fiber was about 1 dBm and the number of OBPFs in the loop was 30. The best performance was obtained with dispersion compensation rate of 85% at 1,558.7 nm and the corresponding average system dispersion was 0.03 ps/km/nm. In the post-transmission dispersion compensation technique[17], optimum dispersion compensation rate for a large number of repeaters is 50% and it is estimated from the formula that this rate becomes close to 100% for a small number of repeaters, considering only jitter reduction. In our cases using periodic dispersion compensation, the optimum compensation rate to give the best Q-factor, calculated by computer simulations, ranges from 80% to 90% and the experimental results almost agree with the results of computer simulations.

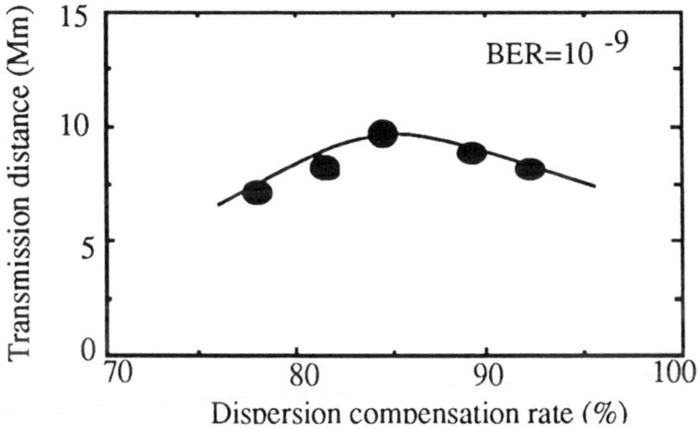

Figure 5.   Transmission distance versus dispersion compensation rate

3.1.2. *Effect of Optical Bandpass Filter*

Figure 6 shows the transmission distance for BER of 10^{-9} as a function of number of OBPFs in the loop for the fixed dispersion compensation rate of 85%. The signal wavelength was 1,558.7 nm for the cases with optical filters and 1,558.4 nm for the case without filter. As shown in Figure 6, 20 Gbit/s over 9,000 km transmission is possible even without inline filters. Although shape of the optical spectrum after 9,000 km transmission without filter was changed, the waveform was not so deformed and the timing jitter observed was small[26]. This result shows that the proposed periodic dispersion compensation technique is quite effective to reduce the Gordon-Haus timing jitter. The transmission performance was improved by employing the optical filters and the best transmission performance was obtained when 12 OBPFs of 5 nm FWHM were used in the loop. Figure 7 shows the BER for the systems without inline filter and with optimum inline filter strength as a function of transmission distance. The transmission distance of 9,000 km for BER of 10^{-9} obtained without inline filter was increased to 14,000 km by using optimum inline filters.

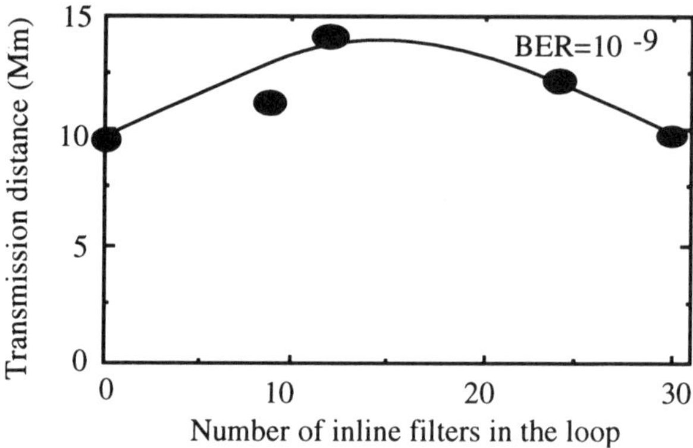

Figure 6. Transmission distance versus the number of inline filters in the loop

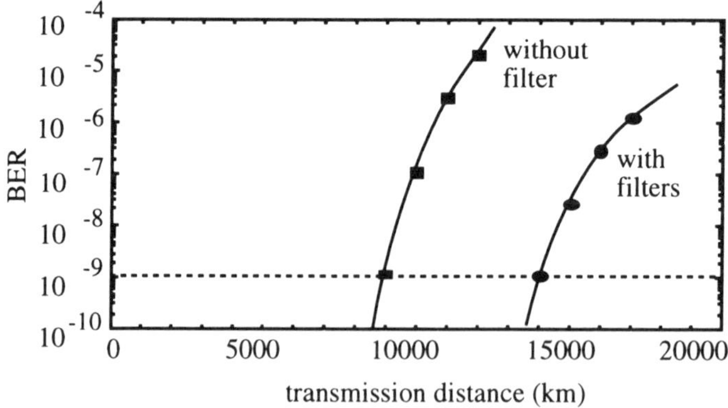

Figure 7. BER versus transmission distance for 20 Gbit/s system with and without inline filters

3.1.3. *Power Window for 20 Gbit/s, 9,000 km System*

Figure 8 shows Q^2 obtained at 9,000 km as a function of nominal repeater output power. Dispersion compensation rate at 1,558.7 nm was 85% and 12 OBPFs was used in the loop. Maximum Q^2 was 19 dB when repeater output power was 1 dBm. Allowing 1dB penalty of Q^2, a large power window of about 2.5 dB was obtained, which leaves an enough margin against repeater output reduction due to aging.

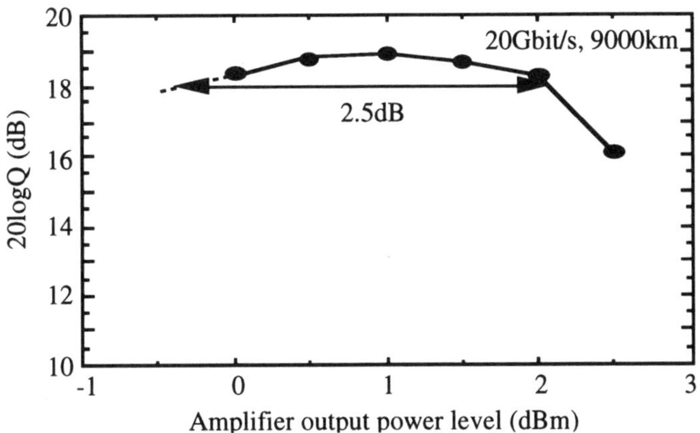

Figure 8. Power window for 20 Gbit/s, 9,000km system

3.2. STRAIGHT-LINE EXPERIMENTS

3.2.1. *8,100 km Test-bed*

We have performed the straight-line transmission experiments[25] to confirm the transmission performance obtained by the loop experiments and computer simulations. The other aspects, which are important for practical system application, such as the robustness of the system against pump failure and supervisory capability, were also evaluated by the straight-line transmission experiments. Figure 9 shows a schematic diagram of the experimental setup. The transmission line comprises 240 repeaters, 8,100 km transmission fiber consisting of dispersion-shifted fiber (DSF) and conventional single-mode fiber (SMF), and 24 spools of dispersion compensation fiber (DCF). More than 40% of the DSF was in the normal dispersion regime at λ_s. The average chromatic dispersion of the DSF and SMF was about 0.2 ps/km/nm at λ_s.

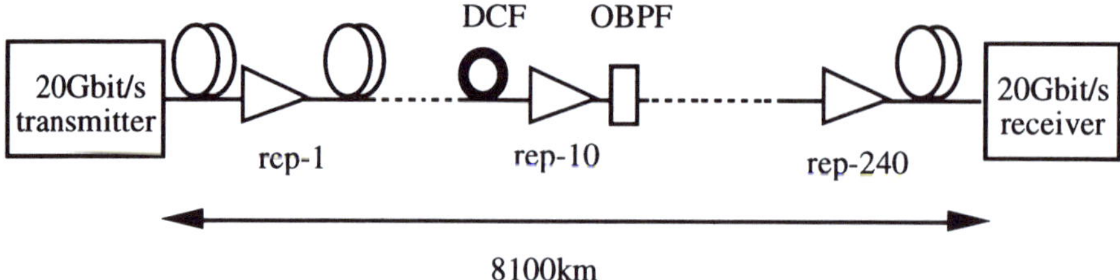

Figure 9. Schematic of experimental setup for 20 Gbit/s, 8,100km transmission

The DCF spools were placed at every 10 repeaters to reduce the timing jitter by compensating for most of the accumulated dispersion in the preceding 10 repeater spans. The repeater spacing ranges from 33 km to 36 km. Optical bandpass filters of 3 nm FWHM were placed at almost every 5 repeaters. We used the same transmitter and receiver as used in loop experiments.

3.2.2. *Transmission Performance*

Figure 10 shows the signal waveform before and after transmission at the nominal repeater output power of 1 dBm. No significant waveform distortion was observed after transmission. To evaluate the transmission performance, the received 10 Gbit/s signals were assessed by Q-factor.

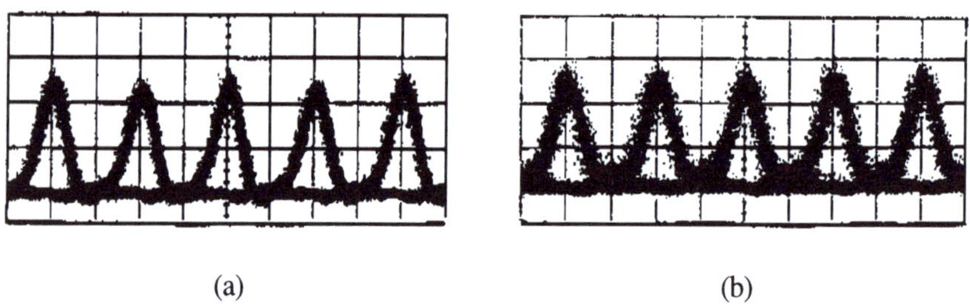

(a)   (b)

Figure 10. `Waveforms (a) before transmission (b) after transmission

Figure 11 shows the Q^2 obtained as a function of transmission distance for both straight-line transmission experiment[25] and 1,000 km-loop transmission experiment[24]. The closed circles represent the results of straight-line experiment and open circles show the results of loop experiments. The Q^2 obtained in straight-line transmission experiment was almost the same as those in the loop transmission experiments. A Q^2 of 19 dB was obtained at 8,100 km. The Q^2 was

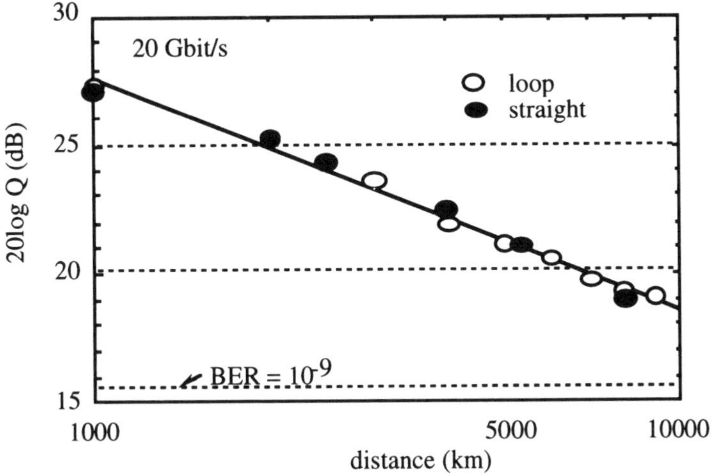

Figure 11. Q-factor versus transmission distance

changed almost linearly with the distance. This tendency indicates that Q^2 degradation is mainly due to signal-to-noise ratio and that the waveform distortion and timing jitter is quite small.

3.2.3. *Feasibility of Supervisory Using the Superimposed-tone Transmission*

In multi-repeater optical amplifier systems, it is commonly used to superimpose intensity modulated low-frequency signals on the main optical signal in order to supervise the systems. So far, such superimposed-tone transmission capability has not been proved for soliton-based systems. To evaluate this, DFB-LD in the transmitter was directly modulated at 10 MHz to superimpose 5%op sinusoidal-tone on the envelope of the 20 Gbit/s optical signal bit stream. The tone level was measured with an O/E converter and a RF spectrum analyzer. The transmitted tone component was clearly observed even at the end of the 8,100 km system. Q^2 degradation due to the tone superimposition was about 0.3 dB. These results show that the tone-superimposition technique is applicable to a soliton-based system[25].

3.2.4. *Robustness against Pump Failure*

Next, to investigate the system robustness to signal level reduction, a failure causing 4 dB repeater-output-reduction was made by reducing the pump power of the repeater by about 50%. The original repeater output was 1 dBm. Firstly, the effect of fault location over the system was evaluated for one, two and three faults. To demonstrate the worst case, two and three faults were made in successive repeaters.

Figure 12 shows the Q^2 degradation versus fault location.

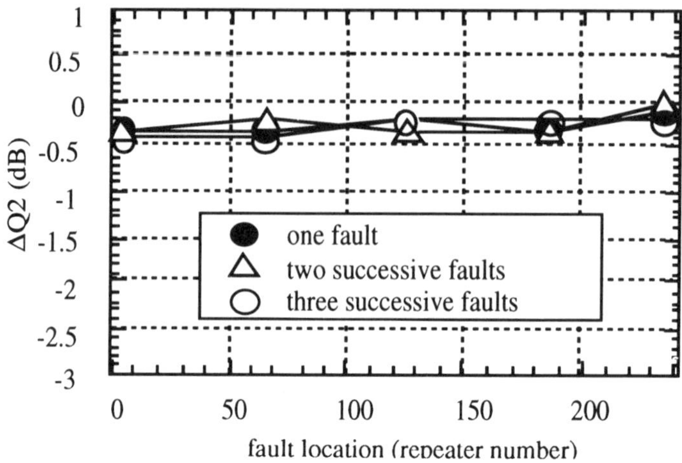

Figure 12. Q^2 degradation versus fault location

Even for three successive-faults, significant degradation was not observed at any locations over the system. Secondly, the effect of number of faults was evaluated. Figure 13 shows the Q^2 degradation versus the number of faults. The each fault location was roughly equally-spaced over the system.

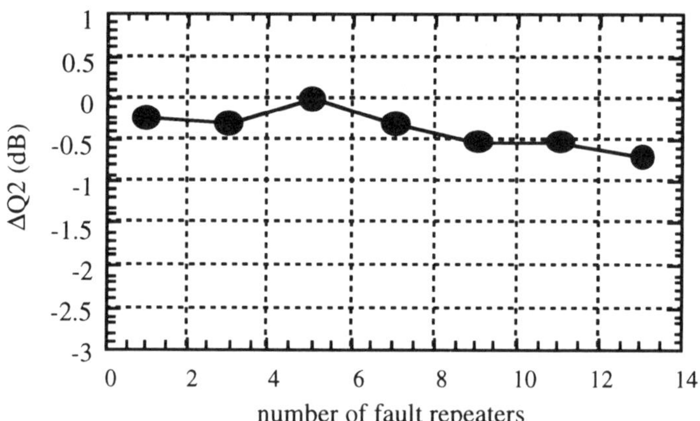

Figure 13. Q2 degradation versus number of faults

No significant Q^2 degradation was observed for up to about 5 faults and no devastating Q^2 degradation occurred even with 13 faults, which corresponds to about 5% pump-LD failure. These results manifest that this system has enough robustness to the failures resulting in repeater output reduction and span loss increase.

4. Toward 100 Gbit/s System

Toward the realization of transoceanic system with the ultra-large capacity around 100 Gbit/s, increase the bit rate by optical time-division-multiplexing (OTDM) and/or the use of more wavelength-division channels should be required. In ultra-high speed soliton transmission systems using OTDM[37], expansion of the transmission distance seems to be difficult, since the impact of the Gordon-Haus jitter and the soliton-soliton interaction on transmission characteristics becomes more severe in high-speed OTDM system with small phase margin of several picoseconds in the receiver. In addition, PMD of the fibers, other

In ultra-high speed soliton transmission systems using OTDM[37], expansion of the transmission distance seems to be difficult, since the impact of the Gordon-Haus jitter and the soliton-soliton interaction on transmission characteristics becomes more severe in high-speed OTDM system with small phase margin of several picoseconds in the receiver. In addition, PMD of the fibers, other effects such as higher-order dispersion and self frequency shift should be taken into account. On the other hand, soliton-WDM transmission seems to be promising to increase the total system capacity[27]-[29], because it offers greater immunity to Gordon-Haus jitter and soliton-soliton interaction than OTDM. 20 Gbit/s over 16,000 km transmission in 2.5 Gbit/s, eight-channel soliton-WDM[28], and 40 Gbit/s over 9,000 km transmission [29] in 5 Gbit/s, eight-channel soliton-WDM have been successfully demonstrated by using frequency guiding filters. As mentioned in the introduction, the major feature of the soliton transmission is the potential of higher bit-rate per one wavelength than NRZ systems. The use of the sliding-frequency guiding filters is more effective for jitter reduction and noise reduction, and it could increase the channel bit rate and the aggregate system capacity. As we saw in the previous section, 20 Gbit/s, 9,000 km transmission systems can be realized with large system margin by using the periodic dispersion compensation technique. Thus, with this technique, dramatic increase in system capacity can be expected by wavelength-division multiplexing 20 Gbit/s signals.

5. Conclusion

The transmission performance, superimposed-tone transmission capability and the robustness to repeater output power reduction for 20 Gbit/s transoceanic system have been evaluated through the 1,000 km-loop transmission experiments and the 8,100 km straight-line transmission experiments. The novel soliton transmission scheme to suppress the accumulation of the Gordon-Haus jitter using periodic dispersion compensation and inline optical filters was adopted in these experiments. The Q^2 of 19 dB, the sufficient power window of 2.5 dB and robustness to the repeater output power reduction were confirmed for 20 Gbit/s, 9,000 km transmission system, in addition to the compatibility of the conventional supervisory scheme. The practicality of the single-channel 20 Gbit/s transoceanic transmission system was proven through these experiments. To realize the systems with an aggregate system capacity of 100 Gbit/s or more, much more efforts to explore the possibility of soliton-WDM transmission with a channel bit rate of more than 10 Gbit/s are expected.

6. Acknowledgment

The authors would like to express their thanks to Mr. H. Tanaka and Dr. Y. Matsushima for providing EA modulators, and Mr. Y. Horiuchi for his help in dispersion measurement and coherent OTDR measurements. They also thank Dr. Y. Urano and Dr. K. Sakai for their continued encouragement.

References

1. for example, Taga H., Edagawa N., Tanaka H., Suzuki M., Yamamoto S. and Wakabayashi H.: 10 Gbit/s, 9000 km IM-DD transmission experiment using 274 Er-doped fiber amplifier repeaters, *OFC'93*, San Jose (1993), **PD-1.**

2. for example, Wakabayashi H., Namihira Y., Akiba S., Yamamoto S., Okawa M. and Yamamoto H.: OS-A optical submarine cable system, *SEE, Suboptics'93*, Versailles (1993), 85-90.

3. Bergano N. S., Davidson C. R., Vengsarkar A. M., Nyman B. N., Evangelides S. G., Darcie J. M., Ma M., Evankow J. D., Corbett P. C., Mills M. A., Ferguson G. A., Pedrazzani J. R., Nagel J.A., Zyskind J. L., Sulhoff J. W. and Lucero A. J.: 100 Gbit/s WDM transmission of twenty 5 Gb/s NRZ data channels over transoceanic distances using a gain flattened amplifier chain, *ECOC'95*, Brussels, post-deadline paper, **Th.A.3.1**

(1995), 967-970.

4. Bergano N. S., Davidson C. R., Nyman B. N., Evangelides S. G., Darcie J. M., Evankow J. D., Corbett P. C., Mills M. A., Ferguson G. A., Pedrazzani J. R., Nagel J. A., Zyskind J. L., Sulhoff J. W., Lucero A. J. and Klein A. A.: 40 Gbit/s WDM transmission of eight 5 Gb/s data channels over transoceanic distances using the conventional NRZ modulation format, *OFC'95*, San Diego, post-deadline paper (1995), **PD19.**

5. Hasegawa A. and Tappert F.: Transmission of stationary nonlinear optical pulses in dispersive dielectric fibers, I Anomalous dispersion", *Appl. Phys. Lett.*, **23** (1973), 142-144.

6. Nakazawa M. , Suzuki K., Kubota H., Yamada E. and Kimura Y.: Dynamic optical soliton communication, *IEEE J. Quantum Electron.*, **QE-26** (1990), 2095-2120.

7. Hasegawa A. and Kodama Y.: Guiding-center solitons in optical fibers, *Opt. Lett.*, **24** (1990), 1443-1445.

8. Mollenauer L. F., Evangelides S. G. and Haus H. A.: Long-distance soliton propagation using lumped amplifiers and dispersion shifted fibers, *IEEE J. Lightwave Tech.*, **LT-9** (1991), 194-197.

9. Blow K. J. and Doran N. J.: Average soliton dynamics and the operation of soliton systems with lumped amplifiers, *IEEE Photon. Technol. Lett.*, **3** (1991), 369-371.

10. Gordon J. P. and Haus H. A.: Random walk of coherently amplified solitons in optical fiber transmission, *Opt. Lett.*, **1 1**, (1986), 665-667.

11. Gordon J. P.: Interaction forces among solitons in optical fibers, *Opt. Lett.*, **8** (1983), 596-598.

12. Chu P. L. and Desem C.: Mutual interaction between solitons of unequal amplitudes in optical fiber, *Electron. Lett.*, **21** (1985), 1133-1134.

13. Mecozzi A. , Moores J. D., Haus H. A. and Lai Y.: Soliton transmission control, *Opt. Lett.*, **16** (1991), 1841-1843.

14. Kodama Y. and Hasegawa A.: Generation of asymptotically stable optical solitons and suppression of the Gordon-Haus effect, *Opt. Lett.*, **17** (1992), 31-33.

15. Mollenauer L. F., Gordon J. P. and Evangelides S. G.: The sliding-frequency guiding filter : an improved form of soliton jitter control, *Opt. Lett.*, **17** (1992), 1575-1577.

16. Nakazawa M., Suzuki K., Yamada E., Kubota H., Kimura Y. and Takaya M.: Experimental demonstration of soliton data transmission over unlimited distance with soliton control in time and frequency domains, *Electron. Lett.*, **29** (1993), 729-730.

17. Forysiak W., Blow K. J. and Doran N. J.: Reduction of Gordon-Haus jitter by post-transmission dispersion compensation, *Electron. Lett.*, **29** (1993), 1225-1226.

18. Suzuki M., Edagawa N., Taga H., Tanaka H., Yamamoto S. and Akiba S.: Feasibility demonstration of 20 Gbit/s single channel soliton transmission over 11500 km using alternating-amplitude solitons, *Electron. Lett.*, **30** (1994), 1083-1084.

19. Suzuki M., Morita I., Yamamoto S., Edagawa N., Taga H. and Akiba S.: Timing jitter reduction by periodic dispersion compensation in soliton transmission, *OFC'95*, San Diego, post-deadline paper (1995), **PD20.**

20. Mollenauer L. F., Lichtman E. , Neubelt M. J. and Harvey G. T.: Demonstration, using

sliding-frequency guiding filters, of error-free soliton transmission over more than 20 Mm at 10 Gbit/s, single-channel, and over more than 13 Mm at 20 Gbit/s in a two-channel WDM, *Electronics Lett.*, **29** (1993), 910-911.

21. Favre F and LeGuen D.: 20 Gbit/s soliton transmission over 19 Mm using sliding-frequency guiding filters, *Electron. Lett.*, 31(1995), 991-992.

22. Aubin G., Jeanney E., Montalant T., Moulu J., Pirio F., Thomine J.-B. and Devaux F.: 20 Gbit/s soliton transmission over transoceanic distances with a 105 km amplifier loop, *Electron. Lett.,* **31** (1995), 1079-1080.

23. King J.P., Hardcastle I., Harvey H.J., Greene P.D., Shaw B.J., Jones M.G., Forbes D.J. and Wright M.C.: Polarisation-independent 20 Gbit/s soliton data transmission over 12500 km using amplitude and phase modulation soliton transmission control, *Electron. Lett.,* **31** (1995), 1090-1091.

24. Suzuki M., Morita I., Edagawa N., Yamamoto S., Taga H. and Akiba S.: Reduction of Gordon-Haus timing jitter by periodic dispersion compensation in soliton transmission, *Electron. Lett.,* **31** (1995), 2027-2029.

25. Edagawa N., Morita I., Suzuki M., Yamamoto S., Taga H. and Akiba S.: 20 Gbit/s, 8100 km straight-line single-channel soliton-based RZ transmission experiment using periodic dispersion compensation, *ECOC'95*, Brussels, post-deadline paper, **Th.A.3.5** (1995), 983-986.

26. Morita I., Suzuki M., Edagawa N., Yamamoto S., Taga H. and Akiba S.: 20 Gbit/s single-channel soliton transmission over 9000 km without inline filters, *Electron. Lett.,* **31** (1995), to be appeared.

27. Mollenauer L. F., Evangelides S. G. and Gordon J. P.: Wavelength division multiplexing with solitons in ultra-long distance transmission using lumped amplifiers, *J. Lightwave Technol.,* **LT-9** (1991), 362-367.

28. Nyman B.M., Evangelides S.G., Harvey G.T., Mollenauer L.F., Mamyshev P.V., Saylors M., Korotky S.K., Koren U., Mizrahi V., Strasser, T.A., Veselka J.J., Evankow J.D., Lucero A.J., Nagel J.A., Sulhoff J.W., Zyskind J.L., Corbett P.C., Mills M.A. and Ferguson G.A.: Soliton WDM transmission of 8 x 2.5 Gbit/s, error free over 10 Mm, *OFC'95,* San Diego, post-deadline paper (1995), **PD21.**

29. Evangelides S. G., Nyman B. M., Mamyshev P.V., Mollenauer L.F. Veselka J. J. and Harvey G. T.: Results of soliton WDM transmission experiments in long circulating loops, *OSA annual meeting*(1995), **PD7.**

30. Edagawa N., Suzuki M., Taga H., Tanaka H., Yamamoto S., Takahashi H. and Akiba S.: Robustness of 20 Gbit/s, 100 km-spacing, 1000 km soliton transmission system, *Electron. Lett.,* **31** (1995), 663-665.

31. Nakazawa M., Suzuki K., Yamada E., Kubota H., and Kimura Y.: Straight-line soliton data transmission over 2000km at 20Gbit/s and 1000km at 40Gbit/s using Erbium-doped fiber amplifiers, Electron. Lett., **29** (1993), 1474-1476.

32. Nakazawa M., Kimura Y., Suzuki K., Kubota H., Komukai T., Yamada E., Sugawa T., Yoshida E., Yamamoto T., Imai T., Sahara A., Nakazawa H., Yamauchi O. and Umezawa M.: Field demonstration of soliton transmission at 10 Gbit/s over 2000 km in Tokyo metropolitan optical loop network, *Electron. Lett.,* **31** (1995), 992-994.

33. Iwatsuki, K., Saito S., Suzuki K., Naka A., Kawai S., Matsuda T. and Nishi S.: Field demonstration of 10 Gb/s-2700 km soliton transmission through commercial submarine optical amplifier system with distributed fiber dispersion and 90 km amplifier spacing, *ECOC'95*, Brussels, post-deadline paper, **Th.A.3.6** (1995), 987-990.

34. Suzuki M.,Tanaka H., Edagawa N., Utaka K. and Matsushima Y.: Transform-limited optical pulse generation up to 20GHz repetition rate by a sinusoidally driven InGaAsP electroabsorption modulator, *J. Lightwave Technol.,* **LT-11** (1993), 468-473.

35. Suzuki M. , Tanaka H., Edagawa N. and Matsushima Y.: New applications of a sinusoidally driven InGaAsP electroabsorption modulator to in-line optical gates with ASE noise reduction effect, *J. Lightwave Technol.,* **LT-10** (1992), 1912-1918.

36. Suzuki M., Tanaka H. and Matsushima Y.: InGaAsP electroabsorption modulator for high-bit-rate EDFA system, *IEEE Photon. Tech. Lett.,* **4** (1992), 586-588.

37. Nakazawa, M.,Yoshida, E., K., Yamada, Suzuki K., Kitoh, H. and Kawachi, M.: 80 Gbit/s soliton data transmission over 500 km with unequal amplitude solitons for timing clock extraction, *Electron. Lett.,* **30** (1994), 1777-1778.

38. Bergano, N. S., Kerfoot, F. W. and Davidson, C. R.: Margin measurements in optical amplifier systems, *IEEE Photonics Technol. Lett.,* **5** (1993), 304-306.

T

U

V

W

List of Contributors (representative)

Abdullaev, F. KH.
Theoretical Div., Physical-Technical Institute
of Uzbek Academy of Sciences
700084, Tashkent-84
G. Mavlyanov Str. 2-B, Uzbekistan
Phone : +7-3712-33-12-71
Fax   : +7-3712-35-43-11
E-mail : fatkh@pti.tashkent.su

Arahira, S.
Oki Electric Industry Co., Ltd.
550-5 Higashiasakawa, Hachioji
Tokyo 193
Japan
Phone : +81-426-62-6765
Fax   : +81-426-67-0545
E-mail : arahira@hlabs.hlabs.oki.co.jp

Chernikov, S. V.
Femtosecond Optics Group
Physics Department, Imperial College
Prince Consort Road, London SW7 2BZ
England
Phone :+44-171-594-7785
Fax   : +44-171-589-9463
E-mail : svch@ic.ac.uk

Chesnoy, J.
Alcatel Corporate Research Centre
33, Rue Emeriau, Paris
France
Phone : +33-1-64-49-10-00
Fax   : +33-1-64-49-06-94
E-mail :

Chi, S.
Institute of Electro-Optical Engineering
National Chiao Tung University
1001 Ta Hsuen Rd. Hsinchu
Taiwan
Phone :
Fax   : +886-35-716631
E-mail : schi@cc.nctu.edu.tw

Doran, N. J.
Electronic Eng. & Applied Physics
Aston University
Aston Triangle, Birmingham B4 7ET
United Kingdom
Phone : +44-121-359-3621, Ext 4973
Fax   : +44-121-359-0156
E-mail : n.j.doran@aston.ac.uk

Elgin, J.
Department of Mathematics
Imperial College
London SW7 2BZ
United Kingdom
Phone :
Fax   : +44-171-594-8517
E-mail : j.elgin@ic.ac.uk

Fujii, Y.
(~ March 1996)
Institute of Industrial Science
University of Tokyo
(April 1996 ~)
College of Science and Technology
Nihon University
7-24-1 Narashinodai, Funabashi 274, Japan
Phone(home) : +81-471-40-7116
Fax(home)   : +81-471-40-7116
E-mail   : MAG00736@niftyserve.or.jp

Gabitov, I. R.
Parmanent Address
L. D. Landau Inst. for Theoretical Physics
Kosygina 2, 117940 Moscow
Russia
Phone : +7-095-137-3244
Fax   : +7-095-938-2077
E-mail : ildar@cpd.landau.ac.ru
Temporary Address(1995-1996)
FZ-241g, FTZ Telekom
Am Kavalleriesand 3
Darmstadt, d-64295
Germany
Phone : +49-(0)6151-83-7667
Fax   : +49-(0)6151-83-4572
E-mail : gabitov@rhrk.uni-kl.de

Grigoryan, V. S.
"IRE-POLUS Co.", Russian Academy
of Sciences
Vvedenskogo sq.1, Fryazino 141120
Moscow region, Russia
Phone : +7-95-330-2456
Fax   : +7-95-202-2767
E-mail : smm@grcc.msk.su

Haelterman, M.
Service d'Optique et Acoustique
Universite Libre de Bruxelles
50, av. F.D. Roosevelt, CP 194/5
B-1050 Brussels
Belgium
Phone : +32-2-650-4494
Fax   : +32-2-650-4496
E-mail : mhaelter@ulb.ac.be

Iwatsuki, K.
NTT Optical Network Systems
1-2356 Take, Yokosuka, Kanagawa 238
Japan
Phone : +81-468-59-3095
Fax   : +81-468-59-3396
E-mail : iwatuki@exa.onlab.ntt.jp

Kamiya, T.
Department of Electronic Engineering
University of Tokyo
7-3-1 Hongo, Bunkyo-ku, Tokyo 113
Japan
Phone : +81-3-3812-2111, Ext 6679
Fax   : +81-3-5684-3984
E-mail : kamiya@ktl.t.u-tokyo.ac.jp

Kivshar, Y. S.
Optical Sciences Centre
The Australian National University
ACT 0200 Canberra
Australia
Phone : +61-6-249-3081
Fax   : +61-6-249-5184
E-mail : ysk124@rsphysse.anu.edu.au

Kodama, Y.
Department of Mathematics
Ohio State University
Columbus, OH 43210
U. S. A.
Phone : +1-614-292-8435
Fax   : +1-614-292-1479
E-mail : kodama@math.ohio-state.edu

Kubota, H.
NTT Access Network System Laboratories
Tokai-mura, Ibarakiken 319-11
Japan
Phone : +81-29-287-7319
Fax   : +81-29-287-7874
E-mail : kubota@lightwave.ntt.jp

Matsumoto, M.
<u>Permanent Address</u>
Osaka University
2-1 Yamadaoka, Suita
Osaka 565
Japan
Phone : +81-6-879-7730
Fax   : +81-6-877-4741
E-mail:matumoto@comm.eng.osaka-u.ac.jp
<u>Temporary Address</u>
c/o Massachusetts Institute of Technology
Phone :
Fax   : +1-617-258-7864
E-mail : mmatsu@mit.edu

Menyuk, C. R.
Dept. of Elec. Eng. & Computer Sciences
University of Maryland Baltimore County
Baltimore, Maryland 21228-5398
U. S. A.
Phone : +1-410-455-3501
Fax   : +1-410-455-6500
E-mail : menyuk@umbc.edu

402

Montes, C.
Lab. de Physique de la Matière Condensée
Univ. de Nice-Sophia Antipolis, Parc Valrose
F-06108 Nice Cedex 2, France
Phone : +33-92-07-67-76
Fax   : +33-92-07-67-54
E-mail : montes@naxos.unice.fr

Nishimura, M.
Sumitomo Electric Industries, Ltd.
1 Taya-cho, Sakae-ku, Yokohama 244
Japan
Phone : +81-45-853-7171
Fax   : +81-45-851-1557
E-mail : nishi@yklab.sumiden.co.jp

Ono, T.
Opto-Electronics Research Laboratories
NEC Corporation
4-1-1 Miyazaki, Minato-ku
Kawasaki-shi, Kanagawa 216
Japan
Phone : +81-44-856-2109
Fax   : +81-44-856-2222
E-mail : ono@optsys.cl.nec.co.jp

Richardson, D. J.
Optoelectronics Research Center
Southampton University
Southampton, S017 1BJ
United Kingdom
Phone : +44-703-59-4524
Fax   : +44-703-59-3142
E-mail : DJR@orc.soton.ac.uk

Runge, P. K.
AT&T Bell Laboratories
AT&T Submarine Systems, Inc.
101 Crawfords Corner Road
P. O. Box 3030, Room 3F-428
Holmdel, NJ 07733-3030
U. S. A.
Phone : +1-908-949-2750
Fax   : +1-908-949-6961
E-mail : 54326PR@HOGPA.HO.ATT.COM

Suzuki, M.
KDD R&D Laboratories
2-1-15 Ohara Kamifukuoka, Saitama 356
Japan
Phone : +81-492-78-7835
Fax   : +81-492-78-7516
E-mail : suzuki@lab.kdd.co.jp

Snyder, A. W.
Optical Sciences Centre
The Australian National University
ACT 0200 Canberra
Australia
Phone :
Fax   : +61-6-249-5184
E-mail : A.Snyder@anu.edu.au

Wabnitz, S.
Fondazione Ugo Bordoni
Via B. Castiglione 59, 00142 Rome
Italy
Phone : +39-6-5480 3206
Fax   : +39-6-5480 4405
E-mail : swab@fub.it

404

Wai, P. K. A.
Deptartment of Electronic Engineering
The Hong Kong Polytechnic University
Hung Hom, Kowloon
Hong Kong
Phone : +852-2766-6231
Fax   : +852-2362-8439
E-mail : wai@encserver.en.polyu.edu.hk

Watanabe, S.
Fujitsu Laboratories Ltd.
1015 Kamikodanaka
Nakahara-ku, Kawasaki 211
Japan
Phone : +81-44-754-2643
Fax   : +81-44-754-2640
E-mail : shigeki@flab.fujitsu.co.jp